Springer
*Tokyo
Berlin
Heidelberg
New York
Barcelona
Hong Kong
London
Milan
Paris
Singapore*

A. Sakurai T. Yokota S. D. Clouse (Eds.)

Brassinosteroids

Steroidal Plant Hormones

With 81 Figures, Including 13 in Color

 Springer

Akira Sakurai
Honorary Scientist of RIKEN
1-33-12 Minami-Ogikubo, Suginami-ku,
Tokyo 167-0052, Japan

Takao Yokota
Professor
Department of Biosciences
Teikyo University
Utsunomiya 320-8551, Japan

Steven D. Clouse
Associate Professor
Department of Horticultural Science, Box 7609
North Carolina State University
Raleigh, NC 27695, USA

ISBN 4-431-70214-8 Springer-Verlag Tokyo Berlin Heidelberg New York

Library of Congress Cataloging-in-Publication Data
Brassinosteroids: steroidal plant hormones / A. Sakurai, T.
Yokota, S.D. Clouse, (eds.).
    p. cm.
    Includes bibliographical references and index.
    ISBN 4-431-70214-8 (alk. paper)
    1. Brassinosteroids. I. Sakurai, A. (Akira), 1936- II. Yokota,
Takao, 1942- III. Clouse, S. D. (Steven D.), 1951-
    QK898.B85 B74 1998
    571.7'42--dc21
                              98-33427
                                  CIP

Printed on acid-free paper
© Springer-Verlag Tokyo 1999
Printed in Tokyo
This work is subject to copyright. All rights are reserved, whether the whole or part of the material is concerned, specifically the rights of translation, reprinting, reuse of illustrations, recitation, broadcasting, reproduction on microfilms or in other ways, and storage in data banks.
The use of registered names, trademarks, etc. in this publication does not imply, even in the absence of a specific statement, that such names are exempt from the relevant protective laws and regulations and therefore free for general use.
Typesetting: From the authors' electronic files
Printing & binding: Obun, Japan
SPIN: 10656073

PREFACE

Brassinosteroids (BRs) are plant-growth-promoting natural products similar in structure to animal steroid hormones, and the accumulated evidence of nearly two decades of research suggests that BRs should be included along with auxins, gibberellins, cytokinins, abscisic acid, and ethylene as essential plant hormones. Since the first discovery of brassinolide from rape pollen in 1979, extensive studies on this notable substance have been undertaken worldwide. Diverse species of plants have been found to contain brassinolide and a variety of structural analogs (which are collectively called BRs), and their characteristic physiological effects on the growth and development of plants as well as their potency in agricultural applications have been examined. Work on BRs from their discovery until 1991 was collected in the book Brassinosteroids, edited by Cutler, Adam, and Yokota, which was the first treatise dedicated solely to these compounds.

From 1991 through 1995, major advances in BR research took place including biochemical studies using a plant cell culture system to elucidate the biosynthetic pathway of brassinolide from the common plant sterol campesterol; the cloning and analysis of the first BR-regulated genes; and the preliminary identification of the first BR-insensitive mutant in Arabidopsis. However, BR research continued to be neglected by the majority of scientists studying plant growth regulation because of uncertainty regarding a unique and essential function for BRs in the growth and development of plants. The turning point in the widespread acceptance of brassinolide as a plant hormone occurred in 1996 and 1997 when several dwarf mutants of Arabidopsis and garden pea were found to be rescued to wild-type by BR treatment. A combined approach including cloning of the genes altered in the mutants, feeding experiments with biosynthetic intermediates, and analysis of endogenous BR levels in the mutant and wild-type plants, led to the identification of the BR biosynthetic step affected in each mutant, and provided convincing genetic evidence that BRs were essential for normal plant development. Moreover, BR-insensitive mutants were characterized from dwarfs of Arabidopsis and garden pea, expanding our knowledge of BR signal transduction. The phenotypic analysis of BR-deficient and insensitive mutants has provided confirmation of many previous studies addressing the importance of BRs in various physiological functions.

Because of the rapid progress in BR research, we decided to present a second treatise devoted solely to this topic by compiling the current works as well as the accumulated results and methodology in the field. This volume addresses the history, natural occurrence, biochemical analysis, chemical synthesis, biosynthesis, metabolism, physiological actions, molecular genetics, structure-activity relationship, and practical applications of BRs. We hope that those readers currently in the field as well as those with a general interest in plant hormone biology will find new aspects of research in plant growth regulation described in the comprehensive reviews by those who have kindly contributed to this volume.

Akira Sakurai
Takao Yokota
Steven D. Clouse

CONTENTS

1. The History of Brassinosteroids: Discovery to Isolation of Biosynthesis and Signal Transduction Mutants 1

    1. Introduction 1
    2. Early History in the United States Department of Agriculture 2
    3. Early History in Japan 5
    4. Advances After Brassinolide 7
    5. Recent Advances 12
    6. Concluding Remarks 14
    References 14

2. Natural Occurrence of Brassinosteroids in the Plant Kingdom 21

    1. Introduction 21
    2. Structural Characteristics of Naturally Occurring Brassinosteroids 26
    3. Distribution of Brassinosteroids in the Plant Kingdom 29
    4. Conclusions 40
    References 41

3. Biochemical Analysis of Natural Brassinosteroids 47

    1. Introduction 47
    2. Bioassays and Immunoassay for Brassinosteroids 48
    3. Purification of Natural Brassinosteroids 50
    4. NMR of Natural Brassinosteroids 52
    5. GC-MS 54
    6. Labeled Brassinosteroids and Related Steroids 56
    7. HPLC 61
    8. LC-MS 63
    9. Conclusions 64
    References 64

4. Chemical Synthesis of Brassinosteroids 69

    1. Introduction 69
    2. Synthesis of Brassinolide 70
    3. Synthesis of 28-Homobrassinolide and 24-Epibrassinolide 85
    4. Synthesis of ^{14}C and ^{3}H-brassinosteroids 87
    5. Conclusions 88
    References 88

5. Biosynthesis 91

    1. Introduction 91
    2. Methods of Metabolic Conversions Using the Cell Culture System 93
    3. Biochemical Analysis of Brassinosteroid Biosynthesis Mutants 97
    4. Early Stage of Biosynthesis 100

5. Downstream Pathways to Brassinolide 103
  6. Regulation of Biosynthesis 107
  7. Conclusion 108
  References 109

6. Uptake, Transport and Metabolism 113

  1. Introduction 113
  2. Uptake and Transport in Plants 113
  3. Metabolic Studies with Explants 115
  4. Metabolism in Cell Cultures 122
  5. Conclusions 134
  References 134

7. Physiological Actions of Brassinosteroids 137

  1. Introduction 137
  2. Sites of Synthesis and Transport of BRs 139
  3. Effects on Vegetative Growth 139
  4. Effects on Cell Division 146
  5. Effects on Development 147
  6. Interactions with Environmental Signals 150
  7. Effects on Insect and Fungal Development 153
  8. Conclusion and Perspectives 154
  References 155

8. Molecular Genetics of Brassinosteroid Action 163

  1. Introduction 163
  2. Cloning of BR-Regulated Genes 164
  3. Molecular Mechanisms of BR-Regulated Cell Expansion 167
  4. Molecular Mechanisms of BR-Modulated Cell Differentiation 168
  5. BR Biosynthesis Mutants 169
  6. BR Signal Transduction 178
  7. Future Prospects 182
  References 184

9. Structure-Activity Relationship 191

  1. Introduction 191
  2. Early Structure-Activity Relationships 192
  3. New Approach for Structure-Activity Relationships Definition 196
  4. Feasibility of Brassinosteroids Hydrogen Bonding with the Receptor 211
  5. Brassinosteroids Inhibitor 218
  6. Conclusions 219
  References 220

10. Practical Application of Brassinosteroids in Agricultural Fields 223

  1. Introduction 223
  2. Application of Brassinosteroids to Crops 225
  3. Practical Application of Brassinosteroids with Long-Lasting Activity 232

4. Conclusions	238
References	238

Index ... 243

1
The History of Brassinosteroids: Discovery to Isolation of Biosynthesis and Signal Transduction Mutants

TAKAO YOKOTA

Department of Biosciences, Teikyo University, Utsunomiya 320-8551, Japan

1. Introduction

It was only recently that brassinosteroids (BRs) were generally accepted as steroidal plant hormones. When brassinolide, the first BR, was isolated from rape (*Brassica napus* L.) pollen by Grove et al in 1979 as a plant growth regulator with remarkable biological activity in the bean second internode assay, little attention was directed to this new growth stimulator. Steroid hormones have been known to play important roles in various organisms. These include sex hormones, androgens, estrogens and gestagen (progesterone), glucocorticoids and mineral corticoids in animals, moulting hormones (ecdysteroids) in insects and crustacea, and an antheridiogen (antheridiol) in a water mold, *Achlya bisexualis*. Therefore, before the isolation of brassinolide, long discussion had been carried out on the question of whether steroidal hormones are present in plants (Heftmann 1971, 1975). There were many papers reporting the occurrence of animal steroid hormones such as estrogens, androgens, progestogens, and a corticosteroid in plants as well as on the their effects on germination, growth, flowering, and sex expression of plants. To my knowledge, the last review on the relation between animal steroid hormones and plant growth appeared in 1978 by Geuns. However, after 1979 when the isolation of brassinolide was reported, such discussion seems to have ceased because brassinolide has a steroidal structure and has apparently hormonal properties.

There was over 10 years' early history in the United States Department of Agriculture (USDA) up to the isolation of brassinolide (Steffens 1991). However, research on BRs had been also carried out concurrently in Nagoya University in Japan

---

Key words: brassins, field test, brassinolide, *Distylium* factors, synthesis, natural occurrence, biosynthesis, biosynthesis mutants, signal transduction mutants

more than 10 years before the first report on brassinolide, but their pioneering attempts to determine the structure terminated at the mid-point without reaching the final goal (Abe and Marumo 1991). After the Grove's report, castasterone was isolated as the second BR at the University of Tokyo (Yokota et al 1982a). Since then, a number of related steroidal compounds, now referred to collectively as brassinosteroids according to the proposal of Mandava (1988), have been isolated from a variety of plant sources. At the same time, physiological activity, field tests, and synthesis of BRs have been investigated, allowing us to conclude that BRs are a group of plant hormones (Sakurai and Fujioka 1993, Adam et al 1996, Fujioka and Sakurai 1997a, Sasse 1997, Yokota 1997). Recently, the major part of the biosynthesis of brassinolide has been established (Fujioka and Sakurai 1997b). Further, simultaneously, mutants of BR biosynthesis and signal transduction have been isolated (Clouse 1996, 1997; Yokota 1997). The molecular genetic and biochemical work on these mutants has furnished conclusive evidence that BRs are a sixth class of plant hormones.

2. Early History in the United States Department of Agriculture

2.1 Brassins Project and the Discovery of Brassinolide

In 1941 Mitchell and Whitehead at the Beltsville Agricultural Research Center (BARC) of USDA reported that *Zea mays* pollen contained plant growth-promoting activity. Mitchell's further interest led to a report in *Nature* in 1970 entitled "Brassins - A New Family of Plant Hormones from Rape Pollen" (Mitchell et al 1970). Brassins-like compounds that are biologically active in the bean second internode bioassay were found in about 20 different kinds of plants (Mandava and Mitchell, 1971) (Table 1). The response of brassins was histologically different from the response induced by gibberellins, which caused only cell elongation. The brassins treatment induced both cell elongation and cell division as well as marked changes in the vascular anatomy (Worley and Mitchell 1971). Brassins were first shown to be a mixture of fatty acid esters of glycerol (glycerides) (Mandava and Mitchell, 1971), and later were reported to be a mixture of fatty acid esters of glucose (Mandava and Mitchell, 1972). The chemical evidence presented by them was criticized because it was premature to support their conclusions (Milborrow and Pryce 1973). There was no formal response from USDA researchers to the published criticism, but they reached a conclusion that other minor components of the brassin complex play a role in its action. In 1974, a cooperative effort to isolate brassins from a large amount of rape pollen (227 kg) was started by USDA scientists at BARC, the Northern Regional Research Center and Eastern Regional Research Center (Steffens 1991). However, the chance that the bioactive natural products might be known gibberellins was seriously considered, relative to the large investment (Cutler 1991). Honeybee-collected

Table 1. Plants in which biological activity was detected by the bean second-internode bioassay or rice-lamina inclination bioassay in the early stage of brassinosteroid history

Bean second-internode bioassay* (Mandava and Mitchell 1971)	Rice-lamina inclination bioassay (Marumo et al 1968, Abe and Marumo 1991)
Aescules hippocastanum	*Arachis hypogaea* (seed)
Alnus glutinosa	*Camellia sasanqua* (leaf)
Brassica napus	*Cocculus trilobus* (leaf)
Carduus nutans	*Corylopsis glabrescens* (leaf)
Crataegus sp.	*Disanthus cercidifolia* (leaf)
Echium vulgare	*Distylium racemosum* (leaf, insect gall)
Pyrus communis	*Dolichos lablab* (seed)
Robiana pseudoacacia	*Glycine max* (seed)
Rhus sp.	*Hamamelis japonica* (leaf)
Secale cereale	*Phaseolus vulgaris* (shoot)
Sinapis arvensis(=*Brassica kaber*)	*Pisum sativum* (seed)
Sisymbrium irio	*Pittosporum tobira* (leaf)
Thea sinensis (=*Camellia sinensis*)	*Thea sinensis* (leaf)
Typha sp.	*Viburnum awabuki* (leaf)
Ulmus sp.	*Vicia faba* (seed)
Zea mays	

*Only pollen was examined by this bioassay.

rape pollen could easily be obtained because pollen was used as a food additive in specialty-type natural food products. The rape pollen was grouped into several batches, and each batch was washed with water and then freeze-dried prior to extraction with isopropanol. The extract was partitioned by counter-current distribution using two-phase solvent mixture composed of methanol, carbon tetrachloride, and water. The upper methanol phases were combined and purified twice by silica gel chromatography (Mandava et al 1978). The enriched brassins fraction derived was delivered to Grove at the Northern Regional Research Center (Steffens 1991). He and his co-workers subjected the extract to column chromatographies using silica gel and C_{18} columns followed by reverse-phase high-performance liquid chromatography (HPLC). After crystallization, they obtained 4 mg of crystals of brassinolide, which represented ca 40 kg of pollen. It was surprising that D.W. Spaulding had conducted about 9000 bean second internode assays at BARC by the time brassinolide was identified (Steffens 1991). The structure of brassinolide was completely established by single crystal X-ray analysis (Grove et al 1979). At the same time, biological activities of a number of synthetic acyl sugars had also been being examined in the bean hypocotyl elongation bioassay but were essentially inactive compared with that of the brassin complex (Grove et al 1978, Bemiller et al 1979). Thus it was concluded that the active principle of brassins is brassinolide, the first steroidal plant growth regulator.

Kishi at Harvard University, who is a synthetic chemist who had moved from

Nagoya University, noticed the announcement of the isolation of brassinolide that appeared in *Chemical & Engineering News* (No 5, p 20, 1979). This news was promptly delivered by a letter to Marumo at Nagoya University in Japan. He was much surprised at the news because brassinolide had chemical properties very similar to those of *Distylium* factors, which he and his co-workers had isolated from plant sources using the rice lamina inclination bioassay (Marumo 1983). Mori (1980) synthesized (22S, 23S)-28-homobrassinolide immediately after being informed of the news by Marumo, and soon after Ikekawa's laboratory at Tokyo Institute of Technology succeeded in the synthesis of brassinolide (Ishiguro et al 1980). As expected, these synthetic compounds were found to be very active in the rice lamina inclination assay, making them conclude that brassinolide or its analogues should be what they had long been seeking from plants. The history thereafter in Japan will be described later.

2.2 Field Test Trials

Mitchell and Gregory (1972) found that seed treatment with brassins increased plant size and/or seed yield and that brassins stimulated the growth of smaller and slow-growing plants to a greater degree than that of vigorously growing ones. Therefore, they carried out large and detailed field experiments in 1974 and 1975 to determine if plants from brassin-treated small seeds grew to the same size and produced as many seeds as plants grown from large seeds (Steffens 1991). Soybean field experiments were done in 1974-1975 in Brazil. Prior to planting, soybean seeds, divided by size, were soaked in dichloromethane solutions containing brassins. No significant effects on seed yield or plant size were found. In 1975, soybean field experiments were repeated using seeds of eight cultivars at seven locations in the U.S.A. Again, no major economic importance was recognized in these trials. In the same year, barley seeds of four cultivars were treated with brassins dissolved in dichloromethane and planted at five locations in the U.S.A. As with the soybean experiments, the differences caused by brassins treatment were relatively small and not considered to be of economic importance.

Because seed treatment did not increase yields, various application techniques were tested in greenhouse and preliminary field experiments in 1977 by Meudt and Gregory (Steffens 1991). Most effective was application of brassins as sprays to very young seedlings. Significant yield increase was observed, for example, for radishes, leafy vegetables, and potatoes. Most of these results were not published because the structure of brassins remained unknown. In 1979, Thompson and his collaborators synthesized 24-epibrassinolide and its (22S, 23S)-epimer. Aqueous solutions of 0.01 ppm of 24-epibrassinolide were applied as sprays to pot-grown seedlings of lettuce, radishes, peppers, beans, tomatoes, and corn (Maugh 1981). The BR treatment increased the yields of lettuce and radishes and that of the larger fruits in the cases of pepper and beans. However, no yield increase was observed for tomatoes and corn. The results were similar to those obtained in the earlier experiments using brassins. Shortly after the success of the brassin project, this American project was coming to a close and no further field work was planned. Some of the scientists involved in this program have died, and most of the others have retired or left the USDA. For a while since then little attention was directed to BRs in the U.S.A., with some exceptions. Further investigation on not only agricultural application of BRs but also the chemistry and physiology was developed largely in other countries. Especially Japanese scientists were enthusiastic in developing BR research because of the historical reasons described in the succeeding section.

3. Early History in Japan

3.1 *Distylium* Factors and the Rice Lamina Inclination Assay

The early history of BRs in Japan has been described in some reviews (Marumo 1983, 1987, Abe and Marumo 1991). Long before the discovery of brassinolide, scientists in Nagoya University had been trying to isolate and determine the structure of new auxin-like substances that were active in the rice lamina inclination bioassay (or rice lamina joint test). This bioassay to observe lamina inclination activity was originally designed to detect synthetic auxins by Maeda (1965). Munakata has been interested in the phenomenon that insect galls in aphid-infested leaves of *Distylium racemosum* rapidly swell, leading to an idea that the gall tissue contains growth-promoting substances, which were later termed *Distylium* factors (Marumo and Wada 1981, Marumo 1983). Marumo subjected the ethyl acetate-soluble neutral fraction separated from the galls to the rice lamina inclination bioassay, resulting in the detection of remarkable activity in the extract. He wrote in his reviews on BRs that they were much impressed upon seeing that the biological activity of this crude extract was far higher than indole-3-acetic acid (Marumo 1983). In this bioassay IAA has only marginal activity. They further found such biological activity in the extracts from leaves and seeds of 14 plant species (Table 1), allowing them to conclude that biologically active substances in this bioassay are distributed in a wide range of plants (Abe and Marumo 1991). Munakata et al (1974) also using this bioassay examined various plant oils from coconut, corn germ, cotton seed, linseed, olive, peanut, poppy, rape, sesame, soya, and turpentine and found biological activity in seven of them, but not in those of linseed oil, poppy oil, sesame oil, and turpentine oil.

Distylium factors were also found to be present in healthy leaves of *D. racemosum*, so a large number of the healthy leaves (430 kg) were collected and extracted with methanol in 1966. A neutral ether-soluble fraction was separated from the methanol extract and purified by a silica gel column. Further chromatography on Sephadex LH-20, silica gel thin-layer chromatography (TLC) and alumina column chromatography resulted in the isolation of three *Distylium* factors A1 (751 µg), A2 (50 µg), and B (236 µg), which were all detected as a single spot on TLC (Marumo et al

1968). They found that these compounds are distinct from indole auxins because of no color reaction characteristic of indoles. However, no further structural information was obtained because technologies of ^{1}H-NMR (nuclear magnetic resonance) and mass spectrometry (MS) at that time were not sensitive enough to analyze such small amounts of compounds. Furthermore *Distylium* factors isolated must have been impure because of the unavailability of HPLC techniques which are currently a prerequisite to purify BRs. Nonetheless, *Distylium* factors A1 and B were much more active than IAA in the rice lamina inclination bioassay. As described in Section 4.2, this chapter, they obtained further important biological evidence supporting the hypothesis that the *Distylium* factors are a new class of plant growth substances (Marumo et al 1968).

As described in Section 2.1, this chapter, immediately after the discovery of brassinolide, brassinolide and synthetic analogues were found to be biologically active in the rice lamina inclination bioassay. This finding as well as the fact that *Distylium* factors have chemical properties similar to BRs suggested that *Distylium* factors were BRs. In 1984, it was reported based on gas chromatography(GC)-MS analysis that *Distylium* factor A1 was a mixture of castasterone and 28-norcastasterone while *Distylium* factor B was a mixture of brassinolide and 28-norbrassinolide (Ikekawa et al 1984). *Distylium* factor A2 has yet to be identified but is supposed to be a 2-deoxy-type BR on the basis of its chromatographic behavior. Munakata et al (1974) also reported the isolation of a corn factor from corn germ oil as an active principle in the rice lamina inclination bioassay. However, the structure of this compound could not be determined.

3.2 Isolation of Castasterone as the Second BR

Chestnut buds parasitized by larvae of a cecidogenetic wasp, *Dryocosmus kuriphilus*, do not grow in a straight manner but swell to form round insect galls. Takahashi at the University of Tokyo was interested in the growth mechanism of this unusual tissue. No clear answer for this problem could be obtained although it was found that the larvae contain compounds to cause swelling and the gall tissues are auxin-rich but have reduced levels of cytokinins (Yokota et al 1974). When brassinolide was isolated from rape pollen in 1979, there was no evidence to suggest that brassinolide widely occurs in plant tissues. Furthermore, it looked like a technically hard task to isolate minute amounts of such compounds. However, triggered by the discovery of brassinolide, Yokota and his co-workers started to isolate brassinolide itself from plant sources. They used chestnut gall tissue because it had been reported that chestnut insect galls contain compounds that are active in the rice lamina inclination assay (Ikeda et al 1977). A large amount (40 kg) of chestnut gall tissue that had been long stored frozen was extracted with methanol prior to solvent partitionings. Successive purification by silica gel chromatography, Sephadex LH-20 chromatography, and HPLC on silica gel and C_{18} silica afforded 95 µg of a new BR termed castasterone, which crystallized from aqueous acetonitrile (Yokota et al 1982a). The structure of castasterone was established by modern techniques such as FAB-mass spectrometry, high-resolution mass spectrometry, and (FT)-NMR. Because castasterone lacked an ether oxygen found in the B-ring lactone of brassinolide, it was considered to be a direct precursor of brassinolide. This hypothesis was later found to be the case (Yokota et al 1990). Further study indicated that castasterone is also present in normal tissues such as shoot, leaf, and flower; and it was concluded that BRs may not be directly involved in the abnormal growth of the insect gall (Arima et al 1984).

4. Advances After Brassinolide

4.1 Occurrence of Numerous BRs in the Plant Kingdom

Like USDA researchers, Japanese scientists involved in the early stages of BR research also believed that BRs are plant hormones. This is well reflected in the title of the paper reporting the isolation of castasterone "Castasterone, a new phytosterol with plant-hormone potency, from chestnut insect gall" (Yokota et al 1982a). In 1982, Ikekawa's laboratory at the Tokyo Institute of Technology developed quite useful GC-MS methods to analyze BRs (Takatsuto et al 1982), in which BRs are converted into bismethaneboronates and analyzed by GC-MS in the electron impact mode or chemical ionization mode. This GC-MS technique together with the rice lamina inclination bioassay must be emphasized because without these techniques BR studies would have been delayed to a large extent. Soon after the isolation of castasterone in 1982, Ikekawa's GC-MS method was used for the identification of brassinolide and castasterone from immature fruits of Chinese cabbage (Abe et al 1982), tea leaves (Morishita et al 1983), and chestnut insect galls (Ikeda et al 1983). Abe in Tokyo University of Agriculture and Technology and his coworkers synthesized some biosynthetically plausible BRs and, using the GC-MS methods, demonstrated the occurrence of 28-norcastasterone, 28-homocastasterone, and/or 28-norbrassinolide in the above three plants (Abe et al 1983). After the isolation of castasterone, Yokota and his coworkers continued to isolate new BRs from plant sources. They isolated dolicholide, dolichosterone, homodolicholide, and homodolichosterone from immature seeds of *Dolichos lablab* (Yokota et al 1982b, 1983, 1984). They established effective purification techniques to isolate small amounts (10-100 µg) of BRs from plant sources throughout these works. They further isolated more than 15 BRs including metabolites and conjugates of BRs from a large amount of *Phaseolus vulgaris* immature seeds (Kim et al 1987, Kim 1991). Details of these findings are summarized in the chapter by S. Fujioka, this volume. Other laboratories have also been involved in the isolation of new BRs. Typhasterol was first isolated from cattail pollen from Suntory Institute for Bioorganic Research, Japan, by Schneider et al (1983). 6-Deoxotyphasterol and 3-dehydro-6-deoxoteasterone were first reported from the University of Melbourne by Griffiths et al (1995). Further, Adam and his co-workers at the Institute of Plant Biochemistry, Halle, Germany, started to analyze naturally occurring BRs around 1990 and isolated secasterone from *Secale cereale*

(Schmidt et al 1995a), 6-deoxo-28-norcastasterone and 6-deoxo-24-epicastasterone from *Ornithopus sativus* (Spengler et al 1995), and 2-deoxybrassinolide from *Apium graveolens* (Schmidt et al 1995b).

Many plant species have now been investigated for the occurrence of BRs using GC-MS techniques. Liquid chromatographic assay for fluorescent boronate derivatives of BRs have also been applied to several plants (Gamoh 1994). Thus, at present, structures of more than 40 BRs including four conjugates (two glucosides and two fatty acid esters) have been elucidated. BRs have been detected in as many as 44 plant species including 37 angiosperms (9 monocots and 28 dicots), 1 alga, and 1 fern (Adam et al 1996, Fujioka and Sakurai 1997a) (see the chapter by S. Fujioka, this volume). Such wide occurrence of BRs in lower to higher plants indicates that BRs are commonly distributed and are essential growth regulators, in other words, plant hormones.

4.2 Unique Physiological Activity

Before the isolation of brassinolide in 1979, the biological activity of brassins and *Distylium* factors had been investigated, and the results obtained allowed the researchers to believe that these compounds were a new class of plant hormones. Physiological activities of brassins were investigated by USDA researchers and Yopp at Southern Illinois University using the bean second-internode bioassay, which was used for the isolation of brassinolide from the brassins extract (Mitchell et al 1970, Mandava and Mitchell 1971, Worley and Mitchell 1971, Yopp et al 1979). Typical activity is to evoke both cell elongation and cell division resulting in elongation, swelling, curvature, and splitting of the second internode: See figures in Grove et al (1979), Thompson et al (1981), and Krizek and Mandava (1983a). Indole auxins and cytokinin (kinetin) never induce both cell division and cell elongation in this assay. Gibberellin causes elongation accompanied by cell enlargement but does not induce cell division (Worley and Mitchell 1971). Yopp et al (1979) found that actions of brassins are distinct from those of gibberellins in bioassays using hook opening and hypocotyl elongation of bean hypocotyls. As described in Section 3.1, this chapter, Marumo et al (1968) found that *Distylium* factors exhibited strong activity in the rice lamina inclination bioassay in which IAA showed marginal activity. Using typical auxin bioassays they found that these compounds are active in the *Avena* straight growth test but inactive in the *Avena* curvature test. In the typical gibberellin bioassays using dwarf maize and rice, these were inactive. Cytokinin-like activity of *Distylium* factors were also observed by them using the radish leaf expansion test and tobacco pith callus test.

Such biological activity was later confirmed using pure BRs after the discovery of brassinolide, as reviewed by Mandava (1988). In many auxin bioassays, BRs showed activity similar to that of auxin. BRs elongated stems (epicotyl, hypocotyl, and mesocotyl) of various plants, inhibited root growth and retarded bean hook opening. BRs were biologically active in many of the gibberellin bioassays using mung bean epicotyls and excised hypocotyls of cucumber and *Phaseolus vulgaris*. Although BRs had cytokinin-like activities such as expansion of cucumber and pea cotyledons as well as callus proliferation, they elicited neither retardation of senescence nor promotion of betacyanin synthesis. Such a wide activity spectrum of BRs was a puzzling question for a long time, and the hormonal status of BRs was accepted only by a limited number of researchers until BR mutants were discovered (Clouse 1996, 1997).

Yopp et al (1979, 1981) observed strong synergistic interactions between BRs and auxin in the bean hypocotyl system and suggested that BR action is mediated through endogenous auxin. Synergism between BRs and auxin was also observed in ethylene production (Arteca et al 1988), but no synergism was observed between BRs and gibberellin. Further physiological studies confirmed the observation made by Yopp et al. However, many findings contradictory to this idea were accumulated by studying stem elongation of pea (Sasse 1990) and cucumber (Katsumi 1991), proton pump activation in root and epicotyls (Cerana et al 1983, Romani et al 1983), and xylem differentiation (Iwasaki and Shibaoka 1991, Yamamoto et al 1997). Furthermore, BRs have unique and specific biological activities such as rice lamina inclination (Wada and Marumo 1981), leaf unrolling (Wada et al 1985), and maize root elongation (Romani et al 1983), which are not explained by an interaction with IAA. Sasse (1990) and Katsumi (1991) stressed clearly that the physiological modes of action of BRs are different from that of auxin and other plant hormones. This was recently beautifully evidenced by gene expression studies with soybean tissue (Clouse et al 1992) and auxin-insensitive mutants of tomato (Zurek et al 1994) and *Arabidopsis* (Clouse et al 1993; see also reviews by Clouse 1996, 1997).

The relationship between photomorphogenesis and BRs had already been pointed out before the isolation of brassinolide by Krizek and Worley (1973) who found that morphological changes induced by brassins are affected by light intensity. Krizek and Mandava (1983a,b) reported that spectral quality also influences the growth response to BR. Light retards ethylene evolution evoked by BR (Arteca and Bachman 1987) but is required for the growth-promoting effects of BR (Kamuro and Inada 1991). However, BRs had no effect on phytochrome-controlled phenomena such as lettuce seed germination and spore germination and cell elongation of protonemata of *Agiantum*; and they are not involved in blue light-controlled cell division of *Agiantum* protonemata (M. Furuya, personal communication; Yokota 1994). However, Furuya's other data indicated that elongation of rice mesocotyl elicited by BRs is likely to be controlled by phytochrome (Yokota 1994). The possibility that BRs are involved in blue light-related photomorphogenesis has also been postulated because both blue light and BRs are quite effective for leaf unrolling (Wada et al 1985). Further, recent reports described that BR-deficient or BR-insensitive *Arabidopsis* mutants show de-etiolation in the dark (Clouse 1996, 1997). Altogether, BRs are likely involved at least indirectly in photomorphogenesis.

Also interesting is that BRs have various anti-stress effects. Such effects were found during the attempts to develop BRs as growth modulators in agriculture and hence will be described in the following section.

4.3 Synthesis and Second-Stage Field Test Trials

Brassinolide attracted a number of synthetic chemists because of its challenging molecular structure, especially the side chain containing four contiguous chiral centers. During the year brassinolide was isolated, 24-epibrassinolide and its (22S, 23S)-isomer were synthesized by Thompson et al (1979). One year later the syntheses of (22S, 23S)-28-homobrassinolide (Mori 1980) and brassinolide (Fung and Siddall 1980, Ishiguro et al 1980) were reported. Since then, a number of BRs were isolated and concurrently various synthetic methods of BRs were developed as described by T.C. McMorris (this volume). Furthermore, a variety of BR analogues were synthesized and evaluated for biological activity in the rice lamina inclination bioassay, bean second internode bioassay, and radish/tomato bioassay, resulting in the establishment of the structure-activity relationship of BRs (Yokota and Mori 1992) (see the chapter by C. Brosa, this volume).

Mitchell and Gregory (1972) found that brassins enhanced the overall growth of some kinds of plants. These findings were later confirmed in many instances (e.g., Gregory 1981, Maugh 1981, Braun and Wild 1984); furthermore USDA researchers obtained promising results in the field tests carried out using synthetic BRs (see Section 2.2, this chapter). These results intrigued Japanese synthetic chemists, agronomists, and horticulturists. Application study of (22S, 23S)-28-homobrassinolide, which was synthesized by Mori (1980), started 3 months after the first report of brassinolide in 1979. The promising results were announced in 1982-1984 from Utsunomiya University, Nissan Chemical Industries Co., and Fruit Tree Research Station (Fujita 1985, Hamada 1985, Takematsu et al 1985). Further, Ikekawa's laboratory developed an economic synthesis of 24-epibrassinolide in 1984 (Takatsuto and Ikekawa 1984). Extensive field trials of 24-epibrassinolide were carried out in Utsunomiya University. Mori et al (1984) also developed an effective synthetic method for brassinolide, which was applied to a large scale synthesis of brassinolide by Aburatani et al (1985). Application trials of brassinolide were extensively carried out in various institutes including Utsunomiya University, Nissan Chemical Industries Co and Agricultural Technology Center of Zen-Noh (the National Federation of Agricultural Co-operatives) (Fujita 1985).

They found that treatment of seeds of grains, legumes, and weeds with BRs enhances seed germination and primary overall growth, often resulting in yield increases. Application of BRs to plants at the time of flowering increased fertilization rates and, especially in fruit trees, prevented flower/fruit drop. Such effects gave rise to yield increases of grains and vegetables such as rice, wheat, corn, soybean, cucumber, and tomato. It was found that, in cases of rice, wheat, and maize, less viable flowers that cannot set fruits under the natural conditions became fertile by the application of BRs. More interesting is that plants treated with BRs acquired resistance or tolerance against a variety of stresses including cold, drought, salt, disease, and herbicide. However, in field tests the effects of BRs were unstable and various promising effects of BRs were not reproduced. The biological activity of these BRs was found to disappear within a few days presumably due to their rapid deactivation in plants (see the chapter by Y. Kamuro and S. Takatsuto, this volume). They also found that the effects of these BRs were largely influenced by environmental conditions. Furthermore, in the natural environments, growth stages of plants are not uniform, prolonging the flowering time compared to that in greenhouses, which makes difficult the timing of BR application. Thus, field trials of these three BRs were suspended in Japan and European countries.

In 1985, Ikekawa launched collaborative studies between Japan and China on the practical application of 24-epibrassinolide in China. These trials revealed that application of 24-epibrassinolide increases yields of wheat, corn, tobacco, watermelon, cucumber, and grape (Ikekawa and Zhao 1991). In 1990, the practical use of 24-epibrassinolide on wheat and corn was tested in areas totaling more than 2664 ha. Although the total area for trials was rapidly expanding (Cutler 1994), further agricultural development of 24-epibrassinolide was suspended for various reasons including the relevant patents (Y. Kamuro, personal communication). China, instead, registered (22S, 23S)-28-homobrassinolide (trade name: BR-120) as a plant growth regulator for tobacco, sugar cane, rape seed, tea, and some fruits (see the chapter by Y. Kamuro and S. Takatsuto, this volume). In Russia and Belarus, state trials (1987-1991) were conducted to exhibit effects of 24-epibrassinolide on yield increases and fungal disease resistance in several crops. As a result, 24-epibrassinolide was registered officially in these countries as a plant growth regulator (trade name: EPIN) for tomato, cucumber, pepper, barley, potato, and apple (Khripach et al 1991, 1997). The safety of 24-epibrassinolide for agricultural use has been established in both Japan and Russia/Belarus (Ikekawa and Zhao 1991, Khripach et al 1997). Success in China and Russia/Belarus may be in part due to their stressful environmental conditions. It has also been stated that the effects of BRs may be more apparent in countries lacking good cultural practice (Cutler 1994).

In Japan, Kamuro and Takatsuto concluded that constant, favorable effects of brassinolide, epibrassinolide, and (22S, 23S)-28-homobrassinolide are difficult to realize in field conditions, mainly because of their extremely short duration of activity, and at the beginning of the 1990s they started trials to develop BR analogues with long-lasting activity. Finally, TS303 which is a dipropionate of homobrassinolide epoxide, was selected as the most promising compound and since 1966 has been subjected to official tests for registration in many countries. Because of synergistic effects between TS303 and PDJ, a jasmonate analogue, a formulation composed of these two chemicals (named TNZ303) is also being officially tested. Seed treatment of rice in Japan, Korea, and China and of wheat, potato and sugar beet in Russia and Poland increased the yields and tolerance against various stresses. Further profitable effects of these growth regulators are described in the chapter by Y. Kamuro and S. Takatsuto, this volume.

5. Recent Advances

5.1 Biosynthesis

A group of BRs such as brassinolide and castasterone are physiologically and biosynthetically most important because they have been detected in all species hitherto examined with the exception of a green alga, *Hydrodictyon reticulatum*. These BRs have the same skeleton as campesterol. From this fact as well as the structures of naturally occurring BRs, it has been hypothesized that the biosynthesis of brassinolide proceeds from campesterol through a pathway of teasterone → typhasterol → castasterone → brassinolide (Yokota et al 1991). Sakurai and his co-workers have investigated the production of BRs in several kinds of cultured cells and found that *Catharanthus roseus* crown gall cells produce brassinolide and castasterone (Park et al 1989). They extensively used *C. roseus* cells to determine the biosynthetic pathway of brassinolide (Fujioka and Sakurai 1997b, Yokota 1997) (see the chapter by A. Sakurai, this volume).

First, the above-mentioned biosynthetic pathway from teasterone to brassinolide was established by feeding labeled intermediates to *C. roseus* cells. In these experiments, 3-dehydroteasterone was newly identified as the intermediate in the conversion of teasterone to typhasterol. Furthermore, the earlier pathway before teasterone was determined to proceed via a pathway of campesterol → campestanol → 6α-hydroxycampestanol → 6-oxocampestanol → cathasterone → teasterone, although the pathway from 6-oxocampestanol from cathasterone remained obscure (Fujioka et al 1995, Suzuki et al 1995) In the course of the analysis of the biosynthetic lesion in the *det2* mutant of Arabidopsis, the pathway from campesterol to campestanol was found to require two intermediates, 3-dehydro-Δ^4-campesterol and 3-dehydrocampestanol (Fujioka et al 1997). The pathway thus established is termed the early C6-oxidation pathway because C6 oxidation occurs prior to hydroxylation of the side chain.

The late C6-oxidation pathway, an alternative one for generating castasterone from campestanol has also been revealed. 6-Deoxocastasterone, an intermediate in this pathway, is usually contained in plant tissues at the highest level. However, 6-deoxoBRs have been considered to be end-pathway products because these BRs elicit very low biological activity when examined by the rice lamina inclination assay (Yokota et al 1987). Griffiths et al (1995) indicated that 3-dehydro-6-deoxoteasterone, 6-deoxotyphasterol, and 6-deoxocastasterone occur in large amounts in the pollen of *Cupressus arizonica* and hypothesized that these 6-deoxoBRs could be precursors of 6-oxoBRs. Their hypothesis turned out to be correct because 6-deoxocastasterone fed to *C. roseus* cells was converted to castasterone and brassinolide. Further study postulated the late C6-oxidation pathway of campestanol → 6-deoxocathasterone → 6-deoxoteasterone → 3-dehydro-6-deoxoteasterone → 6-deoxotyphasterol → 6-deoxocastasterone → castasterone (Choi et al 1997). However, at present, this pathway is incomplete in that no evidence is available for the conversion of campestanol to 6-deoxoteasterone via 6-deoxocathasterone.

Likely pathways cross-linking the late and early C6-oxidation pathways have not yet been recognized. Further, nothing is known about the biosynthesis of other BRs from sitosterol, cholesterol, 24-methylenecholesterol, 24-methylene-25-methylcholesterol, isofucosterol, and 24-epicampesterol. These questions will be quite interesting subjects in future. Metabolism of BRs are not referred to in this chapter but are described by G. Adam and B. Schneider in this volume (Adam et al 1996, Fujioka and Sakurai 1997b).

5.2 Mutants of BR Biosynthesis and Signal Transduction

Because BRs elicit a broad range of biological activities, it has long been difficult to determine their basic functions in plant growth. The clear answer to this question was obtained through the discovery of various mutants since 1996 that cannot synthesize BRs or are not sensitive to BRs (Clouse 1996, 1997, Yokota 1997) (see the chapter by S.D. Clouse and K.A. Feldmann, this volume). There seems to have been three approaches leading to the discoveries of these mutants: 1) studies of Arabidopsis mutants that show de-etiolation in the dark, 2) attempts to isolate BR-insensitive mutants of Arabidopsis whose root growth was not inhibited by BRs, and 3) attempts to select pea mutants of BR biosynthesis and/or sensitivity from known dwarf pea mutants whose dwarfism had not been understood.

The Arabidopsis mutants *det2* and *cpd*, with impaired scotomorphogenesis, are cabbage-like dwarfs in the light. Their phenotypes had been known not to be recovered to wildtype by "classical" plant hormones but were found to be rescued upon treatment with BRs (Kauschmann et al 1996, Li et al 1996, Szekeres et al 1996). These dwarfs were determined to have a lesion in the biosynthesis of brassinolide on the basis of further rescue experiments using biosynthetic precursors of brassinolide as well as functions of *DET2* and *CPD* genes deduced from their structural information. The other BR biosynthesis mutants include the *dim* mutant (Klahre et al 1997) and the *dwf4* mutant (Azpiroz et al 1998, Choe et al 1998). Analysis of the phenotypes of these BR-deficient mutants revealed that BRs are essential steroidal hormones for normal growth and have some role in light-regulated development, either directly or indirectly.

Clouse and his coworkers attempted to identify BR-insensitive mutants of Arabidopsis, resulting in the isolation of the dwarf mutant named *bri1* (Clouse et al 1993, 1996). It was found that the BR-biosynthesis mutants show a phenotype similar to that of the *bri1* mutant. Surprisingly, numerous alleles of *bri1* with an identical phenotype have been identified in independent screens (Kauschmann et al 1996, Li and Chory 1997) (see also the chapter by S.D. Clouse and K.A. Feldmann in this volume). BRI1 has been found to have strong homology to a leucine-rich receptor kinase, although its role as the BR receptor has not yet been confirmed by direct binding studies. However, as discussed in the chapter by S.D. Clouse and K.A. Feldmann, this volume, BRI1 is obviously an important component of the BR signal transduction pathway. Putative involvement of this gene in pathogenic signaling has been discussed (Koncz 1998). It is interesting to note that this idea is in good accord with the facts that BR-treated plants gain resistance to various environmental stresses.

Yokota and co-workers attempted to find BR mutants from known dwarf pea mutants and selected *lka, lkb,* and *lk* mutants as the candidates because these plants have been known not to be responsive to exogenous gibberellins but have a normal status of endogenous gibberellins (Yokota et al 1996, 1997). Analysis of the levels of endogenous BRs and rescue experiments indicated that *lkb* and *lk* mutants have a defect in BR biosynthesis (Nomura et al 1997, Yokota et al 1997). The functions of *LKB* and *LK* were found to be comparable to those of *DIM* and *DET2*, respectively (Yokota et al 1997). The *lka* mutant was not responsive to exogenous BRs and accumulated castasterone, a direct precursor of brassinolide, suggesting that this mutant has a defect in BR-perception/signal transduction (Nomura et al 1997).

Recent analysis of a range of BR biosynthesis and insensitivity mutants of Arabidopsis and pea has been decisive in uncovering the physiological importance of BRs. Although we have a limited knowledge on the mode of action of BRs, we have at least obtained a clear conclusion that BRs function to elongate cells. Mutants will become increasingly important in clarifying the regulation of biosynthesis and action of BRs. At present, there are still some interesting tomato mutants that are likely to be BR-deficient (*d* and *dpy*) or BR-insensitive (*cu-3*) mutants (Bishop et al 1996) (see the chapter by S.D. Clouse and K.A. Feldmann, this volume).

6. Concluding Remarks

From the beginning to the present in the history of BRs, we have seen a number of topics in which a broad range of researchers, plant physiologists, agronomists, natural products chemists, synthetic chemists, biochemists, molecular biologists, and geneticists were involved. In future, research into BRs will become increasingly important in understanding the plant growth. It may be anticipated that interdisciplinary collaborations will give rise to a deeper understanding of the mode of BR action.

References

Abe H, Marumo S (1991) Brassinosteroids in leaves of *Distylium racemosum* Sieb. et Zucc. In: Cutler HG, Yokota T, Adam G (Eds) Brassinosteroids: Chemistry, Bioactivity and Applications. ACS Symp Ser 474. Amer Chem Soc, Washington, DC, pp 18-24

Abe H, Morishita T, Uchiyama M et al (1982) Identification of brassinolide-like substances in Chinese cabbage. Agric Biol Chem 46: 2609-2611

Abe H, Morishita T, Uchiyama M et al (1983) Occurrence of three new brassinosteroids: brassinone, (24S)-24-ethylbrassinone and 28-norbrassinolide, in higher plants. Experientia 39: 351-353

Aburatani M, Takeuchi T, Mori K (1985) Structural revision of the acetal intermediates in brassinolide synthesis. Agric Biol Chem 49: 3557-3562

Adam G, Porzel A, Schmidt J et al (1996) New developments in brassinosteroid research.

Stud Nat Prod Chem 18: 495-549
Arima M, Yokota T, Takahashi N (1984) Identification and quantification of brassinolide-related sterols in the insect gall and healthy tissues of the chestnut plant. Phytochemistry 23: 1587-1591
Arteca RN, Bachman J (1987) Light inhibition of brassinosteroid-induced ethylene production. J Plant Physiol 129: 13-18
Arteca RN, Bachman JM, Mandava NB (1988) Effects of indole-3-acetic acid and brassinosteroid on ethylene biosynthesis in etiolated mung bean hypocotyl segments. J Plant Physiol 133: 430-435
Azpiroz R, Wu Y, LoCascio JC et al (1998) An Arabidopsis brassinosteroid-dependent mutant is blocked in cell elongation. Plant Cell 10: 219-230
Bemiller JN, Leung GLY, Yopp JE (1979) Synthesis and growth regulator activity of fatty acyl derivatives of d-glucose and d-galactose. Phytochemistry 18: 1283-1287
Bishop GJ, Harrison K, Jones JDG (1996) The tomato *Dwarf* gene isolated by heterologous transposon tagging encodes the first member of a new cytochrome P450 family. Plant Cell 8: 959-969
Braun P, Wild A (1984) The influence of brassinosteroid on growth and parameters of photosynthesis of wheat and mustard plants. J Plant Physiol 116: 189-196
Cerana R, Bonetti A, Marre MT et al (1983) Effects of a brassinosteroid on growth and electrogenic proton extrusion in Azuki bean epicotyls. Physiol Plant 59: 23-27
Choe S, Dilkes BP, Fujioka S et al (1998) The *DWF4* gene of Arabidopsis encodes a cytochrome P450 that mediates multiple 22a-hydroxylation steps in brassinosteroid biosynthesis. Plant Cell 10: 231-243
Choi YH, Fujioka S, Nomura T et al (1997) An alternative brassinolide biosynthetic pathway via late C-6 oxidation. Phytochemistry 44: 609-613
Clouse SD (1996) Molecular genetic studies confirm the role of brassinosteroids in plant growth and development. Plant J 10: 1-8
Clouse SD (1997) Molecular genetic analysis of brassinosteroid action. Physiol Plant 100: 702-709
Clouse SD, Zurek DM, McMorris TC et al (1992) Effect of brassinolide on gene expression in elongating soybean epicotyls. Plant Physiol 100: 1377-1383
Clouse SD, Hall AF, Langford M et al (1993) Physiological and molecular effects of brassinosteroids on *Arabidopsis thaliana*. J Plant Growth Regul 12: 61-66
Clouse SD, Langford M, McMorris TC (1996) A brassinosteroid-insensitive mutant in *Arabidopsis thaliana* exhibits multiple defects in growth and development. Plant Physiol 111: 671-678
Cutler HG (1991) Brassinosteroid through the looking glass: an appraisal. In: Cutler HG, Yokota T, Adam G (Eds) Brassinosteroids: Chemistry, Bioactivity and Applications. ACS Symp Ser 474. Amer Chem Soc, Washington, DC, pp 334-345
Cutler HG (1994) Advances in the use of brassinosteroids. In: Hedin PA, Menn JJ, Hollingworth RM (Eds) Natural and Engineered Pest Management Agents. ACS Symp Ser 551. Amer Chem Soc, Washington, DC, pp 85-102
Fujioka S, Sakurai A (1997a) Brassinosteroids. Nat Prod Rep 14: 1-10
Fujioka S, Sakurai A (1997b) Biosynthesis and metabolism of brassinosteroids. Physiol Plant 100: 710-715
Fujioka S, Inoue T, Takatsuto S et al (1995) Identification of a new brassinosteroid, cathasterone, in cultured cells of *Catharanthus roseus* as a biosynthetic precursor of teasterone. Biosci Biotech Biochem 59: 1543-1547

Fujioka S, Li J, Choi YH et al (1997) The Arabidopsis *deetiolated2* mutant is blocked early in brassinosteroid biosynthesis. Plant Cell 9: 1951-1962

Fujita F (1985) Anticipation for agricultural application of brassinolide. Kagaku-to-seibutu 23: 717-725 (in Japanese)

Fung S, Siddall JB (1980) Stereoselective synthesis of brassinolide: A plant growth promoting steroidal lactone. J Am Chem Soc 102: 6580-6581

Gamoh K (1994) Liquid chromatographic assay of brassinosteroids in plants. J Chromatogr 658: 17-25

Geuns JMC (1978) Steroid hormones and plant growth and development. Phytochemistry 17: 1-14

Gregory LS (1981) Acceleration of plant growth through seed treatment with brassins. Amer J Bot 68: 586-588

Griffiths PG, Sasse JM, Yokota T et al (1995) 6-Deoxotyphasterol and 3-dehydro-6-deoxoteasterone, possible precursors to brassinosteroids in the pollen of *Cupressus arizonica*. Biosci Biotech Biochem 59: 956-959

Grove MD, Spencer GF, Pfeffer PE et al (1978) 6-D-Glucopyranosyl fatty acid esters from *Brassica napus* pollen. Phytochemistry 17: 1187-1189

Grove MD, Spencer GF, Rohwedder WK et al (1979) Brassinolide, a plant growth-promoting steroid isolated from *Brassica napus* pollen. Nature 281: 216-217

Hamada K (1985) Brassinolide in crop cultivation. In: McGregor P (Ed) Plant Growth Regulators in Agriculture. FFTC BOOK Ser No. 34, Taipei, pp 188-197

Heftmann E (1971) Functions of sterols in plants. Lipids 6: 128-133

Heftmann E (1975) Functions of steroids in plants. Phytochemistry 14: 891-901

Ikeda M, Sassa T, Miura Y (1977) A new auxin-like active substance from chestnut crown gall. Bull Yamagata Univ Agric Sci 7: 463-466

Ikeda M, Takatsuto S, Sassa T et al (1983) Identification of brassinolide and its analogues in chestnut gall tissue. Agric Biol Chem 47: 655-657

Ikekawa T, Zhao YJ (1991) Application of 24-epibrassinolide in agriculture. In: Cutler HG, Yokota T, Adam G (Eds) Brassinosteroids: Chemistry, Bioactivity and Applications. ACS Symp Ser 474. Amer Chem Soc, Washington, DC, pp 280-291

Ikekawa N, Takatsuto S, Kitsuwa T et al (1984) Analysis of natural brassinosteroids by gas chromatography and gas chromatography-mass spectrometry. J Chromatogr 290: 289-302

Ishiguro M, Takatsuto S, Morisaki M et al (1980) Synthesis of brassinolide, a steroidal lactone with plant-growth promoting activity. J Chem Soc Chem Comm 1980: 962-964

Iwasaki T and Shibaoka H (1991) Brassinosteroids act as regulators of tracheary-element differentiation in isolated *Zinnia* mesophyll cells. Plant Cell Physiol 32: 1007-1014

Kamuro Y, Inada K (1991) The effect of brassinolide on the light-induced growth inhibition in mung bean epicotyl. Plant Growth Regul 10: 37-43

Katsumi M (1991) Physiological mode of brassinolide action in cucumber hypocotyl growth. In: Cutler HG, Yokota T, Adam G (Eds) Brassinosteroids: Chemistry, Bioactivity and Applications. ACS Symp Ser 474. Amer Chem Soc, Washington, DC, pp 246-254

Kauschmann A, Jessop A, Koncz C et al (1996) Genetic evidence for an essential role of brassinosteroids in plant development. Plant J 9: 701-713

Khripach VA, Zhabinskii VN, Litvinovskaya RP (1991) Synthesis and some practical aspects of brassinosteroids. In: Cutler HG, Yokota T, Adam G (Eds) Brassinosteroids: Chemistry, Bioactivity and Applications. ACS Symp Ser 474. Amer Chem Soc, Washington, DC, pp 43-55

Khripach VA, Zhabinskii VN, Malevannaya NN (1997) Recent advances in brassinosteroids study and application. Proc Plant Growth Reg Soc Amer 24: 101-106

Kim SK (1991) Natural occurrences of brassinosteroids. In: Cutler HG, Yokota T, Adam G (Eds) Brassinosteroids: Chemistry, Bioactivity and Applications. ACS Symp Ser 474. Amer Chem Soc, Washington, DC, pp 26-35

Kim SK, Yokota T, Takahashi N (1987) 25-Methyldolichosterone, a new brassinosteroid with a tertiary butyl group from immature seed of *Phaseolus vulgaris*. Agric Biol Chem 51: 2303-2305

Klahre U, Fujioka S, Yokota T et al (1997) Characterization of the *diminuto* mutant and gene regulated by brassinosteroids. Proc Plant Growth Reg Soc Amer 24: 99

Koncz C (1998) Crosstalk between brassinosteroids and pathogenic signaling? Trends Plant Sci 3: 1-2

Krizek DT, Worley JF (1973) The influence of light intensity on the internodal response of intact bean plants to brassins. Bot Gaz 134: 147-150

Krizek DT, Mandava NB (1983a) Influence of spectral quality on the growth response of intact bean plants to brassinosteroid, a growth-promoting steroidal lactone. I. Stem elongation and morphogenesis. Physiol Plant 57: 317-323

Krizek DT, Mandava NB (1983b) Influence of spectral quality on the growth response of intact bean plants to brassinosteroid, a growth-promoting steroidal lactone. II. Chlorophyll content and partitioning of assimilate. Physiol Plant 57: 324-329

Li J, Chory J (1997) A putative leucine-rich repeat receptor kinase involved in brassinosteroid signal transduction. Cell 90: 929-938

Li J, Nagpal P, Vitart V et al (1996) A role for brassinosteroids in light-dependent development of *Arabidopsis*. Science 272: 398-401

Maeda E (1965) Rate of lamina inclination in excised rice leaves. Physiol Plant 18: 813-827

Mandava NB (1988) Plant growth-promoting brassinosteroids. Annu Rev Plant Physiol Plant Mol Biol 39: 23-52

Mandava N, Mitchell JW (1971) New plant hormones: chemical and biological investigations. Indian Agric 15: 19-31

Mandava N, Mitchell JW (1972) Structural elucidation of brassins. Chem Ind 1972: 930-931

Mandava N, Kozempel M, Worley JF et al (1978) Isolation of brassins by extraction of rape (*Brassica napus* L.) pollen. Ind Eng Chem Prod Res Dev 17: 351-354

Marumo S, Wada K (1981) Brassinolide, a new plant growth regulator. Kagaku-to-Seibutu 16: 1-10 (in Japanese)

Marumo S (1983) A new plant growth substance, brassinolide. In: The Agricultural Chemical Society of Japan (Ed) Bioactivity and Bioactive Substances in Organisms. Asakura, Tokyo, pp 102-134 (in Japanese)

Marumo S (1987) The advance of brassinosteroid researches in Japan. Proc Plant Growth Reg Soc Amer 14: 174-185

Marumo S, Hattori H, Abe H et al (1968) The presence of novel plant growth regulators in leaves of *Distylium racemosum* Sieb et Zucc. Agric Biol Chem 32: 528-529

Maugh II TH (1981) New chemicals promise larger crops. Science 212: 33-34

Milborrow BV, Pryce RJ (1973) The brassins. Nature 243: 46

Mitchell JW, Whitehead MR (1941) Responses of vegetative parts of plants following application of extract of pollen from *Zea mays*. Bot Gaz 102: 770-791

Mitchell JW, Gregory LE (1972) Enhancement of overall plant growth, a new response to brassins. Nature New Biol 253: 253-254

Mitchell JW, Mandava N, Worley JF et al (1970) Brassins - a New Family of plant hormones from rape pollen. Nature 225: 1065-1066

Mori K (1980) Synthesis of a brassinolide analog with high plant growth promoting activity. Agric Biol Chem 44: 1211-1212

Mori K, Sakakibara M, Okada K (1984) Synthesis of naturally occurring brassinosteroids employing cleavage of 23,24-epoxides as key reactions: Synthesis of brassinolide, castasterone, dolicholide, dolichosterone, homodolicholide, homodolichosterone, 6-deoxocastasterone and 6-deoxodolichosterone. Tetrahedron 40: 1767-1781

Morishita T, Abe H, Uchiyama M et al (1983) Evidence for plant growth promoting brassinosteroids in leaves of *Thea sinensis*. Phytochemistry 22: 1051-1053

Munakata K, Kato N, Ikeda M (1974) New auxin substance from corn germ oil. In: Sumiki Y (Ed) Plant Growth Substance 1973. Hirokawa Publishing, pp 39-43

Nomura T, Nakayama M, Reid JB et al (1997) Blockage of brassinosteroid biosynthesis and sensitivity causes dwarfism in *Pisum sativum*. Plant Physiol 113: 31-37

Park KH, Saimoto H, Nakagawa S et al (1989) Occurrence of brassinolide and castasterone in crown gall cells of *Catharanthus roseus*. Agric Biol Chem 53: 805-811

Romani G, Marre MT, Bonetti A et al (1983) Effects of a brassinosteroid on growth and electrogenic proton extrusion in maize root segments. Physiol Plant 59: 528-532

Sakurai A, Fujioka S (1993) The current status of physiology and biochemistry of brassinosteroids. Plant Growth Regul 13: 147-159

Sasse JM (1990) Brassinolide-induced elongation and auxin. Physiol Plant 80: 401-408

Sasse JM (1997) Recent progress in brassinosteroid research. Physiol Plant 100: 696-701

Schmidt J, Spengler B, Yokota T et al (1995a) Secasterone, the first naturally occurring 2,3-epoxybrassinosteroid from *Secale cereale*. Phytochemistry 38: 1095-1997.

Schmidt J, Voigt B, Adam G (1995b) 2-Deoxybrassinolide—a naturally occurring brassinosteroid from *Apium graveolens*. Phytochemistry 40: 1041-1043

Schneider JA, Yoshida K, Nakanishi K et al (1983) Typhasterol (2-deoxycastasterone): a new plant growth regulator from cat-tail pollen. Tetrahedron Lett 24: 3859-3860

Spengler B, Schmidt J, Voigt B et al (1995) 6-Deoxo-28-norcastasterone and 6-deoxo-24-epicastasterone-two new brassinosteroids from *Ornithopus sativus*. Phytochemistry 40: 907-910

Steffens GL (1991) U.S. Department of Agriculture Brassins Project: 1970-1980. In: Cutler HG, Yokota T, Adam G (Eds) Brassinosteroids: Chemistry, Bioactivity and Applications. ACS Symp Ser 474. Amer Chem Soc, Washington, DC, pp 2-17

Suzuki H, Inoue T, Fujioka S et al (1995) Conversion of 24-methylcholesterol to 6-oxo-24-methylcholestanol, a putative intermediate of the biosynthesis of brassinosteroids, in cultured cells of *Catharanthus roseus*. Phytochemistry 40: 1391-1397

Szekeres M, Nemeth K, Kalman ZK et al (1996) Brassinosteroids rescue the deficiency of CYP90, a cytochrome P450, controlling cell elongation and de-etiolation in Arabidopsis. Cell 85: 171-182

Takatsuto S, Ikekawa N (1984) Short-step synthesis of plant growth-promoting brassinosteroids. Chem Pharm Bull 32: 2001-2004

Takatsuto S, Ying B, Morisaki M (1982) Microanalysis of brassinolide and its analogues by gas chromatography and gas chromatography-mass spectrometry. J Chromatogr 239: 233-241

Takematsu T, Takeuchi Y, Koguchi M (1985) New plant growth regulators, brassinolide analogs: their biological effects and application to agriculture and bioproduction. Shokucho 18: 2-15 (in Japanese)

Thompson MJ, Mandava N, Anderson JLF et al (1979) Synthesis of brassino steroids: new plant-growth-promoting steroids. J Org Chem 44: 5002-5004

Thompson MJ, Mandava NB, Meudt WJ et al (1981) Synthesis and biological activity of brassinolide and its 22β,23β-isomer: novel plant growth-promoting steroids. Steroids 38: 567-580

Wada K, Marumo S (1981) Synthesis and plant growth-promoting activity of brassinolide analogues. Agric Biol Chem 45: 2579-2585

Wada K, Kondo H, Marumo S (1985) A simple bioassay for brassinosteroids: A wheat leaf-unrolling test. Agric Biol Chem 49: 2249-2251

Worley JF, Mitchell JW (1971) Growth responses induced by brassins (fatty plant hormones) in bean plants. J Amer Soc Hort Sci 96: 270-273

Yamamoto R, Demura T, Fukuda H (1997) Brassinosteroids induce entry into the final stage of tracheary element differentiation in cultured *Zinnia* cells. Plant Cell Physiol 38: 980-983

Yokota T (1994) Brassinosteroids. In: Takahashi N, Masuda Y (Eds) Plant Hormone Handbook. Baifukan, Tokyo, pp 203-240 (in Japanese)

Yokota T (1997) The structure, biosynthesis and function of brassinosteroids. Trends Plant Sci 2: 137-143

Yokota T, Mori K (1992) Molecular structure and biological activity of brassinolide and related brassinosteroids. In: Duax WL, Bohl M (Eds) Molecular Structure and Biological Activity of Steroids. CRC Press, Boca Raton, pp 317-340

Yokota T, Okabayashi M, Takahashi N et al (1974) Plant growth regulators in chestnut gall tissue and wasps. In: Sumiki Y (Ed) Plant Growth Substances 1973. Hirokawa Publishing, pp 28-38

Yokota T, Arima M, Takahashi N (1982a) Castasterone, a new phytosterol with plant-hormone potency, from chestnut insect gall. Tetrahedron Lett 23: 1275-1278

Yokota T, Baba J, Takahashi N (1982b) A new steroidal lactone with plant growth-regulatory activity from *Dolichos lablab* seed. Tetrahedron Lett 23: 4965-4966

Yokota T, Baba J, Takahashi N (1983) Brassinolide-related bioactive sterols in *Dolichos lablab*: brassinolide, castasterone and a new analog, homodolicholide. Agric Biol Chem 47: 1409-1411

Yokota T, Baba J, Koba S et al (1984) Purification and separation of eight steroidal plant-growth regulators from *Dolichos lablab* seed. Agric Biol Chem 48: 2529-2534

Yokota T, Koba S, Kim SK et al (1987) Diverse structural variations of the brassinosteroids in *Phaseolus vulgaris* seed. Agric Biol Chem 51: 1625-1631

Yokota T, Ogino Y, Takahashi N et al (1990) Brassinolide is biosynthesized from castasterone in *Catharanthus roseus* crown gall cells. Agric Biol Chem 54: 1107-1108

Yokota T, Ogino Y, Suzuki H et al (1991) Metabolism and biosynthesis of brassinosteroids. In: Cutler HG, Yokota T, Adam G (Eds) Brassinosteroids: Chemistry, Bioactivity and Applications. ACS Symp Ser 474. Amer Chem Soc, Washington, DC, pp 86-96

Yokota T, Nomura T, Nakayama M et al (1996) Dwarf pea *lka* and *lkb*, possible brassinosteroid biosynthesis and sensitivity mutants. Proc Plant Growth Reg Soc Amer 23: 10

Yokota T, Nomura T, Kitasaka S et al (1997) Biosynthesis lesions in brassinosteroid-deficient pea mutants. Proc Plant Growth Reg Soc Amer 24: 94

Yopp JH, Colclasure GC, Mandava N (1979) Effects of brassin-complex on auxin and gibberellin mediated events in the morphogenesis of the etiolated bean hypocotyl. Physiol Plant 46: 247-254

Yopp JH, Mandava NB, Sasse JM (1981) Brassinolide, a growth-promoting steroidal lactone. I. Activity in selected auxin bioassays. Physiol Plant 53: 445-452

Zurek DM, Rayle DL, McMorris TC et al (1994) Investigation of gene expression, growth kinetics, and wall extensibility during brassinosteroid-regulated stem elongation. Plant Physiol 104: 505-513

2
Natural Occurrence of Brassinosteroids in the Plant Kingdom

SHOZO FUJIOKA

Plant Functions Lab
The Institute of Physical and Chemical Research (RIKEN)
Hirosawa 2-1, Wako-shi, Saitama 351-0198, Japan

1. Introduction

Prior to 1970 Mitchell and coworkers began screening pollen in search of new plant hormones. They screened nearly 60 species of plants and found about one-half caused growth promotion in the bean second internode bioassay. In 1970 they reported a new, naturally occurring group of plant growth-promoting substances, termed brassins, from rape (*Brassica napus* L.) pollen (Mitchell et al 1970). In an effort to isolate the active constitutents, 227 kg of bee-collected rape pollen was extracted and purified. Through extensive efforts an active compound was finally isolated, and the structure of the active component, termed brassinolide (BL), was established to be (22R,23R,24S)-2α,3α,22,23-tetrahydroxy-24-methyl-B-homo-7-oxa-5α-cholestan-6-one by ^{1}H-NMR (nuclear magnetic resonance), mass spectrometry (MS), and X-ray crystallography (Grove et al 1979). BL is the first brassinosteroid (BR) isolated in nature. It has a steroidal skeleton of 5α-cholestane, and possesses a 7-oxalactonic B-ring and two vicinal hydroxyls at the A-ring (C-2α and C-3α) and in the side chain (C-22R and C-23R). In 1982, Yokota et al (1982a) isolated the second naturally occurring BR termed castasterone (CS), from the insect galls of chestnut (*Castanea crenata*) using the rice-lamina inclination assay for monitoring the purification steps. Its structure was determined to be (22R,23R,24S)-2α,3α,22,23-tetrahydroxy-24-methyl-5α-cholestan-6-one by ^{1}H-NMR and MS. The detailed history of brassinosteroid research is described in this volume (see the chapter by Yokota, this volume).

Since the discovery of BL and CS, extensive studies on isolation and identifica-

Key words: brassinolide (BL), castasterone (CS), typhasterol (TY), 3-dehydroteasterone (3DT), teasterone (TE), cathasterone (CT), 6-deoxocastasterone (6-DeoxoCS), BR conjugates, pollen, seeds, shoots, monocots, dicots, gymnosperms, lower plants

tion of BRs from various plant sources have been undertaken. Up to now, 40 free BRs and 4 BR conjugates have been characterized (Figs. 1, 2). BRs can be classified as either C_{27}-BRs, C_{28}-BRs, or C_{29}-BRs according to the number of carbons in the structure (Fig. 1). The bioassay used in the isolation of BL was the bean second-internode assay (Mitchell et al 1970). Later, the rice-lamina inclination assay, which is so sensitive as to detect 0.1 ng/ml BL, has been widely used for the purification of BRs (Wada et al 1981). During earlier studies, some BRs such as BL, CS, dolicholide (DL), dolichosterone (DS), typhasterol (TY), 28-homodolicholide (28-homoDL), and 28-homodolichosterone (28-homoDS) were isolated in pure states, and their structures were rigorously elucidated by ^{1}H-NMR and MS. The other BRs were characterized by gas chromatography(GC)-MS or GC-SIM(selected ion monitoring) without isolation. In these cases synthetic BRs have been used as standard samples. Microanalytical techniques, consisting of methaneboronation of the vicinal hydroxyls, have been developed for the GC-MS analysis of BRs (see the chapter by Takatsuto and Yokota, this volume). A combination of bioassay and GC-MS analysis has greatly contributed to the identification of a number of naturally occurring BRs. This chapter summarizes research on BRs with respect to the structural characteristics of naturally occurring BRs and the distribution of BRs in the plant kingdom. A number of excellent reviews on these subjects are available (Yokota and Takahashi 1986, Adam and Marquardt 1986, Meudt 1987, Mandava 1988, Kim 1991, Marquardt and Adam 1991, Sakurai and Fujioka 1993, Takatsuto 1994, Fujioka and Sakurai 1997a).

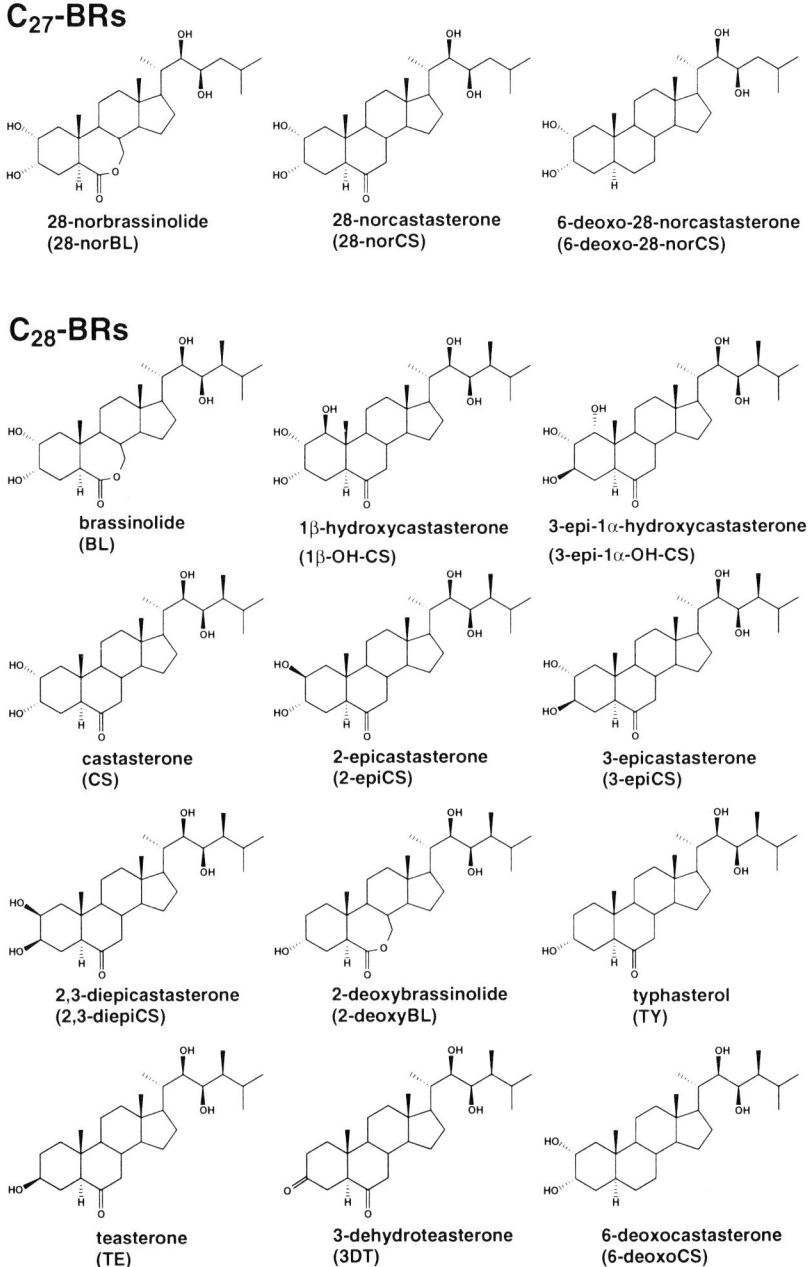

Fig. 1. Structures of free brassinosteroids.

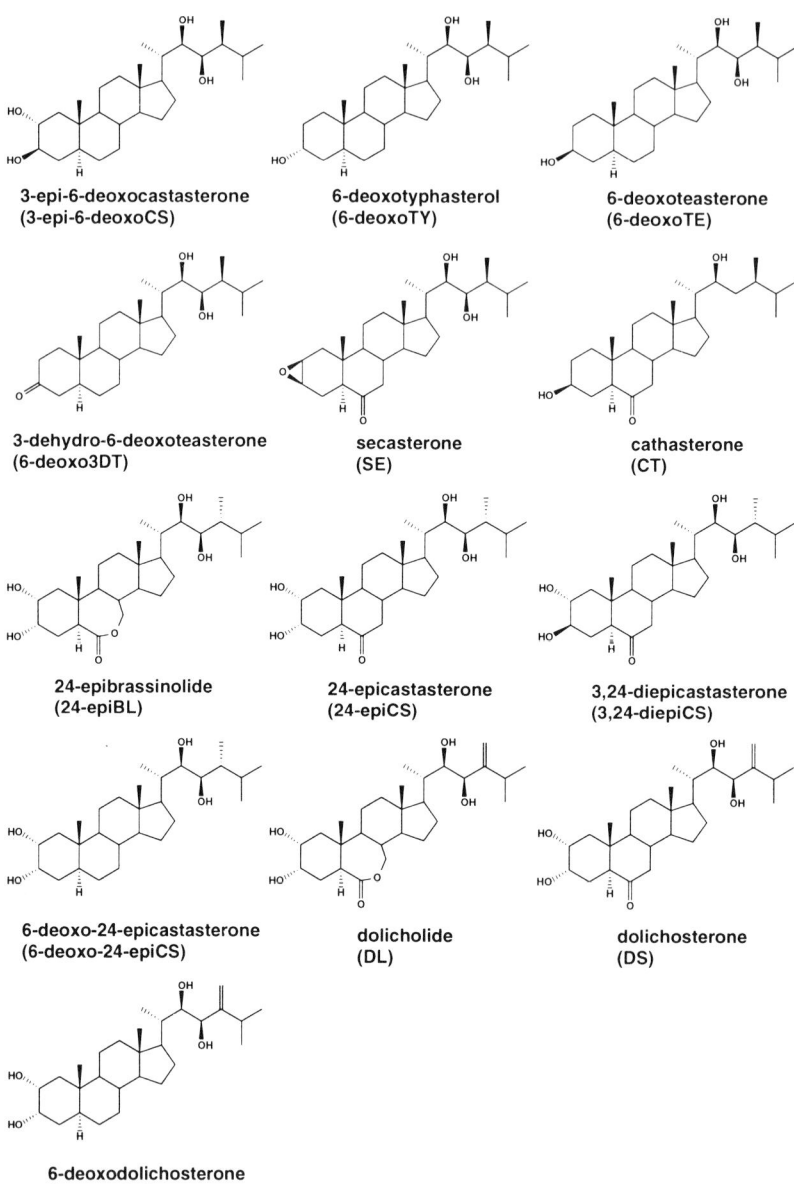

Fig. 1. *Continued.*

C$_{29}$-BRs

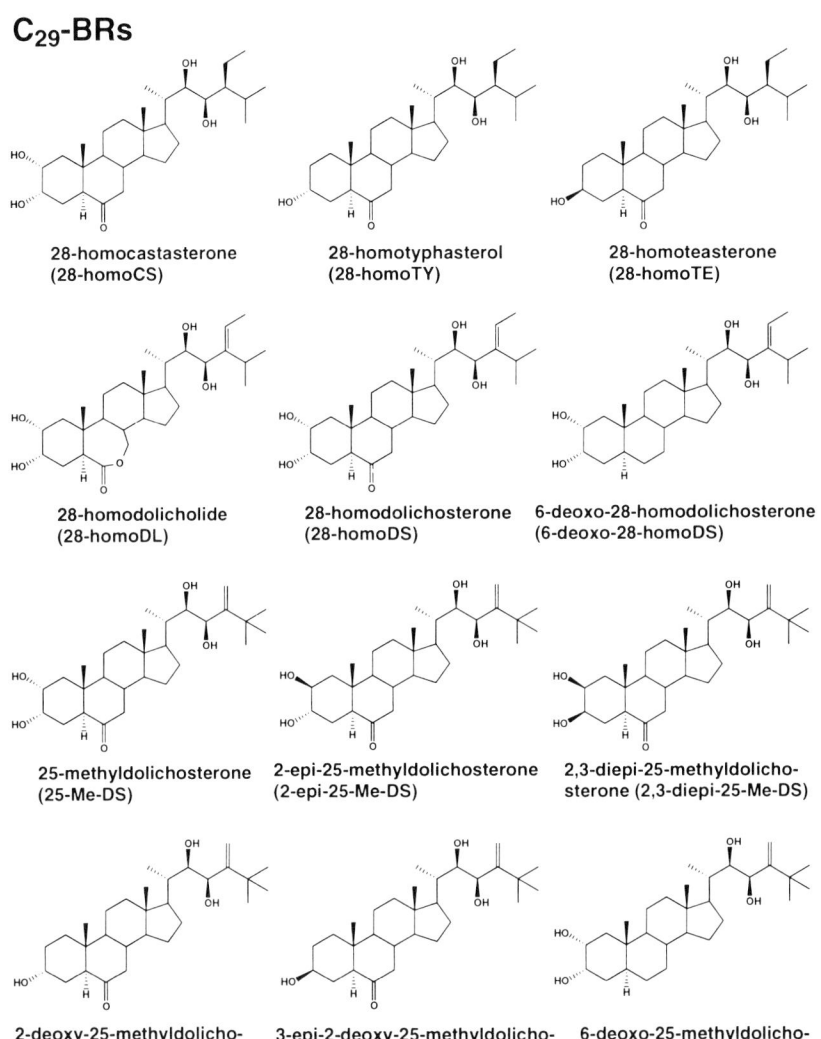

Fig. 1. *Continued.*

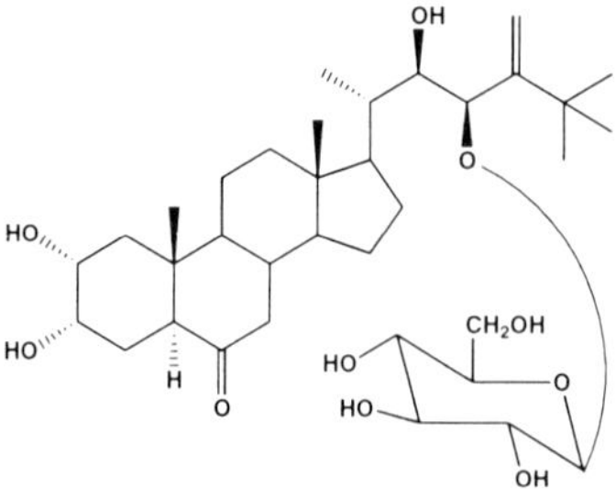

23-O-β-D-glucopyranosyl-25-methyl-dolichosterone (25-Me-DS-Glu)

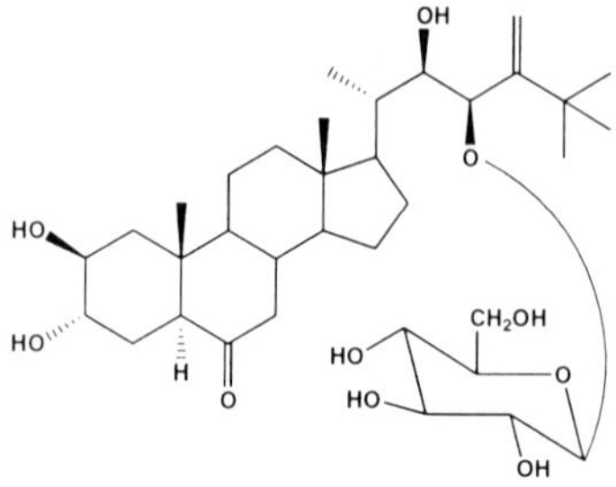

23-O-β-D-glucopyranosyl-2-epi-25methyl-dolichosterone (2-epi-25-Me-DS-Glu)

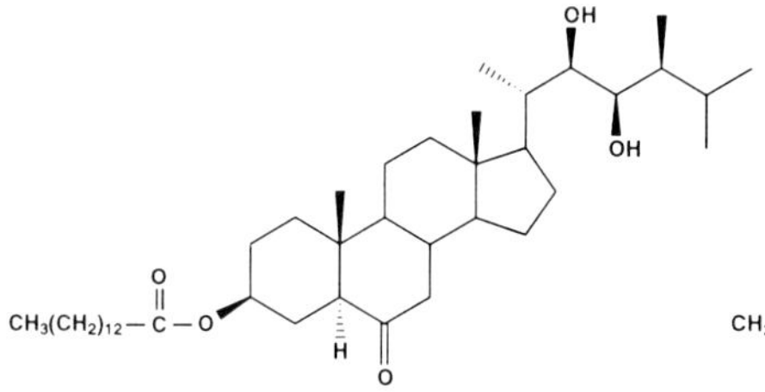

teasterone-3-myristate (TE-3-My)

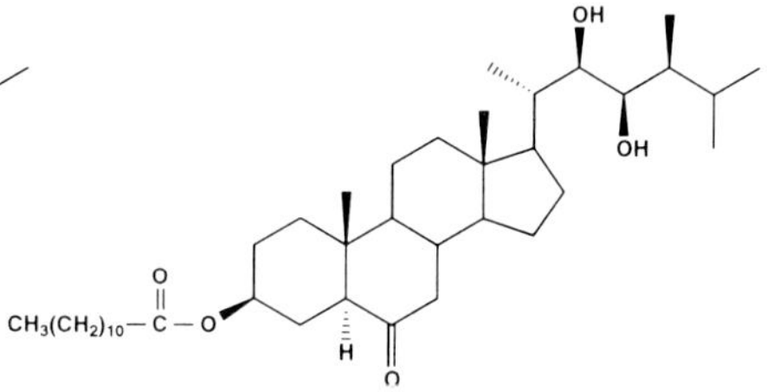

teasterone-3-laurate (TE-3-La)

Fig. 2. Structures of brassinosteroid conjugates.

2. Structural Characteristics of Naturally Occurring Brassinosteroids

Natural BRs so far identified have a common 5α-cholestane skeleton, and their structural variations come from the kinds and orientation of functionalities on the skeleton (Fig. 3).

With respect to the A-ring, there are several variations that come from the number and kinds of functional groups (mainly hydroxyl groups), position, and orientation (Fig. 3). Generally, BRs having vicinal diol hydroxyl groups (at C-2α and C-3α) have strong biological activity (e.g., BL and CS). Exceptionally from the seeds of *Phaseolus vurgaris*, all types of combinations of hydroxyls at C-2 and C-3 (2α,3α; 2α,3β; 2β,3α; 2β,3β) were isolated. There are some BRs having only one hydroxyl in the A-ring (at C-3α or C-3β), such as TY and teasterone (TE). Furthermore, there are two BRs having a 3-keto group such as 3-dehydroteasterone (3DT) and 3-dehydro-6-deoxoteasterone (6-deoxo3DT), one BR having a 2β,3β-epoxy group such as secasterone (SE), and two BRs having three hydroxyl groups in the A-ring.

With respect to the B-ring, 7-oxalactone (7-oxa-6-one), 6-oxo (6-ketone), and 6-deoxo (non-oxidized) types are known (Fig. 3). Generally, 7-oxalactone types show

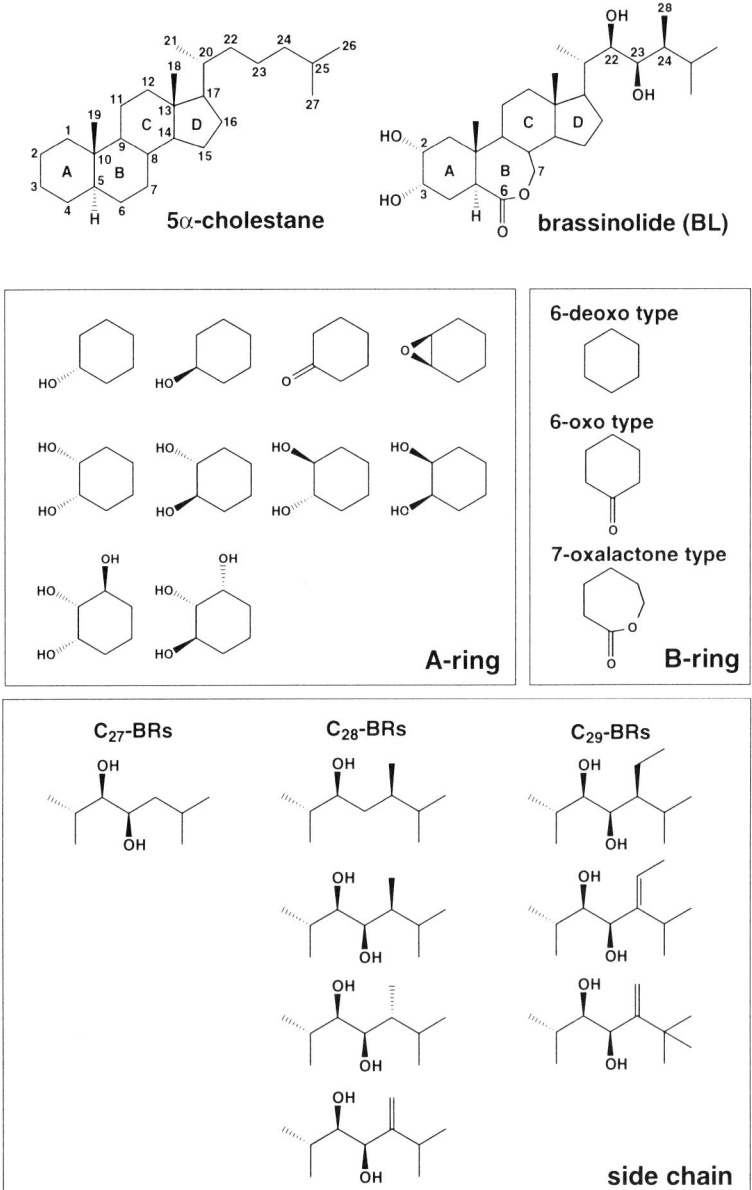

Fig. 3. Structural variations in the A-ring, B-ring, and side chain of naturally occurring brassinosteroids.

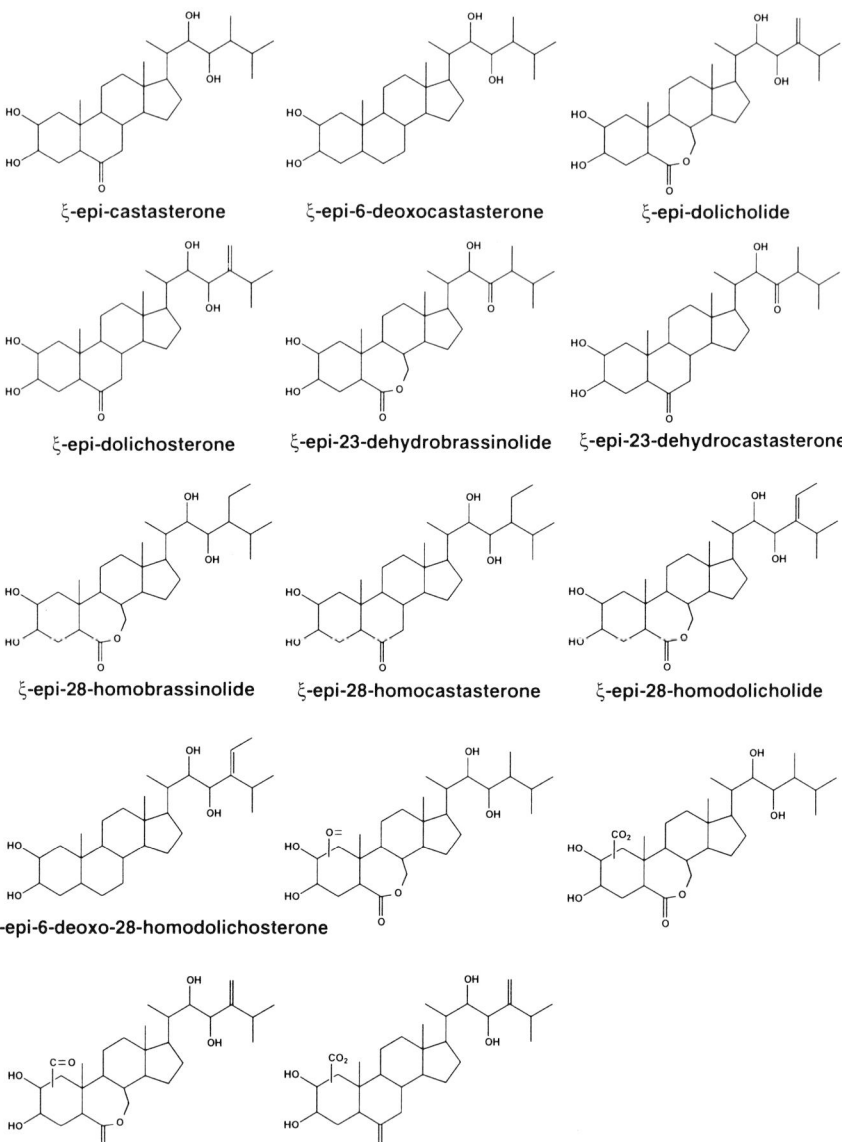

Fig. 4. Possible structures of unknown brassinosteroids.

stronger biological activity than 6-oxo types, and 6-deoxo types have the least biological activity. Among 40 naturally occurring BRs, 6 BRs are 7-oxalactone types, 24 BRs are 6-oxo types, and 10 BRs are 6-deoxo types.

With respect to the side chain skeletons, natural BRs are grouped into seven categories, which reflect those of the parent plant sterols (Fig. 3). Eighteen BRs having a 24α-methyl were found in 42 plant species among 44 plant species so far examined. Exceptions are an alga (*Hydrodictyon reticulatum*) and a dicot (*Gypsophila perfoliata*), which contain only 24β-methyl BR and/or 28-homocastasterone (28-homoCS). Distribution of other BRs is rather limited that have either a 24β-methyl (4 BRs, 8 species), 24-methylene (3 BRs, 3 species), 24α-ethyl (3 BRs, 7 species), 24-ethylidene (3 BRs, 2 species), 24-methylene-25-methyl (6 BRs, 1 species), or no substituent at C-24 (3 BRs, 13 species) (Fig. 1). Generally the most abundant plant sterol in higher plants is sitosterol, which has a 24α-ethyl and accounts for ca. 50-80% of total plant sterol. However, only 7 species among 44 plant species hitherto examined have been known to contain BRs with 24α-ethyl, and the levels in the tissue are low compared to BRs with a 24α-methyl. Almost all BRs so far isolated have vicinal diol functionality at C-22 and C-23, and their orientations are both *R*. An exception is cathasterone (CT), a BR having only one hydroxyl at C-22α, which has been found from cultured cells of *Catharanthus roseus*.

In addtion to known BRs, structural characteristics of unknown BRs have been partially elucidated by GC-MS. Possible structures of unknown BRs are summarized in Fig. 4. Some of them are presumably configurational isomers of known BRs, because their mass spectra are essentially identical, but the retention times [GC and high-performance liquid chromatography (HPLC)] are different from those of the parent compounds. The other unknown BRs might be oxidized during metabolism. The presence of 23-ketoBRs is suggested in *Phaseolus vulgaris* (Kim 1991) and *Crytomeria japonica* (Yokota et al 1998). The presence of BRs bearing carbonyl or BRs bearing an extra carbon atom (CO or COO) on the A ring is also suggested in *Phaseolus vulgaris* (Yokota et al 1987b), although their complete structures remain to be characterized (Fig. 4).

3. Distribution of Brassinosteroids in the Plant Kingdom

Since the discovery of BL, 40 BRs (Fig. 1) and 4 BR conjugates (Fig. 2) have been characterized from 44 plant species including 37 angiosperms (9 monocots and 28 dicots), 5 gymnosperms, 1 alga, and 1 pteridophyte. The distribution of BRs in each plant species is summarized in Table 1 (monocots), Table 2 (dicots), and Table 3 (gymnosperm, alga and pteridophyte).

Brassinosteroids have been found in a wide range of plant species, including higher plants and lower plants, and have been detected in various plant parts such as pollen, seeds, leaves, stems, roots, and flowers. Thus, BRs are widely distributed in the plant kingdom, and they are possibly biosynthesized in all parts of plant organs.

Generally, pollen contains remarkably high concentrations of BRs. Pollen of *Brassica napus*, sunflower (*Helianthus annuus*), and broad bean (*Vicia faba*) contains over 100 ng of BL per gram fresh weight. Pollen in Arizona cypress (*Cupressus arizonica*) contains 1 µg of CS per gram fresh weight, and pollen of corn (*Zea mays*) and *Vicia faba* also contain high concentrations. The highest concentration of BR has been found in *Cupressus arizonica* (6.4 µg of 6-deoxoTY per 1 g pollen). Immature seeds have also high contents of BRs. Compared to the pollen and immature seeds, the other parts of the plant normally contain BRs in only the nanogram or subnanogram levels of BRs per gram fresh weight.

Among BRs, CS is the most widely distributed (37 species) and then BL (25 species). The distribution of TY (20 species), teasterone (TE, 14 species), 6-deoxocastasterone (6-deoxoCS, 13 species), and 28-norcastasterone (28-norCS, 12 species) is relatively wide. The other BRs are distributed in a limited number of plant species. The BRs in more than two species are DS (6 species), 3DT (5 species), 28-homocastasterone (28-homoCS, 5 species), 24-epicastasterone (24-epiCS, 4 species), 6-deoxotyphasterol (6-deoxoTY, 3 species), DL (3 species), 24-epibrassinolide (24-epiBL, 3 species), 28-norbrassinolide (28-norBL, 3 species), 6-deoxodolichosterone (6-deoxoDS, 2 species), 28-homoteasterone (28-homoTE, 2 species), 6-deoxoteasterone (6-deoxoTE, 2 species), and 2-deoxybrassinolide (2-deoxyBL, 2 species). The other 22 BRs and 4 BR conjugates have been found in only one species so far.

Among plant species in which BRs have been found hitherto, *Phaseolus vulgaris* contains the most varied BRs (22 BRs and 2 BR conjugates). A relatively broader spectrum of various BRs is also found in *Cupressus arizonica* (9 BRs), *Lablab purpreus* (8 BRs), *Catharanthus roseus* (8 BRs), *Arabidopsis thaliana* (7 BRs), *Secale cereale* (7 BRs), *Lilium longiflorum* (5 BRs and 2 BR conjugates), *Thea sinensis* (6 BRs), and *Oryza sativa* (6 BRs).

The following sections give a comprehensive survey on the hitherto known BRs isolated from each plant species.

3.1 Brassinosteroids in Monocots

The presence of BRs in monocots has been demonstrated from three families including nine plant species (Table 1).

<u>Gramineae</u>. From rice (*Oryza sativa* cv. Arborio J1) shoots, CS and DS were identified by GC-CI-SIM (where CI is chemical ionization) (Abe et al 1984b). The amounts were roughly estimated to be 13.6 and 8.4 ng/kg, respectively. Trace amounts of BL were also detected (Abe 1991). From rice (*O. sativa* L. cv. Koshihikari) bran, a new BR, $(22R,23R,24S)$-3α,22,23-trihydroxy-24-ethyl-5α-cholestan-6-one, termed 28-homotyphasterol (28-homoTY), was identified by GC-MS, in addition to 28-homoteasterone (28-homoTE) and 6-deoxoCS (Abe et al. 1995a).

From seeds of rye (*Seale cereale*), a new BR termed secasterone (SE), $(22R,23R,24S)$-22,23-dihydroxy-2β,3β-epoxy-24-methyl-5α-cholestan-6-one, was

Table 1. Distribution of brassinosteroids in monocots

Plant species	Plant parts	Brassinosteroids
(Gramineae)		
Oryza sativa	shoot	BL, CS, DS
	bran	6-deoxoCS, 28-homoTY, 28-homoTE
Secale cereale	seed	CS, TY, TE, 6-deoxoCS, SE, 28-homoCS, 28-norCS
Triticum aestivum	bran	CS, TY, 3DT, TE, 6-deoxoCS
	flour	CS, TY, TE, 6-deoxoCS
Zea mays		
dent corn	pollen	CS, TY, TE
sweet corn	pollen	CS, DS, 28-norCS
pioneer hybrid 3358	shoot	CS
(Liliaceae)		
Erythronium japonicum	pollen, anther	TY
Lilium elegans	pollen	BL, CS, TY, TE
Lilium longiflorum	pollen, anther	BL, CS, TY, 3DT, TE, TE-3-My, TE-3-La
Tulipa gesneriana	pollen	TY
(Typhaceae)		
Typha latifolia	pollen	TY, TE

identified by GC-MS (Schmidt et al 1995b). SE represents the first naturally occurring BR with a 2,3-epoxy function. In addition, CS, TY, TE, 6-deoxoCS, 28-homoCS, and 28-norCS were identified from the same plant material.

From wheat (*Triticum aestivum* L.) grains, 3DT, CS, TY, TE, and 6-deoxoCS were identified by GC-MS (Yokota et al 1994). In addition, the presence of a CS epimer and a pair of 6-deoxoCS epimers was indicated in this plant material.

From pollen of *Zea mays* (dent corn), CS, TY, and TE were identified by GC-MS and/or ^{1}H-NMR. The amounts were shown to be about 120 µg/kg (CS), 6.6 µg/kg (TY), and 4.1 µg/kg (TE) (Suzuki et al 1986). It was also found that the anther contained a fairly large amount of BRs by a bioassay. However, in sweet corn pollen, CS, DS, and 28-norCS were found to be present, indicating a variation of the BR constituents, in the different varieties of *Zea mays* (Gamoh et al 1990). The amounts of these BRs was determined by HPLC with fluorimetric detection (DS 16.9 µg/kg; 28-norCS 18.3 µg/kg; CS 27.2 µg/kg). Fom the shoot of *Zea mays* L. (Pioneer Hybrid 3358), CS [ca. 2 µg/kg fresh weight(fw)] was identified (Sekimoto et al 1997).

Liliaceae. Four BRs (BL, CS, TY, TE) in the pollen of *Lilium elegans* were identified by GC-MS or GC-SIM (Suzuki et al 1994b). The amount of the each BR was roughly estimated (CS and TY 10-50 µg/kg, BL and TE 1-5 µg/kg). The main BRs found in lily (*Lilium longiflorum*) pollen were TY, CS, and BL (Abe 1991). 3DT (3-dehydroteasterone or 3-oxoteasterone) was isolated for the first time in the anthers of this plant (Abe et al 1994). In addition, two acyl conjugates of TE [teasterone-3-laurate (TE-3-La) and teasterone-3-myristate (TE-3-My)] were identified by GC-MS and/or liquid chromatography(LC)-MS (Asakawa et al 1994, 1996). This is the first report that acyl conjugates of BR had been discovered in plants. Although TY, CS, and BL were identified in this plant, there were no esterified conjugates of TY, CS, or BL.

Typhasterol was identified by GC-MS in the pollen and anthers of *Erythronium japonicum* Decne (Yasuta et al 1995). The amont of TY was calculated to be ca. 5 µg/kg. In addition, TE, TY, and CS were identified as components of conjugated BRs in the anther.

TY was found in tulip (*Tulipa gesneriana*) pollen (Abe 1991).

Typhaceae. TY (typhasterol or 2-deoxycastasterone) was isolated for the first time from cattail (*Typha latifolia*) pollen (Schneider et al 1983). The structure was determined to be $(22R,23R,24S)$-$3\alpha,22,23$-trihydroxy-24-methyl-5α-cholestan-6-one by ^{1}H-NMR and MS. The yield was 1.7 mg from 25 kg of cattail pollen. TE was also identified from cattail pollen (Abe 1991).

3.2 Brassinosteroids in Dicots

Up to now BRs in dicots have been found in 16 families, including 23 plant species (Table 2).

Apocynaceae. From cultured cells of *Catharanthus roseus*, 8 BRs (BL, CS, TY, TE, CT, 6-deoxoCS, 6-deoxoTY, 6-deoxoTE) were identified (Choi et al 1993, 1996, 1997, Fujioka et al 1995, Park et al 1989, Suzuki et al 1994a). These studies revealed for the first time the production of BRs in cultured cells. Among the above BRs, CT and 6-deoxoTE were new BRs found in the plant kingdom for the first time. $(22S,24R)$-$3\beta,22$-Dihydroxy-24-methyl-5α-cholestan-6-one, named cathasterone (CT) was demonstrated by GC-MS. Its endogenous level was in the range of 2-4 ng/g fw. This study also showed that CT is the first active BR having only one hydroxyl in the side chain, and TE is biosynthesized from CT (Fujioka et al 1995). 6-DeoxoCS was found to be the major BR in *C. roseus*. Its concentration was estimated by GC-SIM using an internal standard to be in the range 5.9-18.9 ng/g (Choi et al 1996). The contents of 6-deoxoTE and 6-deoxoTY in the cells were 47 and 758 pg/g, respectively (Choi et al 1997). The content of CS and BL during cell growth was pursued by GC-SIM using internal standards (Yokota et al 1991). Their concentrations were in the range of 0.4-8.7 ng/g (BL) and 0.6-4.5 ng/g (CS). Furthermore, using this cell culture system, the outlines of BL biosynthetic pathways have been clarified (see the chapter by Sakurai, this volume).

Table 2. Distribution of brassinosteroids in dicots

Plant species	Plant parts	Brassinosteroids
(Apocynaceae)		
Catharanthus roseus	cultured cell	BL, CS, TY, TE, CT, 6-deoxoCS, 6-deoxoTY, 6-deoxoTE
(Betulaceae)		
Alnus glutinosa	pollen	BL, CS
(Caryophyllaceae)		
Gypsophila perfoliata	seed	24-epiBL
(Chenopodiaceae)		
Beta vulgaris	seed	CS, 24-epiCS
(Compositae)		
Helianthus annuus	pollen	BL, CS, 28-norCS
(Convolvulaceae)		
Pharbitis purpurea	seed	CS, 28-norCS
(Cruciferae)		
Arabidopsis thaliana	shoot	CS, TY, 6-deoxoCS, 6-deoxoTY
	seed, silique	BL, CS, TY, 6-deoxoCS, 6-deoxoTY 6-deoxoTE, 24-epiBL
Brassica campestris	seed, sheath	BL, CS, 28-norBL, 28-norCS, 28-homoCS
Brassica napus	pollen	BL, CS
Raphanus sativus	seed	BL, CS, TE, 28-homoTE
(Fagaceae)		
Castanea crenata	gall	BL, CS, 6-deoxoCS, 28-norCS
	leaf	CS, 6-deoxoCS
(Hamamelidaceae)		
Distylium racemosum	gall	CS, 28-norCS
	leaf	BL, CS, 3DT, 28-norBL, 28-norCS
(Leguminosae)		
Cassia tora	seed	BL, CS, TY, TE, 28-norCS
Lablab purpreus (formerly Dolichos lablab)	seed	BL, CS, DL, DS, 6-deoxoCS, 6-deoxoDS, 28-homoDL, 28-homoDS
Ornithopus sativus	seed	CS, 24-epiCS
	shoot	CS, 6-deoxoCS, 24-epiCS, 6-deoxo-24-epiCS 6-deoxo-28-norCS

Phaseolus vulgaris	seed	BL, CS, TY, TE, DS, DL, 6-deoxoCS, 6-deoxoDS, 1β-OH-CS, 3-epi-1α-OH-CS, 2-epiCS, 3-epiCS, 2,3-diepiCS, 3-epi-6-deoxoCS, 3,24-diepiCS, 6-deoxo-28-homoDS, 25-Me-DS, 2-epi-25-Me-DS, 2,3-diepi-25-Me-DS, 2-deoxy-25-Me-DS, 3-epi-2-deoxy-25-Me-DS, 6-deoxo-25-Me-DS, 25-Me-DS-Glu, 2-epi-25-Me-DS-Glu
	stem	BL, CS
Pisum sativum	seed	BL, CS, TY, 6-deoxoCS, 2-DeoxyBL
	shoot	BL, CS, TY, 6-deoxoCS
Robinia pseudo-acacia	pollen	CS, TY, 6-deoxoCS
Vicia faba	seed	BL, CS
	pollen	BL, CS, DS, 24-epiBL, 28-norCS
Vigna radiata		BL, CS
(Moraceae)		
Cannabis sativa	seed	CS, TE
(Polygonaceae)		
Fagopyrum esculentum	pollen	BL, CS
Rheum rhabarbarum	panicle	BL, CS, 24-epiCS
(Rutaceae)		
Citrus sinensis	pollen	BL, CS
Citrus unshiu	pollen	BL, CS, TY, TE
(Solanaceae)		
Lycopersicon esculentum	shoot	CS, 6-deoxoCS, 28-norCS
(Theaceae)		
Thea sinensis	leaf	BL, CS, TY, TE, 28-norCS, 28-homoCS
(Umbelliferae)		
Apium graveolens	seed	2-deoxyBL

Betulaceae. From European alder (*Alnus glutinosa* L.) pollen, BL and CS were identified (Plattner et al 1986). GC-MS-MS has been applied to the identification of BRs in this plant.

Caryophyllaceae. 24-EpiBL was identified from seeds of *Gypsophila perfoliata* L. by GC-MS as the only BR present (Schmidt et al 1996).

Chenopodiaceae. CS and 24-epiCS were identified from seeds of sugar beet (*Beta vulgaris* L.) (Schmidt et al 1994). The ratio of CS to 24-epiCS was ca. 3:1.

Compositae. From the pollen of sunflower (*Helianthus annuus* L.), BL, CS, and 28-norCS were identified by fluorescence HPLC and GC-MS (Takatsuto et al 1989). The amounts were 106, 21, and 65 µg/kg, respectively.

Convolvulaceae. In immature seeds of *Pharbitis purpurea*, CS and 28-norCS were identified in quantities of about 1.1 and 0.2 µg/kg, respectively (Suzuki et al 1985).

Cruciferae. BRs have been identified from four plant species of the Cruciferae family.

As mentioned above, the first naturally occurring BR, brassinolide, was found from the pollen of rape (*Brassica napus*) (Grove et al 1979). Rape pollen was also shown to contain CS (Mandava 1988).

From immature seeds and sheaths of Chinese cabbage (*Brassica campestris*), C_{27}-BRs, C_{28}-BRs, and C_{29}-BRs of BL, CS, 28-homoCS, 28-norBL, and 28-norCS were identified (Abe et al 1982, 1983, Ikekawa et al 1984). These BRs in seeds and sheaths were estimated to be present in extremely small amounts (BL 9.4 ng/kg, 28-norBL 1.3 ng/kg, CS 1.6 ng/kg, 28-norCS 0.78 ng/kg, 28-homoCS 0.13 ng/kg).

The CS and BL were identified from seeds of radish (*Raphanus sativus*) (Schmidt et al 1991). The amounts were in the range of 300-800 ng/kg. Later 28-homoTE and TE were identified from the seeds (Schmidt et al 1993b). 28-HomoTE was found as a natural product in this plant for the first time.

Brassinosteroids in *Arabidopsis thaliana* have been recently reported. From the shoots, CS, TY, 6-deoxoCS and 6-deoxoTY were identified by GC-MS (Fujioka et al 1996), and the amounts of BR were quantified by GC-SIM using internal standards. In wild-type (ecotype Columbia) shoots, BR contents were as follows: CS 0.75 ng/g fw, 6-deoxoCS 0.71 ng/g, TY 0.11 ng/g, 6-deoxoTY 0.95 ng/g (Fujioka et al 1997). The study also revealed that *det2* mutant is BR-deficient (see the chapter by Sakurai, this volume). From seeds and siliques of the same ecotype, BL (0.5-1.9 ng/g), CS (0.4-5 ng/g), TY (1.3 ng/g), 6-deoxoCS (1.5-3 ng/g), 6-deoxoTY (0.5-5.4 ng/g), and 6-deoxoTE (0.5-1 ng/g) were identified (Fujioka et al 1998). From the seeds of *A. thaliana* (ecotype 24), 24-epiBL (0.22 ng/g) and CS (0.36 ng/g) were identified (Schmidt et al 1997).

Fagaceae. As a second naturally occurring BR, CS (95 μg) was isolated from 40 kg of insect galls of chestnut (*Castanea crenata*) using the rice-lamina inclination assay to verify bioactivity (Yokota et al 1982a). Later, the presence of BL and CS was reported by another group (Ikeda et al 1983). Their amounts in 3.5 kg of chestnut galls were reported to be about 2 and 40 μg, respectively. 28-NorCS was also identified from the same material, and the amount was 40 ng (Abe et al 1983). Subsequent study revealed that CS and 6-deoxoCS were identified not only in the insect gall of chestnut but also in the healthy tissues, including the shoot, leaf, and flower bud. The amounts were reported to be in the range of 2-12 μg/kg (CS) and 9-30 μg/kg (6-deoxoCS) (Arima et al 1984). However, the presence of 28-norCS could not be confirmed by them.

Hamamelidaceae. BL, CS, 28-norBL, and 28-norCS were identified from leaves of *Distylium racemosum*, while CS and 28-norCS were found in the insect galls (Ikekawa et al 1984). It should be noted that the amounts of CS and 28-norCS in the insect galls are higher (ca. 20-300 times) than those in the leaves of the same plant. 3DT was also detected in the leaves of this plant (Abe et al 1994).

Leguminosae. In the Leguminosae family, endogenous BRs have been extensively investigated. Especially, immature seeds of *Phaseolus vulgaris* cv. Kentucky Wonder have been the most rigorously investigated. It was found that the seeds of this plant contain 6-deoxoCS, 6-deoxoDS, CS, DL, 6-deoxo-28-homodolichosterone (6-deoxo-28-homoDS), DS, and BL (Yokota et al 1983b, 1987b). In addition, the presence of a number of unidentified BRs has been suggested by HPLC and GC-MS analysis. From a large scale experiment, 13 BRs [2-epicastasterone (2-epiCS), 3-epicastasterone (3-epiCS), 2,3-diepicastasterone (2,3-diepiCS), 3,24-diepicastasterone (3,24-diepiCS), 1β-hydroxycastasterone (1β-OH-CS), 3-epi-1α-hydroxycastasterone (3-epi-1α-OH-CS), 3-epi-6-deoxocastatseorne (3-epi-6-deoxoCS), 25-methyldolichosterone (25-Me-DS), 2-epi-25-methyldolichosterone (2-epi-25-Me-DS), 2,3-diepi-25-methyldolichosterone (2,3-diepi-25-Me-DS), 2-deoxy-25-methyldolichosterone (2-deoxy-25-Me-DS), 3-epi-2-deoxy-25-methyldolichosterone (3-epi-2-deoxy-25-Me-DS), and 6-deoxo-25-methyldolichosterone (6-deoxo-25-Me-DS)], and 2 BR conjugates [23-O-β-D-glucopyranosyl-25-methyldolichosterone and 23-O-β-D-glucopyranosyl-2-epi-25-methydolichosterone] were isolated in pure states (Kim et al 1987, Kim 1991, Yokota et al 1987c). The structures were characterized by ^{1}H-NMR and MS. In addition, TE and TY were also found. Among BRs found in this plant, 25-Me-DS is the first 25-methylated BR, and interestingly its biological activity in the rice lamina inclination test is ca. 10 times higher than that of its non-25-methylated counterpart, DS. From the same plant material, in addition to the above BRs, the presence of over 30 kinds of unknown BRs was shown by GC-MS, although their complete structures remain to be characterized. Analysis of BRs in shoots and seeds of this plant by radioimmunoassay has demonstrated that numerous BRs occur in the seeds but only CS and BL are found in the shoots (Yokota et al 1990). This study suggests that the biosynthesis of BRs might be differently controlled in each organ.

From immature seeds of broad bean (*Vicia faba*), CS and BL were identified (Park et al 1987). From bee-collected pollen of the broad bean, 24-epiBL was identified, together with BL, CS, and 28-norCS (Ikekawa et al 1988). This work demonstrated for the first time the occurrence of 24-epiBL in the plant kingdom. The amounts of BL and 24-epiBL in the pollen were at least 190 and 5 µg/kg, respectively. Gamoh et al. (1989) analyzed BRs in bee-collected pollen of the broad bean. After derivatization, the biologically active fraction was analyzed by HPLC. BL, DS, 28-norCS, and CS were identified. The amounts were calculated to be 181, 537, 628, and 134 ng/g, respectively.

From immature seeds of *Lablab purpreus* (formerly *Dolichos lablab*), DL was isolated for the first time (Yokota et al 1982b). The structure was elucidated as $(22R,23R)$-$2\alpha,3\alpha,22,23$-tetrahydroxy-B-homo-7-oxa-24-methyl-5α-cholest-24(28)-en-6-one by ^{1}H-NMR and MS. The yield was 160 µg from 34 kg of the seeds. In the same plant material, other new BRs [DS (50 µg), 28-homoDS (20 µg), 28-homoDL (12 µg)] were isolated (Baba et al 1983, Yokota et al 1983a), and their structures were characterized by ^{1}H-NMR and MS. BL, CS, 6-deoxoCS, and 6-deoxoDS were also identified (Yokota et al 1983a, 1984).

The co-occurrence of two epimeric BRs, CS and 24-epiCS, was reported from the seeds of *Ornithopus sativus* (Schmidt et al 1993a). The contents of two epimeric BRs per 1 kg seeds were ca. 5 and 25 µg, respectively. Two new BRs were detected from shoots of *Ornithopus sativus* and identified as 6-deoxo-24-epicastasterone (6-deoxo-24-epiCS) and 6-deoxo-28-norcastasterone (6-deoxo-28-norCS) by GC-MS. Furthermore, CS, 24-epiCS, and 6-deoxoCS were found in the same plant material (Spengler et al 1995).

From fully grown seed of garden pea (*Pisum sativum*), 2-deoxyBL was identified, together with BL, CS, TY, and 6-deoxoCS (Yokota et al 1996). In addition, the presence of a BL with an extra hydroxy group and an epimer of 6-deoxoCS was indicated, although their complete structures could not be clarified. Endogenous BRs in the shoots of dwarf mutants and wildtype garden pea (*Pisum sativum*) were investigated. BL, CS, TY, and 6-deoxoCS were identified and quantified by GC-SIM using internal standards (Nomura et al 1997). In wild-type plants, the levels of BRs were as follows: BL (164-840 ng/kg), CS (355-2360 ng/kg), TY (995 ng/kg), and 6-deoxoCS (3133 ng/kg). This study also revealed that the *lkb* mutant is BR-deficient (see the chapter by Sakurai, this volume).

From immature seeds of *Cassia tora*, BL, CS, TY, TE, and 28-norCS were identified (Park et al 1994). From mung bean (*Vigna radiata*), BL and CS were identified (Abe 1991). From the pollen of *Robinia pseudo-acacia* L., CS, 6-deoxoCS, and TY were identified (Abe et al 1995b).

Moraceae. Two BRs, CS and TE, were identified from the seeds of *Cannabis sativa* L. (Takatsuto et al 1996). Subsequent study strongly suggested that TE occurs as both free and ester forms (Takatsuto et al 1997).

Polygonaceae. From pollen of buckwheat (*Fagopyrum esculentum* Moench) BL and CS were identified by fluorescence HPLC and GC-HR-SIM microanalyses (where HR is high resolution) (Takatsuto et al 1990b). The amounts of the BRs were calculated to be 5.0 and 7.1 ng/g, respectively. From the panicles of rhubarb (*Rheum rhabarbarum* L.), BL, CS, and 24-epiCS were identified (Schmidt et al 1995a).

Rutaceae. BL, CS, TY, and TE were identified from pollen of *Citrus unshiu* (Abe 1991). In pollen of *Citrus sinensis* Osbeck, BL and CS were identified by HPLC with fluorimetric detection (Motegi et al 1994). The amounts were calculated to be 36.2 and 29.4 µg kg of pollen, respectively.

Solanaceae. CS, 6-deoxoCS, and 28-norCS were identified from shoots of tomato (*Lycopersicon esculentum*) (Yokota et al 1997). Endogenous levels were estimated to be 210, 1690, and 29 ng/kg fw, respectively.

Theaceae. 28-NorCS, 28-homoCS, BL, CS, TY, and TE were identified from leaves of green tea (*Thea sinensis*) (Abe et al 1983, 1984a, Ikekawa et al 1984, Morishita et al 1983). The respective amounts of the above BRs were roughly estimated to be 130 ng, 30 ng, 390 ng, 7.2 µg, 4 µg, and 1 µg in 65 kg of the leaves.

Umbellifererae. A new BR was detected in seeds of celery (*Apium graveolens*). It was shown to be 2-deoxyBL by GC-MS (Schmidt et al 1995c). This BR is the sixth BR identified to contain an oxalactone functionality.

3.3 Brassinosteroids in Gymnosperms

The occurrence of BRs in gymnosperms has been reported from five conifer species (Table 3).

Cupressaceae. Griffiths et al (1995) reported that two new BRs [$(22R,23R,24S)$-22,23-dihydroxy-24-methyl-5α-cholestan-3α-ol and $(22R,23R,24S)$-22,23-dihydroxy-24-methyl-5α-cholestan-3-one], which are termed 6-deoxotyphasterol (6-deoxoTY) and 3-dehydro-6-deoxoteasterone (6-deoxo3DT), occur in very large amounts (6.4 and 2.3 µg/g, respectively) in the mature pollen of Arizona cypress (*Cupressus arizonica*). They also found that 6-deoxoCS (1.2 µg/g), CS (1.0 µg/g), TY (0.46 µg/g), TE (0.005 µg/g), 28-homoCS (0.004 µg/g), 3-DT (0.002 µg/g), and BL (< 0.001 µg/g) were present in the same pollen. Furthermore, the presence of an epiCS-like compound was shown in this study. Among plant species so far reported, the contents of BRs in this plant are the highest.

Pinaceae. Yokota et al (1983c) isolated 1.7 mg of TY from 19 kg of pollen of Japanese black pine (*Pinus thunbergii*). The structure was established to be $(22R,23R,24S)$-3α,22,23-trihydroxy-24-methyl-5α-cholestan-6-one by ^{1}H-NMR and MS. Later they isolated CS from the same pollen. They also identified CS and TY

Table 3. Distribution of brassinosteroids in gymnosperm, alga and pteridophyte

Plant species	Plant parts	Brassinosteroids
Gymnosperm		
(Cupressaceae)		
Cupressus arizonica	pollen	BL, CS, TY, 3DT, TE, 6-deoxoCS, 6-deoxoTY, 6-deoxo3DT, 28-homoCS
(Pinaceae)		
Picea sitchensis	shoot	CS, TY
Pinus silverstris	cambial region	BL, CS
Pinus thunbergii	pollen	CS, TY
(Taxodiaceae)		
Cryptomeria japonica	pollen, anther	TY, 3DT, DL
Alga		
(Chlorophyceae)		
Hydrodictyon reticulatum	colony	24-epiCS, 28-homoCS
Pteridophyte		
(Equisetaceae)		
Equisetum arvense	strobilus	CS, DS, 28-norBL, 28-norCS

from shoots of Sitka spruce (*Picea sitchensis*) by GC-MS (Yokota et al 1985). The amounts of CS and TY from 1 kg of shoots were estimated to be ca. 5 and 7 µg, respectively. Kim et al (1990) identified CS (major) and BL (minor) from cambial region of Scots pine (*Pinus silverstris*) by GC-MS after detection using a modified dwarf rice lamina inclination assay.

Taxodiaceae. From pollen and anther of Japanese cedar (*Cryptomeria japonica*), 3DT, TY, and DL were identified by GC-MS. Occurrence of several new BRs, four stereoisomers of 23-dehydrobrassinolide (23-dehydroBL), three stereoisomers of 28-homoBL, and an isomer of 28-homoDL were revealed (Yokota et al 1998). *C. japonica* is unique in that its pollen and anther contain a variety of BRs with either a 24-methyl (C_{28}-BRs), 24-methylene (C_{28}-BRs), 24-ethyl (C_{29}-BRs), or 24-ethylidene (C_{29}-BRs). In contrast, in other conifers, C_{28}-BRs having a 24α-methyl are predominant.

3.4 Brassinosteroids in Lower Plants

The presence of BRs in lower plants has been shown in one pteridophyte and one green alga (Table 3).

Pteridophyte. Takatsuto et al (1990a) investigated BRs in pteridophyte. From the strobilus of *Equisetum arvense* L., 28-norBL, DS, 28-norCS, and CS were detected by a newly developed HPLC microanalytical method. The amounts of identified BRs were calculated as 152, 746, 349, and 171 ng/kg fw, respectively. This work is the first report of BRs in pteridophytes.

Green alga. 24-EpiCS and 28-homoCS have been identified in green alga, *Hydrodictyon reticulatum* (Yokota et al 1987a). The amounts of 24-epiCS and 28-homoCS were estimated to be 0.3 and 4.0 μg/kg fw, respectively. This study demonstrates for the first time that BRs occur in algae and that a 24β-methyl type BR is present in plants.

4. Conclusions

Since the discovery of BL, extensive studies on the isolation and identification of BRs from various plant species have been undertaken. To date, 40 free BRs and 4 BR conjugates have been found and fully characterized by spectrometric methods. A number of new BRs and BR conjugates are also suggested to be present in plants, although their structures have not been fully characterized. BRs have been found in at least 44 plant species including 37 angiosperms (9 monocots and 28 dicots), 5 gymnosperms, 1 alga, and 1 pteridophyte. Bioassay data not mentioned here suggest that BRs are more widely distributed than presently established. Thus, it is clear that BRs are ubiquitously distributed in the plant kingdom.

Among BRs hitherto found, BL, CS, TY, TE, and 6-deoxoCS, which are C_{28}-BRs having a 24α-methyl group in the side chain, are widely found in the plant kingdom. These BRs have been recently established to be biosynthesized from campesterol (Fujioka and Sakurai 1997b, Sakurai and Fujioka 1997, Yokota 1997). Among these BRs, BL is the most biologically active BR. Therefore, BL is considered to be the most important BR and plays pivotal roles in the hormonal regulation of plant growth and development.

Judging from the diverse structural variations of the A-ring, B-ring, and side chain, more than 100 BRs might be present in the plant kingdom. In future, a number of BRs and BR conjugates will be isolated and characterized. Among such BRs, an immediate catabolite of BL would be one of the most intriguing BRs worth surveying, because it is important to know how endogenous levels of BL are controlled in intact plants. Up to the present, 23-dehydrobrassinolide was tentatively identified as an endogenous BR from Japanese black pine, and 23-glucosylbrassinolide was identified as the metabolite of BL in mung bean explants (see the chapter of Adam and Schneider, this volume). However, these catabolites do not seem to be involved in general deactivation processes of BL.

During the past two decades, microanalytical methods including GC-MS, HPLC, and radioimmunoassay, in combination with bioassays, have greatly contributed to the isolation and identification of naturally occurring BRs. They will continue to widen our knowledge of the natural occurrence and distribution of BRs in the plant kingdom. However, current interest seems to have shifted to elucidation of the physiological roles, biosynthesis and its regulation, and signal transduction of BRs in plants. One of the important approaches to defining the roles of BRs in the regulation of plant growth and development is to know the endogenous levels in plant tissues quantitatively. Because BRs in plant tissues are present in very low concentrations except in pollen and immature seeds, more sophisticated analytical methodology should be developed to address the above.

References

Abe H (1991) Rice-lamina inclination, endogenous levels in plant tissues and accumulation during pollen development of brassinosteroids. In: Cutler HG, Yokota T, Adam G (Eds) Brassinosteroids; Chemistry, Bioactivity, and Applications. ACS Symp Ser 474, Amer Chem Soc, Washington, DC, pp 200-207

Abe H, Morishita T, Uchiyama M, et al (1982) Identification of brassinolide-like substances in Chinese cabbage. Agric Biol Chem 46: 2609-2611

Abe H, Morishita T, Uchiyama M, et al (1983) Occurrence of three new brassinosteroids: brassinone, (24S)-24-ethylbrassinone and 28-norbrassinolide, in higher plants. Experientia 39: 351-353

Abe H, Morishita T, Uchiyama M, et al (1984a) A new brassinolide-related steroid in the leaves of *Thea sinensis*. Agric Biol Chem 48: 2171-2172

Abe H, Nakamura K, Morishita T, et al (1984b) Endogenous brassinosteroids of the rice plant: Castasterone and dolichosterone. Agric Biol Chem 48: 1103-1104

Abe H, Honjo C, Kyokawa Y, et al (1994) 3-Oxoteasterone and the epimerization of teasterone: Identification in lily anthers and *Distylium racemosum* leaves and its biotransformation into typhasterol. Biosci Biotech Biochem 58: 986-989

Abe H, Takatsuto S, Nakayama M, et al (1995a) 28-Homotyphasterol, a new natural brassinosteroid from rice (*Oryza sativa* L.) bran. Biosci Biotech Biochem 59: 176-178

Abe H, Takatsuto S, Okuda R, et al (1995b) Identification of castasterone, 6-deoxocastasterone, and typhasterol in the pollen of *Robinia pseudo-acacia* L. Biosci Biotech Biochem 59: 309-310

Adam G, Marquardt V (1986) Brassinosteroids. Phytochemistry 25: 1787-1799

Arima M, Yokota T, Takahashi N (1984) Identification and quantification of brassinolide-related steroids in the insect gall and healthy tissues of the chestnut plant. Phytochemistry 23: 1587-1591

Asakawa S, Abe H, Kyokawa Y, et al (1994) Teasterone 3-myristate: A new type of brassinosteroid derivative in *Lilium longiflorum* anthers. Biosci Biotech Biochem 58: 219-220

Asakawa S, Abe H, Nishikawa N, et al (1996) Purification and identification of new acyl-conjugated teasterones in lily pollen. Biosci Biotech Biochem 60: 1416-1420

Baba J, Yokota T, Takahashi N (1983) Brassinolide-related new bioactive steroids from *Dolichos lablab* seed. Agric Biol Chem 47: 659-661

Choi YH, Inoue T, Fujioka S, et al (1993) Identification of brassinosteroid-like active substances in plant-cell cultures. Biosci Biotech Biochem 57: 860-861

Choi YH, Fujioka S, Harada A, et al (1996) A brassinolide biosynthetic pathway via 6-deoxocastasterone. Phytochemistry 43: 593-596

Choi YH, Fujioka S, Nomura T, et al (1997) An alternative brassinolide biosynthetic pathway via late C-6 oxidation. Phytochemistry 44: 609-613

Fujioka S, Sakurai A (1997a) Brassinosteroids. Nat Prod Rep 14: 1-10

Fujioka S, Sakurai A (1997b) Biosynthesis and metabolism of brassinosteroids. Physiol Plant 100: 710-715

Fujioka S, Inoue T, Takatsuto S, et al (1995) Identification of a new brassinosteroid, cathasterone, in cultured cells of *Catharanthus roseus* as a biosynthetic precursor of teasterone. Biosci Biotech Biochem 59: 1543-1547

Fujioka S, Choi YH, Takatsuto S, et al (1996) Identification of castasterone, 6-deoxocastatsterone, typhasterol and 6-deoxotyphasterol from the shoots of *Arabidopsis thaliana*. Plant Cell Physiol 37: 1201-1203

Fujioka S, Li J, Choi YH, et al (1997) The Arabidopsis *deetiolated2* mutant is blocked early in brassinosteroid biosynthsis. Plant Cell 9: 1951-1962

Fujioka S, Noguchi T, Yokota T, et al (1998) Brassinosteroids in *Arabidopsis thaliana*. Phytochemistry 48: 595-599

Gamoh K, Omote K, Okamoto N, et al (1989) High-performance liquid chromatography of brassinosteroids in plants with derivatization using 9-phenanthreneboronic acid. J Chromatogr 469: 424-428

Gamoh K, Okamoto N, Takatsuto S, et al (1990) Determination of traces of natural brassinosteroids as dansylaminophenylboronates by liquid chromatography with fluorimetric detection. Anal Chim Acta 228: 101-105

Griffiths PG, Sasse JM, Yokota T, et al (1995) 6-Deoxytyphasterol and 3-dehydro-6-deoxoteasterone, possible precursors to brassinosteroids in the pollen of *Cupressus arizonica*. Biosci Biotech Biochem 59: 956-959

Grove MD, Spencer GF, Rohwedder WK, et al (1979) Brassinolide, a plant growth-promoting steroid isolated from *Brassica napus* pollen. Nature 281: 216-217

Ikeda M, Takatsuto S, Sassa T, et al (1983) Identification of brassinolide and its analogues in chestnut gall tissue. Agric Biol Chem 47: 655-657

Ikekawa N, Takatsuto S, Kitsuwa T, et al (1984) Analysis of natural brassinosteroids by gas chromatography and gas chromatography-mass spectrometry. J Chromatogr 290: 289-302

Ikekawa N, Nishiyama F, Fujimoto Y (1988) Identification of 24-epibrassinolide in bee pollen of the broad bean, *Vicia faba* L. Chem Pharm Bull 36: 405-407

Kim SK (1991) Natural occurrences of brassinosteroids. In: Cutler HG, Yokota T, Adam G (Eds) Brassinosteroids; Chemistry, Bioactivity, and Applications. ACS Symp Ser 474, Amer Chem Soc, Washington, DC, pp 26-35

Kim SK, T. Yokota T, Takahashi N (1987) 25-Methyldolichosterone, a new brassinosteroid with a tertiary butyl group from immature seed of *Phaseolus vulgaris*. Agric Biol Chem 51: 2303-2305

Kim SK, Abe H, Little CHA, et al (1990) Identification of two brassinosteroids from the cambial region of Scots pine (*Pinus silverstris*) by gas chromatography-mass spectrometry, after detection using a dwarf rice lamina inclination bioassay. Plant Physiol 94: 1709-1713

Mandava NB (1988) Plant growth-promoting brassinosteroids. Annu Rev Plant Physiol Plant Mol Biol 39: 23-52

Marquardt V, Adam G (1991) Recent advances in brassinosteroid research. In: Ebing W (Ed) Chemistry of Plant Protection, Vol 7. Springer-Verlag, Berlin Heidelberg, pp 103-139

Meudt WJ (1987) Chemical and biological aspects of brassinolide. In: Fuller G, Nes WD (Eds) Ecology and Metabolism of Plant Lipids. Amer Chem Soc, Washington, DC, pp

53-75
Mitchell JW, Mandava N, Worley JF, et al (1970) Brassins - a new family of plant hormones from rape pollen. Nature 225: 1065-1066
Morishita T, Abe H, Uchiyama M, et al (1983) Evidence for plant growth promoting brassinosteroids in leaves of *Thea sinensis*. Phytochemistry 22: 1051-1053
Motegi C, Takatsuto S, Gamoh K (1994) Identification of brassinolide and castasterone in the pollen of orange (*Citrus sinensis* Osbeck) by high-performance liquid chromatography. J Chromatogr 658: 27-30
Nomura T, Nakayama M, Reid JB, et al (1997) Blockage of brassinosteroid biosynthesis and sensitivity causes dwarfism in garden pea. Plant Physiol 113: 31-37
Park KH, Yokota T, Sakurai A, et al (1987) Occurrence of castasterone, brassinolide and methyl 4-chloroindole-3-acetate in immature *Vicia faba* seeds. Agric Biol Chem 51: 3081-3086
Park KH, Saimoto H, Nakagawa S, et al (1989) Occurrence of brassinolide and castasterone in crown gall cells of *Catharanthus roseus*. Agric Biol Chem 53: 805-811
Park KH, Park JD, Hyun KH, et al (1994) Brassinosteroids and monoglycerides in immature seeds of *Cassia tora* as the active principles in the rice lamina inclination bioassay. Biosci Biotech Biochem 58: 1343-1344
Plattner RD, Taylor SL, Grove MD (1986) Detection of brassinolide and castasterone in *Alnus glutinosa* (European alder) pollen by mass spectrometry/mass spectrometry. J Nat Prod 49: 540-545
Sakurai A, Fujioka S (1993) The current status of physiology and biochemistry of brassinosteroids. Plant Growth Regul 13: 147-159
Sakurai A, Fujioka S (1997) Studies on biosynthesis of brassinosteroids. Biosci Biotech Biochem 61: 757-762
Schmidt J, Yokota T, Adam G, et al (1991) Castasterone and brassinolide in *Raphanus sativus* seeds. Phytochemistry 30: 364-365
Schmidt J, Spengler B, Yokota T, et al (1993a) The co-occurrence of 24-epi-castasterone and castasterone in seeds of *Ornithopus sativus*. Phytochemistry 32: 1614-1615
Schmidt J, Yokota T, Spengler B, et al (1993b) 28-Homoteasterone, a naturally occurring brassinosteroid from seeds of *Raphanus sativus*. Phytochemistry 34: 391-392
Schmidt J, Kuhnt C, Adam G (1994) Brassinosteroids and sterols from seeds of *Beta vulgaris*. Phytochemistry 36: 175-177
Schmidt J, Himmelreich U, Adam G (1995a) Brassinosteroids, sterols and lup-20(29)-en-2α,3β,28-triol from *Rheum rhabarbarum*. Phytochemistry 40: 527-531
Schmidt J, Spengler B, Yokota T, et al (1995b) Secasterone, the first naturally occurring 2,3-epoxybrassinosteroid from *Secale cereale*. Phytochemistry 38: 1095-1097
Schmidt J, Voigt B, Adam G (1995c) 2-Deoxybrassinolide - a naturally occurring brassinosteroid from *Apium graveolens*. Phytochemistry 40: 1041-1043
Schmidt J, Bohme F, Adam G (1996) 24-Epibrassinolide from *Gypsophila perfoliata*. Z Naturforsch 51c: 897-899
Schmidt J, Altmann T, Adam G (1997) Brassinosteroids from seeds of *Arabidopsis thaliana*. Phytochemistry 45: 1325-1327
Schneider JA, Yoshihara K, Nakanishi K, et al (1983) Typhasterol (2-deoxycastasterone): A new plant growth regulator from cat-tail pollen. Tetrahedron Lett 24: 3859-3860
Sekimoto H, Hoshi M, Nomura T, et al (1997) Zinc deficiency affects the levels of endogenous gibberellins in *Zea mays* L. Plant Cell Physiol 38: 1087-1090
Spengler B, Schmidt J, Voigt B, et al (1995) 6-Deoxo-28-norcastasterone and 6-deoxo-24-

epicastasterone - two new brassinosteroids from *Ornithopus sativus*. Phytochemistry 40: 907-910

Suzuki Y, Yamaguchi I, Takahashi N (1985) Identification of castasterone and brassinone from immature seeds of *Pharbitis purpurea*. Agric Biol Chem 49: 49-54

Suzuki Y, Yamaguchi I, Yokota T, et al (1986) Identification of castasterone, typhasterol and teasterone from the pollen of *Zea mays*. Agric Biol Chem 50: 3133-3138

Suzuki H, Fujioka S, Takatsuto S, et al (1994a) Biosynthesis of brassinolide from teasterone via typhasterol and castasterone in cultured cells of *Catharanthus roseus*. J Plant Growth Regul 13: 21-26

Suzuki H, Fujioka S, Yokota T, et al (1994b) Identification of brassinolide, castasterone, typhasterol, and teasterone from the pollen of *Lilium elegans*. Biosci Biotech Biochem 58: 2075-2076

Takatsuto S (1994) Brassinosteroids: distribution in plants, bioassays and microanalysis by gas chromatography-mass spectrometry. J Chromatogr A 658: 3-15

Takatsuto S, Yokota T, Omote K, et al (1989) Identification of brassinolide, castasterone and norcastasterone (brassinone) in sunflower (*Helianthus annuus* L.) pollen. Agric Biol Chem 53: 2177-2180

Takatsuto S, Abe H, Gamoh K (1990a) Evidence for brassinosteroids in strobilus of *Equisetum arvense* L. Agric Biol Chem 54: 1057-1059

Takatsuto S, Omote K, Gamoh K, et al (1990b) Identification of brassinolide and castasterone in buckwheat (*Fagopyrum esculentum* Moench) pollen. Agric Biol Chem 54: 757-762

Takatsuto S, Abe H, Yokota T, et al (1996) Identification of castasterone and teasterone in seeds of *Cannabis sativa* L. J Jpn Oil Chem Soc 45: 871-873

Takatsuto S, Kawashima T, Noguchi T, et al (1997) Identification of teasterone and phytosterols in the lipid fraction from seeds of *Cannabis sativa* L. J Jpn Oil Chem Soc 46: 1499-1504

Wada K, Marumo S, Ikekawa N, et al (1981) Brassinolide and homobrassinolide promotion of lamina inclination of rice seedlings. Plant Cell Physiol. 22: 323-325

Yasuta E, Terahara T, Nakayama M, et al (1995) Free and conjugated brassinosteroids in the pollen and anthers of *Erythronium japonicum* Decne. Biosci Biotech Biochem 59: 2156-2158

Yokota T (1997) The structure, biosynthesis and function of brassinosteroids. Trends Plant Sci 2: 137-143

Yokota T, Takahashi N (1986) Chemistry, physiology and agricultural application of brassinolide and related steroids. In: Bopp M (Ed) Plant Growth Substances 1985. Springer-Verlag, Berlin Heidelberg New York, pp 129-138

Yokota T, Arima M, Takahashi N (1982a) Castasterone, a new phytosterol with plant-hormone potency, from chestnut insect gall. Tetrahedon Lett 23: 1275-1278

Yokota T, Baba J, Takahashi N (1982b) A new steroidal lactone with plant growth-regulatory activity from *Dolichos lablab* seed. Tetrahedron Lett 23: 4965-4966

Yokota T, Baba J, Takahashi N (1983a) Brassinolide-related bioactive sterols in *Dolichos lablab*: Brassinolide, castasterone and a new analog, homodolicholide. Agric Biol Chem 47: 1409-1411

Yokota T, Morita M, Takahashi N (1983b) 6-Deoxocastasterone and 6-deoxodolichosterone: Putative precursors for brassinolide-related steroids from *Phaseolus vulgaris*. Agric Biol Chem 47: 2149-2151

Yokota T, Arima M, Takahashi N, et al (1983c) 2-Deoxycastasterone, a new brassinolide-related bioactive steroid from *Pinus* pollen. Agric Biol Chem 47: 2419-2420

Yokota T, Baba J, Koba S, et al (1984) Purification and separation of eight steroidal plant-growth regulators from *Dolichos lablab* seed. Agric Biol Chem 48: 2529-2534

Yokota T, Arima M, Takahashi N, et al (1985) Steroidal plant growth regulators, castasterone and typhasterol (2-deoxycastasterone) from the shoots of Sitka spruce (*Picea sitchensis*). Phytochemistry 24: 1333-1335

Yokota T, Kim SK, Fukui Y, et al (1987a) Brassinosteroids and sterols from a green alga, *Hydrodictyon reticulatum*: Configuration at C-24. Phytochemistry 26: 503-506

Yokota T, Koba S, Kim SK, et al (1987b) Diverse structural variations of the brassinosteroids in *Phaseolus vulgaris* seed. Agric Biol Chem 51: 1625-1631

Yokota T, Kim SK, Kosaka Y, et al (1987c) Conjugation of brassinosteroids. In: Schreiber K, Schutte HR, Sembdner G (Eds) Conjugated plant hormones, structure, metabolism and function. VEG Deuchter Verlag der Wissenschaften, Berlin, pp 288-296

Yokota T, Watanabe S, Ogino Y, et al (1990) Radioimmunoassay for brassinosteroids and its use for comparative analysis of brassinosteroids in stems and seeds of *Phaseolus vulgaris*. J Plant Growth Regul 9: 151-159

Yokota T, Ogino Y, Suzuki H, et al (1991) Metabolism and biosynthesis of brassinosteroids. In: Cutler HG, Yokota T, Adam G (Eds) Brassinosteroids; Chemistry, Bioactivity, and Applications. ACS Symp Ser 474, Amer Chem Soc, Washington, DC, pp 86-96

Yokota T, Nakayama M, Wakisaka T, et al (1994) 3-Dehydroteasterone, a 3,6-diketobrassinosteroid as a possible biosynthetic intermediate of brassinolide from wheat grain. Biosci Biotech Biochem 58: 1183-1185

Yokota T, Matsuoka T, Koarai T, et al (1996) 2-Deoxybrassinolide, a brassinosteroid from *Pisum sativum* seed. Phytochemistry 42: 509-511

Yokota T, Nomura T, Nakayama M (1997) Identification of brassinosteroids that appear to be derived from campesterol and cholesterol in tomato shoots. Plant Cell Physiol 38: 1291-1294

Yokota T, Higuchi K, Takahashi N, et al (1998) Identification of various brassinosteroids with epimerized substituents and/or the 23-oxo group in pollen and anther of Japanese cedar. Biosci Biotech Biochem 62: 526-531

3
Biochemical Analysis of Natural Brassinosteroids

SUGURU TAKATSUTO[1] AND TAKAO YOKOTA[2]

[1] Department of Chemistry, Joetsu University of Education, Joetsu, Niigata 943-0815, Japan
[2] Department of Biosciences, Teikyo University, Utsunomiya 320-8551, Japan

1. Introduction

Brassinolide (BL) was first isolated in 1979 in a crystalline form as a new plant growth promoter from rape (*Brassica napus* L.) pollen, and its structure was determined by spectroscopic analysis and X-ray diffraction (Grove et al 1979). Subsequently, another new BL-related compound termed castasterone (CS), was isolated in pure form from the insect galls of chestnut (*Castanea crenata*), and its structure was determined by spectroscopic methods (Yokota et al 1982). In isolating these brassinosteroids (BRs) from the plant sources, bioassays sensitive and specific to BRs were indispensable. Further, the development of gas chromatography-mass spectrometry (GC-MS) enabled us to investigate natural BRs in the plant kingdom. Nowadays, more than 40 natural BRs have been characterized from a number of plant families and species (Fujioka and Sakurai 1997, see the chapter by S. Fujioka, this volume), revealing that BRs are ubiquitously distributed from higher to lower plants.

In addition to the GC-MS microanalysis, other microanalytical methods for BRs including high performance liquid chromatography (HPLC) and liquid chromatography-mass spectrometry (LC-MS) have also been developed. Some reviews on microanalytical methods of BRs are available (Ikekawa and Takatsuto 1984, Takatsuto 1991 and 1994, Gamoh and Takatsuto 1994). Coupled with GC-MS microanalysis, labeled BRs have played important roles in the recent clarification of the biosynthesis of BRs and the quantitative analysis of endogenous BRs (Sakurai and Fujioka 1997, Fujioka and Sakurai 1997, see the chapter by A. Sakurai, this volume).

In this chapter, we summarize analytical methods for evaluating BRs, including

Key words: bioassay, radioimmunoassay, purification, NMR, GC-MS, HPLC, LC-MS, labeled brassinosteroids, sterols

bioassays and immunoassays, purification techniques for plant extracts, nuclear magnetic resonance (NMR) spectroscopy, GC-MS microanalysis, labeling BRs, HPLC microanalysis, and LC-MS microanalysis.

2. Bioassays and Immunoassay for Brassinosteroids

2.1 Bioassays

For the isolation and purification of BRs from plant sources, highly sensitive and specific bioassays are required, especially because of the very low concentrations of BRs in plants. Until now the rice lamina inclination assay and the bean second-internode bioassays have been used for this purpose. These bioassays as well as the wheat leaf-unrolling test, which have also been used for the structure-activity relationship study of BRs, are explained here. BRs have been known to be biologically active in other various bioassays systems (Mandava 1988, Sasse 1997), which are described in the chapter by J. Sasse, this volume.

The bean second-internode bioassay has been developed by Mitchell and Livingston (1968). This bioassay was used to monitor isolation of BL from rape pollen (Grove et al 1979) and also to evaluate the biological activity of BL and analogs by Thompson et al (1981, 1982). As a lanolin paste, BRs are applied to 6-day-old light-grown bean seedlings (*Phaseolus vulgaris* L., pinto variety), which showed not only elongation but also curvature, swelling, and splitting of the internodes, depending on the amount of BRs. Application of BL at as low as 10 ng per plant causes remarkable elongation with curvature and swelling. Synthetic stereoisomers of BL as well as 6-oxo-type BRs were also found to be active, although their activities were decreased drastically. In this bioassay BRs evoke both cell elongation and cell division, resulting in characteristic morphological changes of the second internode. In contrast, gibberellins cause only elongation of the treated and upper internodes. Auxins and cytokinins are not biologically active in this bioassay.

The rice-lamina inclination bioassay was developed by Maeda (1965) to examine the biological activity of synthetic auxins. In this bioassay, explants comprising leaf blade, sheath, and the joint between them are excised from etiolated rice seedlings grown under intermittent red light and floated on test solutions. Cells aligned in the adaxial side of the joint are swollen by BRs, causing lamina bending (Maeda and Saka 1968, Cao and Chen 1995). It was found earlier than the isolation of brassinolide that some plants contain biologically active substances in this bioassay (Marumo et al 1968, Munakata et al 1973) that later turned out to be BRs. This bioassay was found to be highly sensitive and specific for BRs (Wada et al 1981) and was successfully used to isolate CS from chestnut insect galls (Yokota et al 1982, Arima et al 1984). Later on, this bioassay was widely employed to isolate many BRs from a number of plant sources because of its simplicity. A further advantage for use of this bioassay for isolation of BRs from plant sources is that it does not show stringent structure-activity relationships as observed in the bean second-internode bioassay (Yokota and Mori 1992). Thus, this bioassay easily can detect 7-oxalactone-, 6-oxo-, and 2-deoxy-type BRs at relatively lower concentrations. The detection limit of BL is as low as 0.05 ng/ml. The response to BL and CS is linear to the logarithm of the concentrations from 5.00 to 0.05 ng/ml (Wada et al 1984), enabling quantitative bioassay of BRs. Although some cultivars, including Arborio J-1, Kinmaze, Koshihikari, and Nihonbare, have been used for this bioassay, the sensitivity and specificity of these cultivars to BRs seem to be similar.

Dipotassium maleate has been known to enhance lamina bending (Maeda 1965). However, dipotassium maleate is not essential for the bioassay because it never raises the detection limit of BRs. Indole-3-acetic acid (IAA) (Wada et al 1984) and monoglycerides (especially monoolein) (Park et al 1994) are biologically active in this bioassay although five orders of magnitude less active than BL (Wada et al 1984). Cytokinins were inactive and actually counteracted the effect of BL. Abscisic acid (ABA) also counteracted the effect of BL (Wada et al 1984).

Application of BRs to intact dwarf rice seedlings also causes lamina inclination (Takeno and Pharis 1982). This inclination effect is synergized by IAA. Based on these findings, a very sensitive dwarf rice lamina inclination assay for detection and semi-quantification of BRs has been developed (Kim et al 1990). The assay uses 3-day-old seedlings, and the detection of BRs is sensitized by co-application of IAA. The minimum detectable amount of BL is less than 0.1 ng/plant when BL is applied to the plants pretreated with 1000 ng of IAA. The dwarf rice assay is just as sensitive as the excised rice lamina inclination test. This bioassay has advantages that it needs only 1 week for the entire assay, and red light treatment is not needed to prepare the assay seedlings.

The wheat leaf unrolling test using explants from etiolated wheat seedlings is a convenient bioassay for BRs (Wada et al 1985). This assay is about 10-fold less sensitive than the rice-lamina inclination test. At 0.5 ng/ml BL and CS markedly stimulate unrolling, and at 0.01 µg/ml or higher BL produced complete unrolling of the leaf segments. Gibberellic acid (GA_3) produces only slight unrolling at 0.1 to 10 µg/ml, as does the cytokinin 6-(3-methyl-2-butenyl)aminopurine. However, zeatin causes complete unrolling at 1 µg/ml and has a measurable effect at 0.001 µg/ml. ABA, IAA, and indoleacetonitrile inhibit unrolling of leaf segments. This bioassay is advantageous because it needs simpler manipulation than the rice-lamina inclination test.

2.2 Radioimmunoassay

Radioimmunoassays (RIAs) for BRs have been developed by two groups. Horgen et al (1984) used synthetic 24-epibrassinolide (24-epiBL) as the hapten that was noncovalently bound to fetal calf serum. Hybridoma clones were generated from CAF1 mice and employed in an enzyme-linked immunosorbent assay (ELISA) for examination of BL distribution in *Brassica napus* tissues. It has been claimed that pollen-producing tissues gave the highest ELISA values. Although 24-epiBL could be detected by this ELISA, this assay was almost equally responsive to ecdysone and sitosterol (a very common phytosterol). This unusual response indicates that this method is unreliable and cannot be used for the analysis of BRs

Yokota et al (1990b) have reported a useful RIA for BRs. Antiserum against CS was produced by immunizing a rabbit with CS-carboxymethoxylamine oxime conjugated with bovine serum albumin. Detection limits of CS and BL were approximately 0.3 pmol, which are superior to those of the rice bioassay and the GC-SIM. It was difficult to attain higher sensitivity because of high background due to nonspecific binding of BRs to assay tubes. Cross-reactivities of 27 compounds including the natural and synthetic BRs were investigated, revealing the specificities of the antiserum in the recognition of various functional groups. Although there is an exception that the cross-reactivities of BRs with a 24-methylene group were quite low, it is of interest that the functional group recognition observed in the RIA correlates to high biological activities obtained by structure-activity relationship studies. This RIA system was successfully used for analyzing endogenous BRs in seeds and stems of *Phaseolus vulgaris* L (Yokota et al 1990b). The antiserum obtained has been utilized to localize BRs in pollen tissue by immunocytochemistry (Taylor et al 1993).

3. Purification of Natural Brassinosteroids

The level of BRs in vegetative tissues of plants is only few micrograms to submicrograms per kilogram fresh weight, so that purification techniques are quite important before qualitative and quantitative analysis of BRs by GC-MS. Figure 1 shows a typical procedure that is frequently applied to the analysis of endogenous BRs (Yokota et al 1984, Nomura et al 1997, Yokota et al 1997b).

The solvent used for extraction of BRs from plant tissues is methanol in many cases. Other solvents such as 2-propanol have been used to extract brassinolide from rape pollen. The aqueous residue obtained by evaporating the organic solvent is subjected to solvent partitionings. BRs can be transferred to the organic phase when partitioned against chloroform or ethyl acetate. Acidic substances are removed by partitioning the organic phase against an alkaline buffer. Partitionings between hexane and 80% methanol are also effective for removing nonpolar lipids. BRs are partitioned into the 80% methanol fraction, while lipids go into the hexane phase.

Endogenous BRs are then subjected to conventional column chromatography. Because there are a number of naturally occurring BRs with different chromatographic behaviors, they may be separated into a number of fractions during purification. However, this makes very tedious and difficult further purification and instrumental analysis of BRs. Thus, all endogenous BRs should be purified as a group before being subjected to HPLC. For this purpose, a combination of silica gel and Sephadex LH-20 chromatography has been frequently used. Among various solvent systems used for silica gel chromatography (Arima et al 1984, Yokota et al 1984), mixtures of methanol-chloroform seem to be most successful. Note that elution pro-

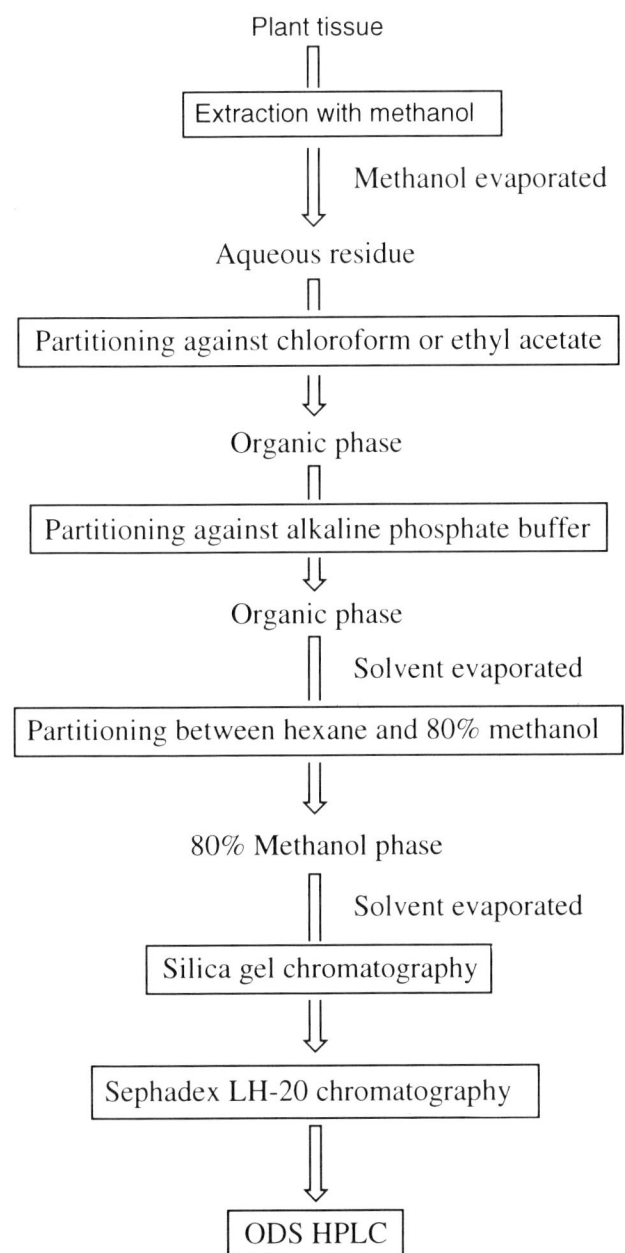

Fig. 1. Typical purification procedure of natural brassinosteroids.

Table 1. Effects of methanol concentrations on the elution of BRs in silica gel chromatography

	Methanol % in chloroform	
	Alcohol-stabilized chloroform	Alcohol-free chloroform
6-Deoxo3DT	0	1
6-DeoxoTE	0	1
3-DehydroTE	0, 0.5	2.5
6-DeoxoTY	0.5	2.5
TE	1, 2.5	5
TY	1, 2.5	5
6-DeoxoCS	2.5	5
CS	2.5, 5	5
BL	5	7

3DT, 3-dehydroteasterone; TE, teasterone; TY, typhasterol; CS, castasterone; BL, brassinolide.

files of BRs are different depending on whether commercial (alcohol-stabilized) chloroform or alcohol-free chloroform is used (Table 1). Sephadex LH-20 chromatography is generally a very effective purification technique to collect BRs as a group when proper mobile phase such as a 4:1 mixture of methanol-chloroform is used (Yokota et al 1984). Charcoal chromatography using methanol-chloroform mixtures as eluting solvents has been frequently used as a useful technique to enrich BRs from crude extracts (Park et al 1994, Yokota et al 1996, Nomura et al 1997). A short column of diethylaminosilica is often used to purify BRs prior to HPLC (Schmidt et al 1991).

Respective BRs are finally separated by HPLC with ODS columns, and then subjected to GC-MS. Separation of endogenous BRs in pea seed by ODS HPLC (Yokota et al 1996) is shown in Fig. 2, in which retention times of typical BRs are also indicated. Silica gel HPLC is also useful for extensively purifying BRs from plant extracts (Yokota et al 1984) but is less important in the GC-MS-based qualitative and quantitative analysis of BRs.

4. NMR of Natural Brassinosteroids

^{1}H NMR has been greatly advanced and thus is a powerful tool to determine the structures of minute amounts of BRs obtained from plants as well as of metabolites obtained in feeding experiments. This technique has been successfully used for the structure determination of 95 µg of CS isolated from chestnut insect galls (Yokota et al 1982), and of 160 µg of dolicholide, 50 µg of dolichosterone, 20 µg of homodolichosterone and 12 µg of homodolicholide isolated from *Dolichos lablab* seed (Yokota et al 1984). However, BRs to be examined should be purified to homogeneity prior to ^{1}H NMR. Thus this technique is not applicable for analysis of partially purified plant extracts. Details of ^{1}H and ^{13}C NMR are given in some references (Porzel et al 1992, Ando et al 1993, Adam et al 1996).

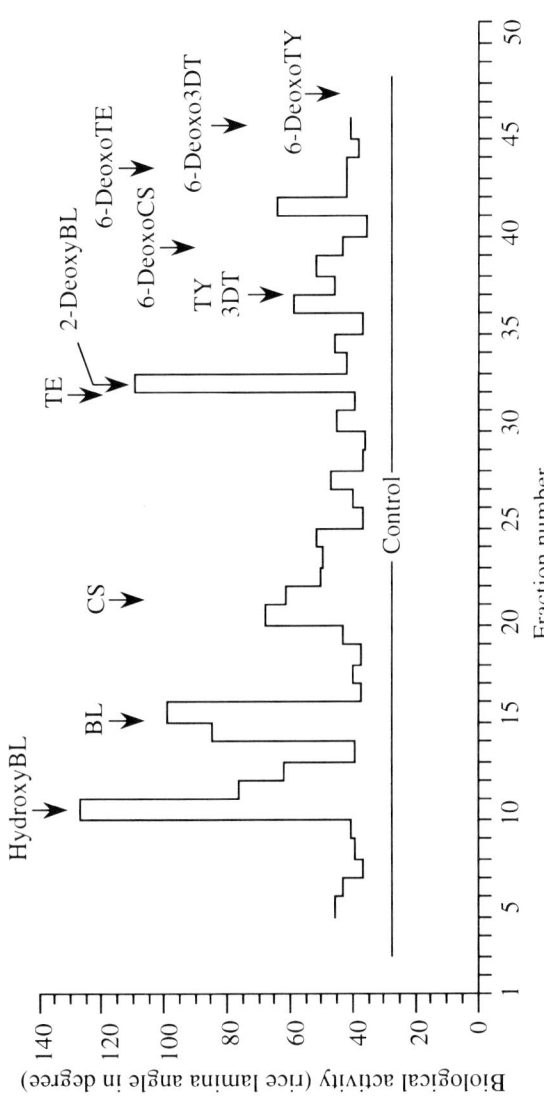

Fig. 2. Separation of endogenous brassinosteroids in pea seed by reverse-phase (ODS) HPLC (modified from Yokota et al 1996). Elution positions of authentic brassinosteroids are also indicated by arrows. BL, brassinolide; CS, castasterone; TE, teasterone; TY, typhasterol; 3DT, 3-dehydroteasterone. An ODS column (250 x 8 mm) was used. Elution solvent (flow rate 2.5 ml/min) was 0-20 min, 45% acetonitrile; 20-40 min, gradient to pure acetonitrile; 40-50 min, pure acetonitrile. Fractions were collected every minute. HydroxyBL (unknown structure), BL, CS, 2-deoxyBL, TY, 6-deoxoCS, and an epimer of 6-deoxoCS have been identified from the pea seed by GC-MS.

5. GC-MS

GC-MS and GC-MS-selected ion monitoring (SIM) are highly sensitive and selective analytical methods that can be applied to partially purified BR fractions from a limited amount of plant materials (Takatsuto et al 1982, Ikekawa and Takatsuto 1984, Takatsuto and Ikekawa 1986a, Takatsuto 1991 and 1994).

Brassinosteroids are involatile compounds because of the presence of two sets of vicinal diol groups in the A-ring and side chain. Hence they must be converted into volatile derivatives prior to GC. Considering its applicability to the microanalysis of natural BRs in plant extracts, the desired derivatives of BRs are bismethaneboronates (BMBs) which can be prepared with methaneboronic acid (Fig. 3). Because methaneboronic acid reacts specifically with vicinal diols, BRs are easily separated from plant constituents having no vicinal diols by GC after methaneboronation. A typical GC profile of authentic BRs is presented in Fig. 4. A pair of 24-epimers of BL and CS were also completely separated by GC using capillary columns (Takatsuto et al 1982, Ikekawa et al 1984). In the case of 2-deoxy BRs such as teasterone (TE) and typhasterol (TY), which have a vicinal diol function in the side chain and an isolated hydroxyl group in the A-ring, the diol moiety is first methaneboronated and then the remaining 3-hydroxyl group is trimethylsilylated with appropriate silylating reagents to give a methaneboronate-trimethylsilyl (MB-TMS) derivative (Takatsuto

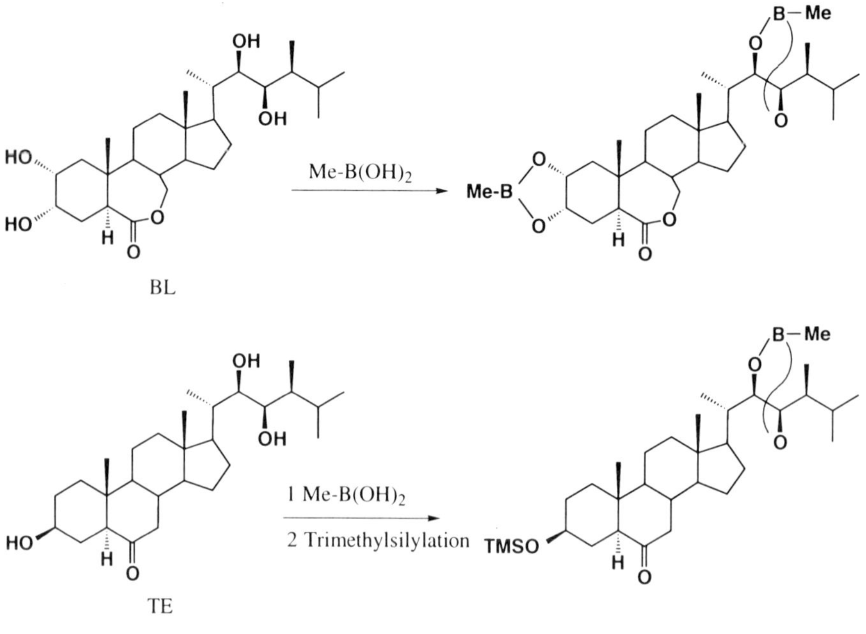

Fig. 3. Derivatization of brassinolide to a bismethaneboronate and of teasterone to a methaneboronate-trimethylsilyl ether.

3 Analysis of Brassinosteroids   55

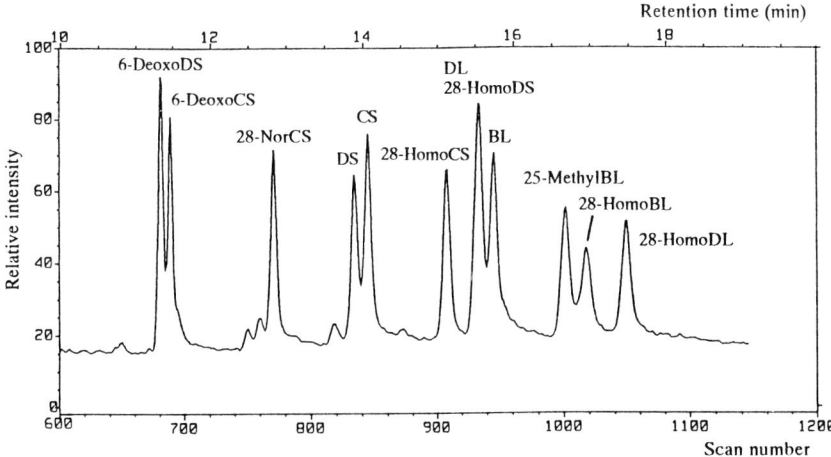

Fig. 4. Total ion current trace of bismethnaboroantes of authentic brassinosteroids in gas chromatography-mass spectrometry (T Yokota unpublished results). A J & W DB-1 capillary column (15 m x 0.25 mm; 0.25 μm film thickness) was used. The column oven temperature was maintained at 170° for the first 2 min, elevated to 270° at 32°/min and then to 290° at 2°/min, and finally maintained at 290°.

and Ikekawa 1986a) (Fig. 3). The MB-TMS derivatives of TE, TY, and their 28-homo and 28-nor analogs are well separated by GC using capillary columns (Takatsuto and Ikekawa 1986a).

The detection limits of these derivatives are the levels of nanograms for full-scan EI-MS and the levels of subnanograms for GC-EI-SIM. Fragmentation patterns of these BR derivatives have been summarized (Ikekawa and Takatsuto 1984, Takatsuto and Ikekawa 1986a, Takatsuto 1991 and 1994). These derivatives show several characteristic fragment ions that are useful for characterizing or identifying BRs (Takatsuto et al 1982, Ikekawa and Takatsuto 1984). Characteristic fragment ions and Kovats retention time indices of the biosynthetically important BRs are summarized in Table 2.

The GC-MS microanalysis techniques have been employed to determine the structures of new BRs and have also contributed greatly to revealing the wide occurrence of BRs in the plant kingdom (Takatsuto 1994, Fujioka and Sakurai 1997, see the chapter by S. Fujioka, this volume). Further, in combination with the use of deuterium-labeled BRs, the GC-MS techniques have also played crucial roles in the quantitation of BRs in plants as well as in the biosynthetic studies of BRs (Sakurai and Fujioka 1997, Fujioka and Sakurai 1997, Sasse 1997, see the chapter by A. Sakurai, this volume).

Table 2. GC-EIMS data for biosynthetically important brassinosteroids

Compound	Kovats retention index	Prominent ions (relative intensity)	Ref
BL[a]	3724	528 (M$^+$, 11), 457 (13), 397 (14), 374 (47), 345 (36), 332 (34), 319 (20), 177 (88), 163 (36), 155 (100)	e
CS[a]	3619	512 (M$^+$, 81), 441 (17), 358 (37), 287 (26), 155 (100)	e, f, g
TY[b]	3545	544 (M$^+$, 100), 529 (46), 526 (41), 515 (92), 454 (82), 155 (19)	f
3DT[c]	3607	470 (M$^+$, 49), 399 (6), 339 (4), 316 (20), 287 (11), 260 (8), 245 (15), 155 (100)	f
TE[b]	3656	544 (M$^+$, 51), 529 (100), 515 (92), 454 (32), 155 (12)	f
CT[d]	-	561 (M$^+$-15, 6), 491 (4), 462 (100), 187 (61)	h
6-DeoxoCS[a]	3429	498 (M$^+$, 77), 483 (22), 288 (21), 273 (100), 155 (26)	e, f
6-DeoxoTY[b]	-	530 (M$^+$, 23), 440 (31), 425 (29), 369 (7), 305 (17), 285 (17), 230 (34), 215 (100), 155 (29)	g
6-Deoxo3DT[c]	-	456 (M$^+$, 64), 385 (8), 301 (24), 246 (23), 231 (100), 155 (64)	g
6-DeoxoTE[b]	-	530 (M$^+$, 100), 515 (48), 473 (14), 440 (41), 425 (31), 369 (9), 305 (31), 285 (17), 257 (15), 230 (28), 215 (82), 155 (21)	h

CT, cathasterone.
Mass pectral data were obtained by a magnetic mass spectrometer (70 eV) equipped with a capillary column DB-1 or DB-5 (15 m x 0.25 mm, 0.25 μm film thickness). Kovats retention indices were obtained by the DB-1 column. The column temperature was held at 170° for the first 1.5 min, elevated to 280° at the rate of 37°/min, then to 300° at the rate of 1.5°/min, and finally held at 300°.

a) Bismethaneboronate derivative, b) methaneboronate-trimethylsilyl derivative, c) methaneboronate derivative, d) bistrimethylsilyl derivative, e) Yokota et al 1996 f) Yokota et al 1994, g) Griffiths et al 1995, h) unpublished results (as for the quadrupole spectrum of cathasterone, see Fujioka et al 1995).

6. Labeled Brassinosteroids and Related Steroids

Labeled compounds are essential for clarifying the quantitation, biosynthesis, metabolism and transport of BRs. This section summarizes the syntheses and applications of compounds labeled with radioactive or stable isotopes.

Several radiolabeled BRs have been synthesized. The structures and the labeled positions with tritium or ^{14}C are shown in Fig. 5. Yokota et al (1990a, 1991) synthesized [24,28-^{3}H$_2$]CS and [24,28-^{3}H$_2$]BL by catalytic reduction of the C-24(28) double bond of dolichosterone and dolicholide with tritium gas, respectively, followed by separation of the resulting 24-epimers by means of ODS-HPLC (see the chapter by T.C. McMorris, this volume). Using these radio-labeled BRs, they have developed a radioimmunoassay (Yokota et al 1990b) and also investigated the metabolism of CS

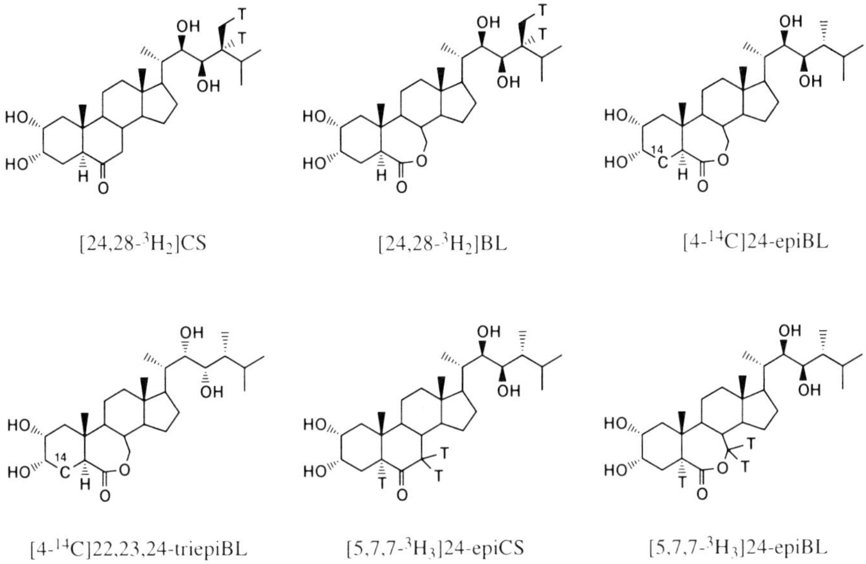

[24,28-^{3}H$_2$]CS   [24,28-^{3}H$_2$]BL   [4-^{14}C]24-epiBL

[4-^{14}C]22,23,24-triepiBL   [5,7,7-^{3}H$_3$]24-epiCS   [5,7,7-^{3}H$_3$]24-epiBL

Fig. 5. Radio-labeled brassinosteroids.

and BL in mung bean seedlings (Suzuki et al 1993). The first evidence that BL is biosynthesized from CS was obtained by feeding ^{3}H-labeled CS to cultured crown gall cells of *Catharansus roseus* (Yokota et al 1990a).

Seo et al (1989) have synthesized ^{14}C-labeled 24-epiBL during the course of development of 24-epiBL as a practical candidate for agricultural use. They introduced a ^{14}C atom into the C-4 position of brassicasterol, and then the labeled brassicasterol was transformed into ^{14}C-labeled 24-epiBL and 22,23,24-triepiBL (see the chapter by T.C. McMorris, this volume).

Kolbe et al (1995) synthesized ^{3}H- and ^{2}H-labeled 24-epiCS by base-catalyzed exchange reaction. These were converted to ^{3}H- and ^{2}H-labeled 24-epiBL (see the chapter by T.C. McMorris, this volume). ^{3}H-labeled CS and BL were used to clarify the metabolism of 24-epiCS and 24-epiBL in cell suspension cultures of *Lycopersicon esculentum* and *Ornithopus sativus* (Adam et al 1996, see the chapter by G. Adam and B. Schneider, this volume).

Several ^{2}H-labeled BRs have been synthesized (Fig. 6). [26,28-^{2}H$_6$]BL, [26,28-^{2}H$_6$]CS, [26,28-^{2}H$_6$]TY, and [26,28-^{2}H$_6$]TE have been synthesized from [26,28-^{2}H$_6$]-crinosterol, which was prepared by introducing two deuterated-methyl groups stepwise into the side chain of crinosterol (Takatsuto and Ikekawa, 1986b,c). Later, using these ^{2}H-labeled-BRs and synthetic intermediates as starting materials, several ^{2}H-labeled BRs including cathasterone (CT), 3-dehydroTE (3DT), 6-deoxoTE, 6-deoxo3DT, 6-deoxoTY, and 6-deoxoCS were synthesized (Suzuki et al 1994, Fujioka et al 1995, Choi et al 1997). These ^{2}H-labeled BRs were used as internal

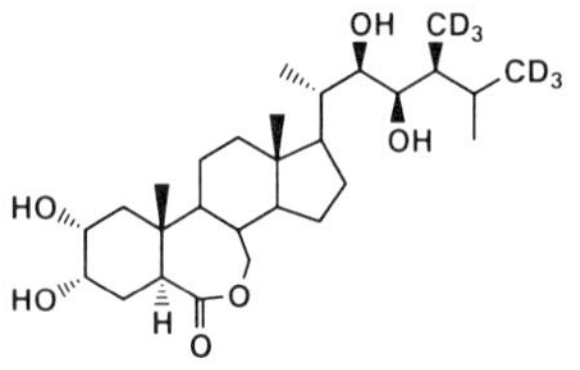

[26,28-$^{2}H_{6}$]BL

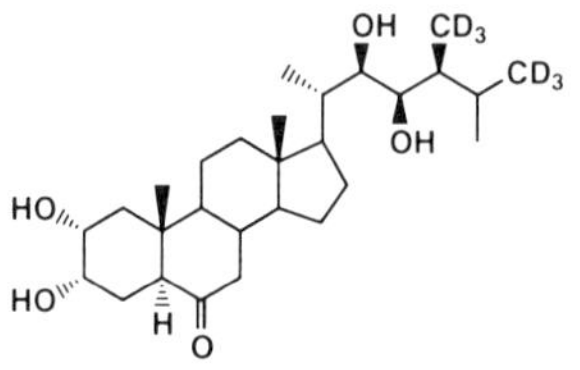

[26,28-$^{2}H_{6}$]CS

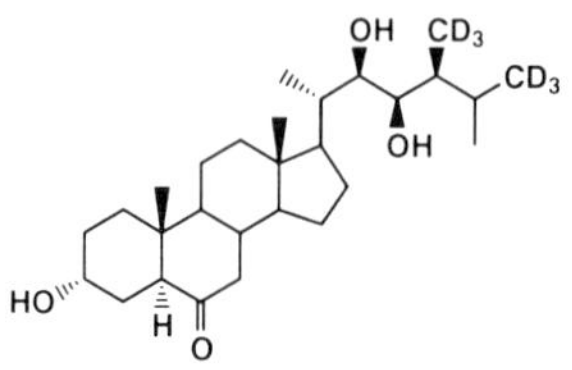

[26,28-$^{2}H_{6}$]TY

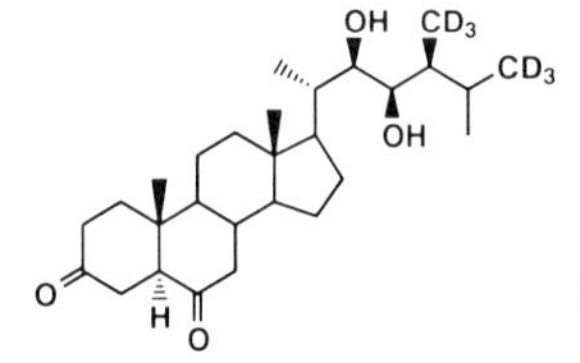

[26,28-$^{2}H_{6}$]3DT

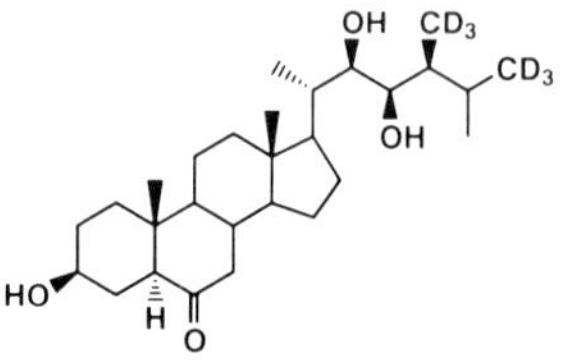

[26,28-$^{2}H_{6}$]TE

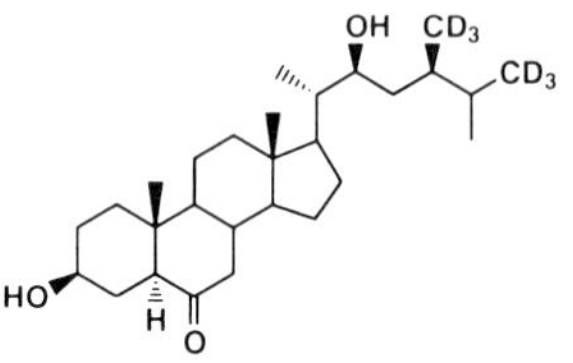

[26,28-$^{2}H_{6}$]CT

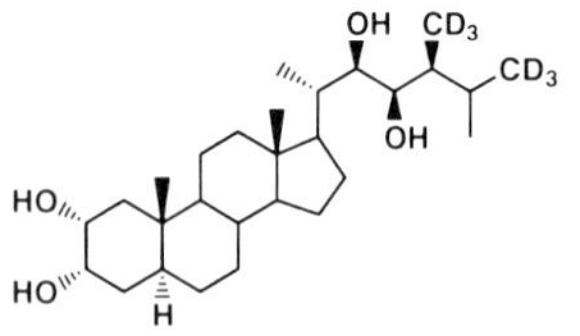

[26,28-$^{2}H_{6}$]6-deoxoCS

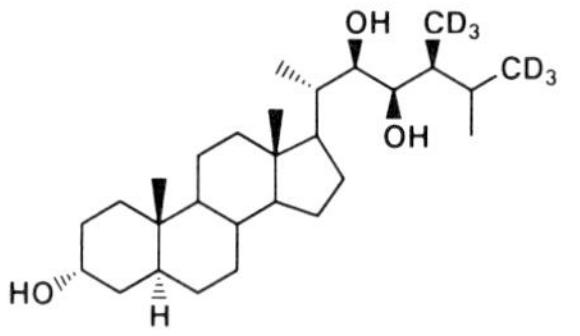

[26,28-$^{2}H_{6}$]6-deoxoTY

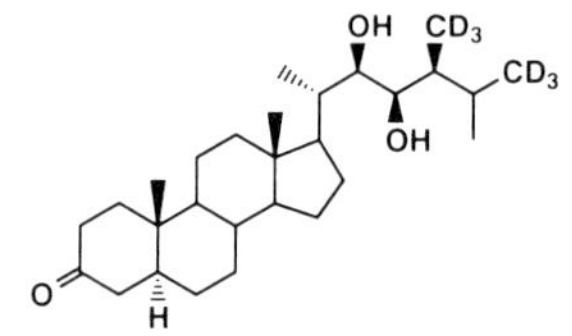

[26,28-$^{2}H_{6}$]6-deoxo3DT

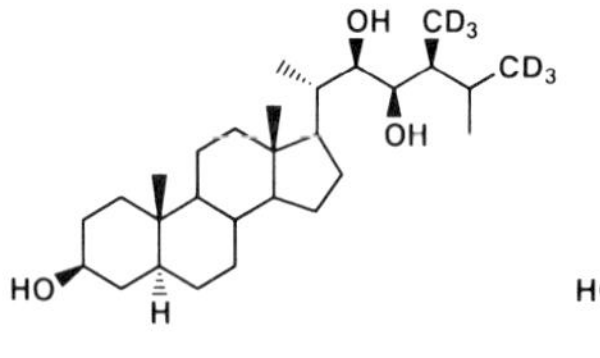

[26,28-$^{2}H_{6}$]6-deoxoTE

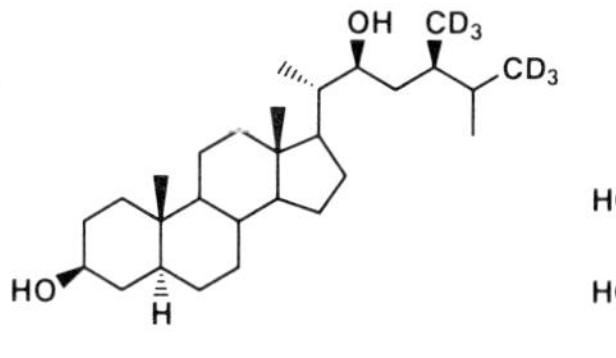

[26,28-$^{2}H_{6}$]6-deoxoCT

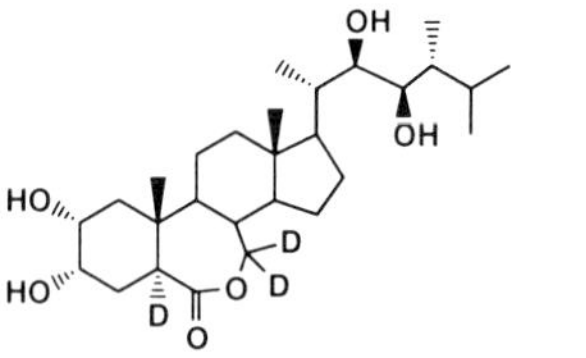

[5,7,7-$^{2}H_{3}$]24-epiBL

Fig. 6. Deuterio-labeled brassinosteroids.

standards for quantitative analysis of endogenous BRs of several plants. They have also been used for the biosynthetic study of BRs in cultured cells of *Catharanthus roseus* and several plant seedlings, establishing two biosynthetic sequences, the early C6-oxidation pathway: CT → TE → 3-dehydroTE → TY → CS and the late C6-oxidation pathway: 6-deoxoTE → 6-deoxo3DT → 6-deoxoTY → 6-deoxoCS → CS, as well as the final pathway from CS to BL (Fujioka and Sakurai 1997, Sakurai and Fujioka 1997, see the chapter by A. Sakurai, this volume).

[5,7,7-$^{2}H_{3}$]24-epiBL was synthesized by Allevi et al (1988). Three deuterium atoms were incorporated into the C-5 and C-7 positions of 24-epiCS by base-catalyzed enolization, which was then transformed into [5,7,7-^{2}H$_3$]24-epiBL. They used this compound for evaluation of the uptake of 24-epiBL in growing maize root segments by GC-SIM.

Campesterol, a common phytosterol, has been hypothesized to be an early precursor of C$_{28}$ BRs such as CS and BL, the most frequently occurring BRs, because campesterol has the same carbon skeleton as these BRs, including the stereochemistry of the methyl group at the C-24 position (Yokota et al 1991). Thus, for biosynthetic purposes, labeled campesterol, and a number of related labeled sterols have been prepared by synthesis or biotransformation (Fig. 7). [^{13}C$_5$]Campesterol and [^{13}C$_5$]campestanol were prepared by feeding [(R)-2-^{13}C] mevalonic acid to the cultured cells of *C. roseus* in the presence of compactin, a mevalonic acid biosynthetic inhibitor (Suzuki et al 1995, see the chapter by A. Sakurai, this volume). [^{14}C]Campesterol and [^{14}C]campestanol were also prepared from [(R)-2-^{14}C] mevalonic acid using the same method (Suzuki et al 1995). Chemical synthesis of [26,28-^{2}H$_6$]campesterol, [26,28-^{2}H$_6$]campestanol, [26,28-^{2}H$_6$]6α-hydroxycampesterol and [26,28-^{2}H$_6$]6-oxocampesterol were elaborated from [26,28-^{2}H$_6$]crinosterol mentioned above. Feeding experiments using these sterols established the early BL biosynthetic pathway of campesterol → campestanol → 6α-hydroxycampesterol → 6-oxocampesterol (Suzuki et al 1995, Fujioka et al 1997a, see the chapter by A. Sakurai, this volume).

Conversion of campesterol to campestanol has been found to be impaired in the BR-deficient dwarf mutant *det2* of *Arabidopsis thaliana* (Li et al 1996) and *lk* of pea (Yokota et al 1997a, Kitasaka 1997). To clarify multistep reactions between campesterol and campestanol and specify the blocked point in the mutant, several tentative candidates including [26,28-^{2}H$_6$](24R)-24-methylcholest-4-en-3β-ol, [26,28-^{2}H$_6$](24R)-24-methylcholest-4-en-3-one, and [26,28-^{2}H$_6$](24R)-5α-24-methylcholestan-3-one were synthesized from [26,28-^{2}H$_6$]campesterol (S. Takatsuto and S. Fujioka, unpublished results) and fed to the *det2* plants. Thus, a pathway of campesterol → (24R)-24-methylcholest-4-en-3β-ol → (24R)-24-methylcholest-4-en-3-one → (24R)-5α-24-methylcholestan-3-one → campestanol was established in the wild type plants of *A. thaliana*, while conversion of (24R)-24-methylcholest-4-en-3-one to (24R)-5α-24-methylcholestan-3-one was found to be blocked in the *det2* mutant (Fujioka et al 1997a, Noguchi et al 1997, see the chapter by A. Sakurai, this volume). The same biosynthetic pathway was also established in the cultured cells of *Catharanthus roseus* (Noguchi et al 1998).

Biosynthesis of campesterol from 24-methylenecholesterol has been found to be blocked in the dwarf mutant *lkb* of pea (Yokota et al 1997a) and the dwarf mutant *dim* of *A. thaliana* (Klahre et al 1997) because these mutants are deficient in campesterol but accumulate 24-methylenecholesterol. Thus, [25,26,27-^{2}H$_7$]24-methylenecholesterol and its related sterols were synthesized (Takatsuto et al 1998). Using these labeled compounds, it has recently been found that these dwarf mutants have a defect in the isomerization-reduction step of 24-methylenecholesterol into campesterol via 24-methyldesmosterol (Yokota et al 1997a, Klahre et al 1997, Fujioka et al 1997b).

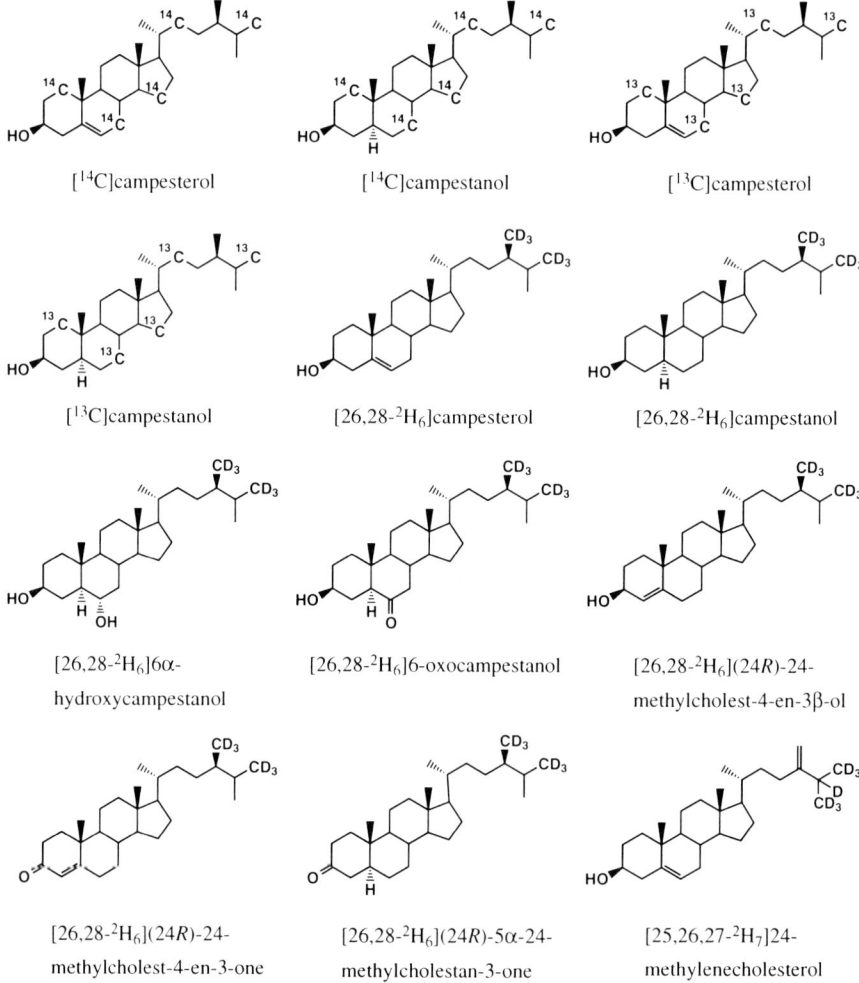

Fig. 7. Labeled sterols for biosynthetic study of BRs.

Biosyntheses from campesterol to CT in the early C6-oxidation pathway and to 6-deoxoTE in the late C6-oxidation pathway are still not fully understood. Among a number of presumable intermediates, [26,28-^{2}H$_6$]6-deoxoCT was synthesized as a precursor of CT and 6-deoxoTE, and [26,28-^{2}H$_6$]6α-hydroxyCT as a precursor of CT (H Kuriyama, S Takatsuto, T Noguchi, S Fujioka unpublished results).

7. HPLC

Highly sensitive HPLC analytical methods for BRs have been developed (Gamoh and Takatsuto 1994). Because BRs have no suitable chromophore for detection, they are derivatized with pre-labeling reagents so as to be responsive to either ultraviolet (UV), fluorimetric, or electrochemical detectors. Several boronic acid reagents that react with the vicinal diol group(s) of BRs have been developed as the prelabeling reagents (Fig. 8), including naphthaleneboronic acid for UV detection (Gamoh et al 1988), 9-phenanthreneboronic acid (Gamoh et al 1989), 1-cyanoisoindole-2-*m*-phenylboronic acid (Gamoh and Takatsuto 1989), and dansylaminophenylboronic acid (Gamoh et al 1990a) for fluorimetric detection and ferroceneboronic acid for electrochemical detection (Gamoh et al 1990b). The detection limits depend on the derivatives, ranging from 25 to 100 pg per injection. Among these reagents, dansylaminophenylboronic acid seems to be most effective, because BR dansylaminophenylboronates can be detected at longer wavelength (excitation 345 nm/emission 515 nm) than the other boronates and hence are least interfered with by the matrix. The derivatized BRs are effectively separable by HPLC using ODS columns with acetonitrile-water as the mobile phase. A typical chromatogram of authentic BR dansylaminophenylboronates is presented in Fig. 9.

Although BL and CS have an α-oriented methyl at C-24, their epimers with a β-methyl also occur in plants, though less frequently. As described in the section on GC-MS, a pair of 24-epimers of BL and CS are easily separated by GC using capillary columns. The 24-epimers can also be separated by HPLC but, after derivatizing

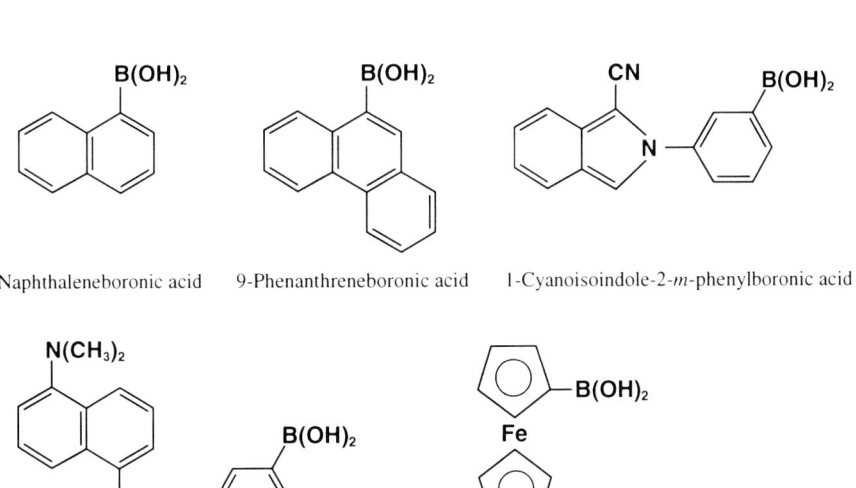

Fig. 8. Boronic acid reagents used for pre-labeling of brassinosteroids in high-performance liquid chromatography.

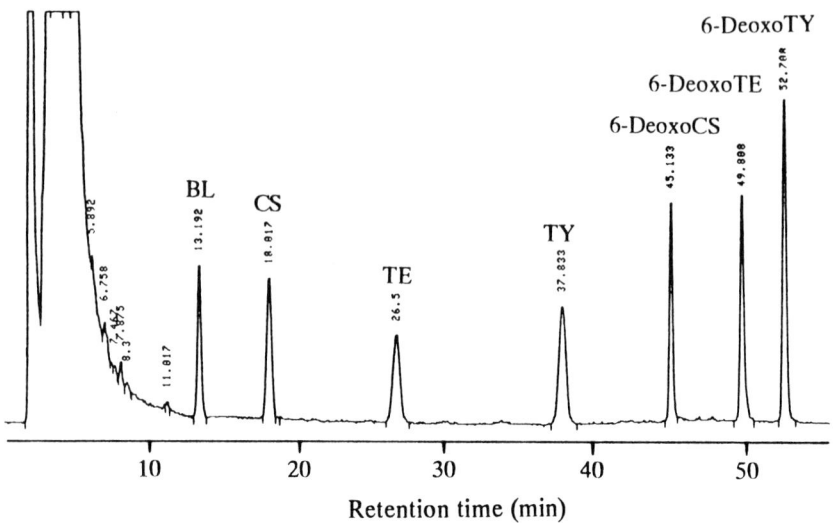

Fig. 9. Chromatographic profile of dansylaminophenylboronates of authentic brassinosteroids (each 1 ng) in reverse-phase (ODS) high-performance liquid chromatography (T Yokota unpublished results). An ODS column (150 x 6 mm) was used. Elution solvent (flow rate 1 ml/min) was 0-30 min, 80% acetonitrile; 30-40 min, gradient to pure actonitrile; 40 -60 min, pure acetonitrile.

with the above boronic acid reagents, cannot be resolved. Gamoh et al (1992) succeeded in the separation and detection of the 24-epimeric pairs of BL and CS by HPLC using a combination of a pre-column labeling and a post-column fluorescence detection. 24-Epimeric mixtures of BL or CS are derivatized to m-aminophenylboronates, which are easily separable using an ODS column with a mobile phase of acetonitrile-1% acetic acid (75:25, v/v) containing 20 mM 18-crown-6. These boronates can be detected using a post-column reaction system where m-aminophenylboronates are reacted with o-phthalaldehyde in the presence of cyanide ions to give 1-cyanoisoindole derivatives, which fluoresce at 400 nm (excitation at 330 nm). BL amounts as low as 40 ng (S/N = 3) are detectable by this technique.

The HPLC methods stated above have been successfully applied to analyze BRs in the pollen of broad bean, corn, sunflower, buckwheat, and orange (Takatsuto 1991, Gamoh and Takatsuto 1994, Motegi et al 1994). Recently, the HPLC method using dansylaminophenylboronates has been employed for the identification of BRs in the pollen of *Cupressus arizonica* (Griffiths et al 1995). Shim et al (1996) have identified BL and examined its fluctuation during development of rice by HPLC with 1-cyanoisoindole-2-m-phenylboronic acid as a prelabeling reagent, resulting in the detection of BL at the heading and panicle formation stages but not at the blooming and elongation stages.

A solid-phase extraction technique of BRs using an immobilized phenylboronic acid gel that has high affinity for BRs carrying a C-22,23-diol has been developed (Gamoh et al 1994). BRs having a C-22,23-diol group are trapped as their cyclic boronates by the gel in a mixture of pyridine and acetonitrile. Then the parent BRs are easily released from their boronates by a mixed solution of hydrogen peroxide and acetonitrile. A highly selective and sensitive analytical method of BRs based on the molecular recognition ability of boronic acid derivatives has been studied (Gamoh 1996).

8. LC-MS

Microanalytical LC-MS methods to analyze BRs as their boronates have been developed using either atmospheric pressure chemical ionization (APCI) or electrospray ionization (ESI). The APCI-based LC-MS was applied to naphthaleneboronate derivatives of BRs and optimum results were obtained by the use of reverse-phase HPLC with a mobile phase of a mixture of acetonitrile and water (Gamoh et al 1996a). Typical ions observed in the positive-ion spectra of the naphthaleneboronates are a pseudo-molecular ion, $[M + H]^+$, and a fragment ion, $[M + H - H_2O]^+$. The $[M + H]^+$ ion is the most abundant in BL while the $[M + H - H_2O]^+$ ion is the most abundant in CS, TE, and TY. Full-scan mass spectra were readily obtained from 400 ng of free BRs, whereas a limit of detection in the SIM mode is 2 ng (S/N = 3). This method was applied to the analysis of *Cannabis sativa* L seeds. Both the ESI and APCI methods have been applied to the reversed-phase HPLC analysis of BR boronates including naphthaleneboronates, *m*-aminophenylboronates, and dansylaminophenylboronates (Gamoh et al 1996b). The effects of additives to the mobile phase on the detectability of the boronates were examined. The most effective mobile phase for naphthaleneboronates was found to be a 9:1 mixture of acetonitrile and water containing 0.5% formic acid. When LC-MS analysis of BL naphthaleneboronate was conducted using this mobile phase in combination with the ESI mode, detection limits of 25 pg (S/N = 3) and 2.5 pg (S/N = 4) were obtained in the SIM mode and multiple reaction monitoring mode, respectively.

By means of LC-MS analysis, TE-3-myristate has been identified as the first BR fatty acid ester from the lily anther (Asakawa et al 1994). Detection was done by negative ion fast atom bombardment (FAB)-MS. A column-switching LC system was used, consisting of a trapping C_8-LC column and an analytical ODS column. The mobile phase was 70% methanol in water in the trapping column and 10% chloroform in methanol in the analytical column. Introduction of the eluate to the flit-FAB-MS system was done using a splitter to reduce the flow rate.

9. Conclusions

With respect to the biological methods to detect BRs, highly sensitive and specific bioassays and radioimmunoassays have been developed. Concerning the instrumental microanalysis of BRs, GC-MS (SIM), HPLC, and LC-MS have been developed. Investigation of endogenous BRs using a combination of bioassays, purification methods, and instrumental microanalyses has revealed that BRs occur unexceptionally in plants, allowing us to conclude that BRs are plant hormones. In particular, GC-MS, coupled with the labeled compounds, has played a decisive role in the analysis of the metabolites from feeding experiments, leading to the clarification of BR biosynthesis. Furthermore, these techniques were successfully used to pinpoint the biosynthetic lesions in BR-deficient mutants of pea and Arabidopsis. Thus, during almost the past two decades, we have greatly widened our knowledge on the chemistry, physiology, and molecular biology of BRs and are still envisaging their practical applications in agriculture. The microanalyses will continue to contribute to deeper understanding of the regulation of the biosynthesis and the physiological mechanism of BRs.

References

Adam G, Porzel A, Schmidt J, et al (1996) New developments in brassinosteroid research. In: Atta-ur-Rahman (Ed) Studies in Natural Products Chemistry, vol 18 Elsevier Science BV, Amsterdam, pp 495-549

Allevi P, Anastasia M, Cerana R, et al (1988) 24-Epibrassinolide uptake in growing maize root segments evaluated by multiple-selected ion monitoring. Phytochemistry 27:1309-1313

Ando T, Aburatani M, Koseki N, et al (1993) ^{13}C NMR assignments of brassinosteroids by two-dimensional techniques. Magn Reson Chem 30:651-657

Arima M, Yokota T, Takahashi N (1984) Identification and quantification of brassinolide-related steroids in the insect gall and healthy tissues of the chestnut plant. Phytochemistry 23:1587-1591

Asakawa S, Abe H, Kyokawa Y, et al (1994) Teasterone 3-myristate: a new type of brassinosteroid derivatives in *Lilium longiflorum* anthers. Biosci Biotech Biochem 58:219-220

Cao H, Chen S (1995) Brassinosteroid-induced rice lamina inclination and its relation to indole-acetic acid and ethylene. Plant Growth Regul 16:189-196

Choi Y-H, Fujioka S, Nomura T, et al (1997) An alternative brassinolide biosynthetic pathway via late C-6 oxidation. Phytochemistry 44:609-613

Fujioka S, Sakurai A (1997) Brassinosteroids. Nat Prod Rep 14:1-10

Fujioka S, Inoue T, Takatsuto S, et al (1995) Identification of a new brassinosteroid, cathasterone, in cultured cells of *Catharanthus roseus* as a biosynthetic precursor of teasterone. Biosci Biotech Biochem 59:1543-1547.

Fujioka S, Li J, Choi Y-H, et al (1997a) The Arabidopsis *det2* mutant is blocked early in brassinosteroid biosynthesis. Plant Cell 9:1951-1962

Fujioka S, Noguchi T, Takatsuto S, et al (1997b) Dwarf Arabidopsis mutant *dim* is brassinosteroid-deficient due to blocked biosynthesis of campesterol. Abstract paper of the 32nd Annual Meeting of Jap Soc for Chem Regul Plants, Tokyo, pp 91-92

Gamoh K (1996) Study of a highly selective and sensitive analytical method based on the molecular recognition ability of boronic acid derivatives. Bunseki Kagaku 45:19-30 (in Japanese)

Gamoh K, Takatsuto S (1989) A new boronic acid derivative as a highly sensitive fluorescence derivatization reagent for brassinosteroids in liquid chromatography. Anal Chim Acta 222:201-204

Gamoh K, Takatsuto S (1994) Liquid chromatographic assay of brassinosteroids in plants. J Chromatogr A 658:17-25

Gamoh K, Kitsuwa T, Takatsuto S, et al (1988) Determination of trace brassinosteroids by high performance liquid chromatography. Anal Sci 4:533-535

Gamoh K, Omote K, Okamoto N, et al (1989) High performance liquid chromatography of brassinosteroids in plants with derivatization using 9-phenanthreneboronic acid. J Chromatogr 469:424-428

Gamoh K, Okamoto N, Takatsuto S, et al (1990a) Determination of traces of natural brassinosteroids as dansylaminophenylboronates by liquid chromatography with fluorimetric detection. Anal Chim Acta 228:101-105

Gamoh K, Sawamoto H, Takatsuto S, et al (1990b) Ferroceneboronic acid as a derivatization reagent for the determination of brassinosteroids by high-performance liquid chromatography with electrochemical detection. J Chromatogr 515:227-231

Gamoh K, Takatsuto S, Ikekawa N (1992) Effective separation of C-24 epimeric brassinosteroids by liquid chromatography. Anal Chim Acta 256:319-322

Gamoh K, Yamaguchi I, Takatsuto S (1994) Rapid and selective sample preparation for the chromatographic determination of brassinosteroids from plant material using solid-phase extraction method. Anal Sci 10:913-917

Gamoh K, Abe H, Shimada K, et al (1996a) Liquid chromatography/mass spectrometry with atmospheric pressure chemical ionization of free brassinosteroids. Rapid Commun Mass Spectrom 10:903-906

Gamoh K, Prescott MC, Goad LJ, et al (1996b) Analysis of brassinosteroids by liquid chromatography/mass spectrometry Bunseki Kagaku 45:523-527 (in Japanese)

Griffiths PG, Sasse JM, Yokota T, et al (1995) 6-Deoxotyphasterol and 3-dehydro-6-deoxoteasterone, possible precursors to brassinosteroids in the pollen of *Cupressus arizonica*. Biosci Biotech Biochem 59:956-959

Grove MD, Spencer GF, Rohwedder WK, et al (1979) Brassinolide, a plant growth-promoting steroid isolated from *Brassica napus* pollen. Nature 281:216-217

Horgen PA, Nakagawa CH, Irvin RT (1984) Production of monoclonal antibodies to a steroid plant growth regulator. Can J Biochem Cell Biol 62:715-721

Ikekawa N, Takatsuto S (1984) Microanalysis of brassinosteroids in plants by gas chromatography/mass spectrometry. Mass Spectrosc 32:55-70

Ikekawa N, Takatsuto S, Kitsuwa T, et al (1984) Analysis of natural brassinosteroids by gas chromatography and gas chromatography-mass spectrometry. J Chromatogr 290:289-302

Kim S-K, Abe H, Little CHA, et al (1990) Identification of two brassinosteroids from cambial region of Scots pine (*Pinus silverstris*) by gas chromatography-mass spectrometry, after detection using a dwarf rice lamina inclination bioassay. Plant Physiol 94:1709-1713

Kitasaka Y, Nomura T, Takatsuto S, et al (1997) New brassinosteroid-deficient pea mutant *lk* with blocked synthesis of campesterol. Abstract paper of the 32nd Annual Meeting of

Jap Soc Chem Regul Plants, Tokyo, pp 89-90

Klahre U, Fujioka S, Yokota T, et al (1997) Characterization of the *diminuto* mutant and genes regulated by brassinosteroids. Proc Plant Growth Reg Soc Amer 24: 99

Kolbe A, Schneider B, Porzel A, et al (1995) Investigations on the metabolism of ^{3}H labeled brassinosteroids in cell suspension cultures of *Ornithopus sativus*. In: Allen J, Voges R (Eds) Proc 5th Int Symp on Synth Appl Isot Labelled Compd 1994, Chichester, pp 331-334

Li J, Nagpal P, Vitart V, et al (1996) A role of brassinosteroids in light-dependent development of Arabidopsis. Science 272:398-401

Maeda E (1965) Rate of lamina inclination in excised rice leaves. Physiol Plant 18: 813-827

Maeda E, Saka H (1968) Establishment of a new method for study of the inclination rate of lamina in the excised rice leaves. Proc Crop Sci Soc Japan 37:37-44

Mandava NB (1988) Plant growth-promoting brassinosteroids. Annu Rev Plant Physiol Plant Mol Biol 39:23-52

Marumo S, Hattori H, Abe H, et al (1968) The presence of novel plant growth regulators in leaves of *Distylium racemosum* Sieb et Zucc. Agric Biol Chem 32:528-529

Mitchell JW, Livingston GA (1968) Methods of studying plant hormones and growth-regulating substances. Agricultural Handbook No. 336. Agric Res Ser USDA

Motegi C, Takatsuto S, Gamoh K (1994) Identification of brassinolide and castasterone in the pollen of orange (*Citrus sinensis* Osbeck) by high-performance liquid chromatography. J Chromatogr A 658:27-30

Munakata K, Kato N, Ikeda M (1973) New auxin substance from corn germ oil In: Sumiki Y (Ed) Plant Growth Substance 1973. Hirokawa, Tokyo, pp 39-43

Noguchi T, Fujioka S, Takatsuto S, et al (1997) Clarification of blocking site of brassinosteroid biosynthesis in Arabidopsis dwarf mutant *det2*. Abstract paper of the 32nd Annual Meeting of Jap Soc Chem Regul Plants, Tokyo, pp 93-94

Noguchi T, Fujioka S, Takatsuto S, et al (1998) Biosynthetic pathway from campesterol to campestanol in cultured cells of *Catharanthus roseus* cells. Abstract paper of the Annual Meeting of Jap Soc Biosci Biotech Agrochem, Nagoya, p 70

Nomura T, Nakayama M, Reid JB, et al (1997) Blockage of brassinosteroid biosynthesis and sensitivity causes dwarfism in garden pea. Plant Physiol 113:31-37

Park KH, Park JD, Hyun KH, et al (1994) Brassinosteroids and monoglycerides in immature seeds of *Cassia tora* as the active principles in the rice lamina inclination bioassay. Biosci Biotech Biochem 58:1343-1344

Porzel A, Marquardt V, Adam G, et al (1992) ^{1}H and ^{13}C NMR analysis of brassinosteroids. Magn Reson Chem 30:651-657

Sakurai A, Fujioka S (1997) Studies on biosynthesis of brassinosteroids. Biosci Biotech Biochem 61:757-762

Sasse JM (1997) Recent progress in brassinosteroid research. Physiol Plant 100:696-701

Schmidt J, Adam G, Yokota T, et al (1991) Castasterone and brassinolide in *Raphanus sativus* seeds. Phytochemistry 30:364-365

Seo S, Nagasaki T, Katsuyama Y, et al (1989) Synthesis of (22R,23R)- and (22S,23S)-[4-^{14}C]-24-epibrassinolide. J Labelled Compd Radiopharm 27:1383-1393

Shim JH, Kim IS, Lee KB, et al (1996) Determination of brassinolide by HPLC equipped with fluorescence detector in rice (*Oryza sativa* L.) Han'guk Nonghwa Hakhoechi 39:84-88 (in Korean)

Suzuki H, Kim SK, Takahashi N, et al (1993) Metabolism of castasterone and brassinolide in mung bean explant. Phytochemistry 33:1361-1367

Suzuki H, Inoue T, Fujioka S, et al (1994) Possible involvement of 3-dehydroteasterone in the conversion of teasterone and typhasterol in cultured cells of *Catharanthus roseus*. Biosci Biotech Biochem 58:1186-1188

Suzuki H, Inoue T, Fujioka S, et al (1995) Conversion of 24-methylcholesterol to 6-oxo-24-methylcholesterol, a putative intermediate of the biosynthesis of brassinosteroids, in cultured cells of *Catharanthus roseus*. Phytochemistry 40:1391-1397

Takatsuto S (1991) Microanalysis of naturally occurring brassinosteroids. In: Cutler HG, Yokota T, Adam G (Eds) Brassinosteroids: Chemistry, Bioactivity, and Applications, ACS Symp Ser 474. Amer Chem Soc, Washington DC, pp 107-120

Takatsuto S (1994) Brassinosteroids: distribution in plants, bioassays and microanalysis by gas chromatography-mass spectrometry. J Chromatogr A 658:3-15

Takatsuto S, Ikekawa N (1986a) Analysis of 2-deoxybrassinosteroids by gas chromatography-mass spectrometry. Chem Pharm Bull 34:3435-3439

Takatsuto S, Ikekawa N (1986b) Synthesis of deuterio-labeled brassinosteroids, [26,28-$^2H_6$]brassinolide, [26,28-$^2H_6$]castasterone, [26,28-$^2H_6$]typhasterol and [26,28-$^2H_6$]teasterone. Chem Pharm Bull 34:4045-4049

Takatsuto S, Ikekawa N (1986c) Synthesis of [26,28-$^2H_6$]crinosterol, a synthetic intermediate of [26,28-$^2H_6$]brassinolide and [26,28-$^2H_6$]castasterone. J Chem Soc Perkin Trans 1 1986:591-593

Takatsuto S, Ying B, Morisaki M, et al (1982) Microanalysis of brassinolide and its analogs by gas chromatography and gas chromatography-mass spectrometry. J Chromatogr 239:233-239

Takatsuto S, Gotoh C, Noguchi T, et al (1998) Synthesis of deuterio-labeled 24-methylenecholesterol and related steroids. J Chem Res 1998:206-207

Takeno K, Pharis RP (1982) Brassinosteroid-induced bending of the leaf lamina of dwarf rice seedlings: an auxin-mediated phenomenon. Plant Cell Physiol 23:1275-1281

Taylor PE, Spuck K, Smith PM, et al (1993) Detection of brassinosteroids in pollen of *Lolium perenne* L by immunocytochemistry. Planta 189:91-100

Thompson MJ, Mandava N B, Meudt WJ, et al (1981) Synthesis and biological activity of brassinolide and its 22,23-isomer: novel plant growth promoting steroids. Steroids 38:567-680

Thompson MJ, Meudt WJ, Mandava NB, et al (1982) Synthesis of brassinosteroids and relationship of structure to plant growth-promoting effects. Steroids 39:89-95

Wada K, Marumo S, Ikekawa N, et al (1981) Brassinolide and homobrassinolide promotion of lamina inclination of rice seedlings. Plant Cell Physiol 22:323-325

Wada K, Marumo S, Abe H, et al (1984) A rice lamina inclination test — a micro-quantitative bioassay for brassinosteroids. Agric Biol Chem 48:719-726

Wada K, Kondo H, Marumo S (1985) A simple bioassay for brassinosteroids: a wheat leaf-unrolling test. Agric Biol Chem 49:2249-2251

Yokota T, Mori K (1992) Molecular structure and biological activity of brassinolide and related brassinosteroids. In: Bohl M, Duax WL (Eds) Molecular Structure and Biological Activity of Steroids. CRC Press, Boca Raton, pp 317-340

Yokota T, Arima M, Takahashi N (1982) Castasterone, a new phytosterol with plant-hormone potency, from chestnut insect gall. Tetrahedron Lett 23:1275-1278

Yokota T, Baba J, Koba S, et al (1984) Purification and separation of eight steroidal plant-growth regulators from *Dolichos lablab* seed. Agric Biol Chem 48:2529-2534

Yokota T, Ogino Y, Takahashi N, et al (1990a) Brassinolide is biosynthesized from castasterone in *Catharanthus roseus* crown gall cells. Agric Biol Chem 54:1107-1108

Yokota T, Watanabe S, Ogino Y, et al (1990b) Radioimmunoassay for brassinosteroids and its use for comparative analysis of brassinosteroids in stems and seeds of *Phaseolus vulgaris*. J Plant Growth Regul 9:151-159

Yokota T, Ogino Y, Suzuki H, et al (1991) Metabolism and biosynthesis of brassinosteroids. In: Cutler HG, Yokota T, Adam G (Eds) Brassinosteroids: Chemistry, Bioactivity and Applications. ACS Symp Ser 474. Amer Chem Soc, Washington DC, pp 86-96

Yokota T, Nakayama M, Wakisaka T, et al (1994) 3-Dehydroteasterone, a 3,6-diketobrassinosteroid as a possible biosynthetic intermediate of brassinolide from wheat grain. Biosci Biotech Biochem 58:1183-1185

Yokota T, Matsuoka T, Koarai T, et al (1996) 2-Deoxybrassinolide, a brassinosteroid from *Pisum sativum* seed. Phytochemistry 42:509-511

Yokota T, Nomura T, Kitasaka Y, et al (1997a) Biosynthetic lesion in brassinosteroid-deficient pea mutants. Proc Plant Growth Reg Soc Amer 24: 94

Yokota T, Nomura T, Nakayama M (1997b) Identification of brassinosteroids that appear to be derived from campesterol and cholesterol in tomato shoots. Plant Cell Physiol 38:1291-1294

4
Chemical Synthesis of Brassinosteroids

TREVOR C. MCMORRIS

Department of Chemistry and Biochemistry
University of California, San Diego, 9500 Gilman Drive, La Jolla, CA 92093-0506, USA

1. Introduction

The discovery of brassinolide [$2\alpha,3\alpha,22(R),23(R)$-tetrahydroxy-$24(S)$-methyl-B-homo-7-oxa-$5\alpha$-cholestan-6-one; **1**] (Fig. 1), first reported in 1979 (Grove et al 1979), aroused intense interest among steroid synthetic chemists. This was the first steroid to be identified with specific hormonal functions in plants. Its polyhydroxy structure, somewhat reminiscent of the ecdysones (insect moulting hormones) and its unprecedented B-ring lactone, presented a challenging synthetic target. The low abundance of brassinolide in natural sources gave added impetus to synthetic efforts.

Besides brassinolide, many closely related steroids have been identified in or isolated from a wide variety of plants (see chapter by S. Fujioka, this volume).

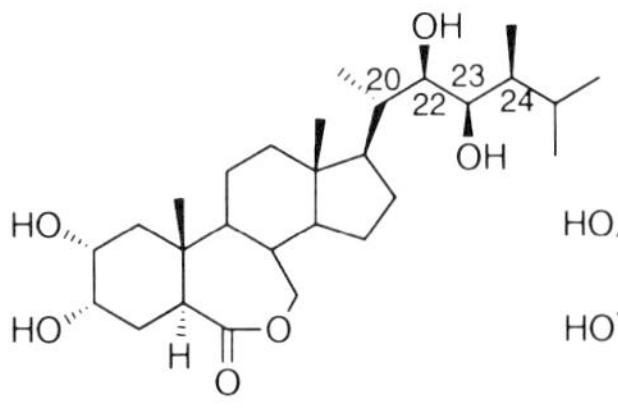

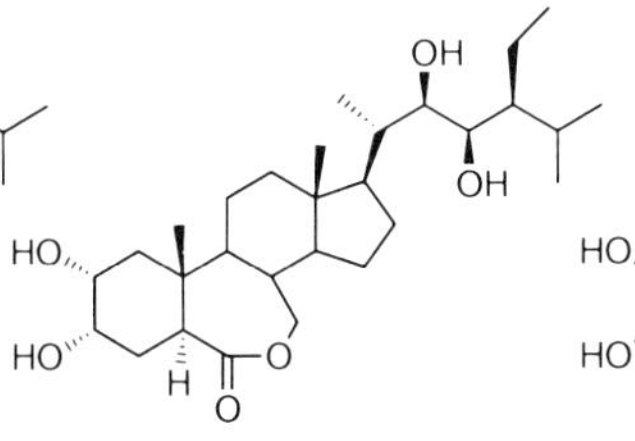

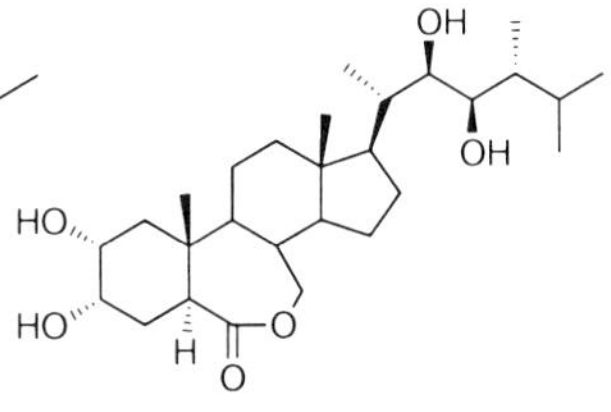

Fig. 1. Structures of brassinolide (1) 28-homobrassinolide (2) and 24-epibrassinolide (3).

Key words: brassicasterol, brassinolide, brassinosteroids, castasterone, crinosterol, dolichosterone, 24-epibrassinolide, ergosterol, 28-homobrassinolide, plant-growth hormones, stigmasterol, synthesis

Brassinolide is the most biologically active of the more than 40 known brassinosteroids, although 28-homobrassinolide (**2**) and 24-epibrassinolide (**3**) possess high biological activity (see chapter by Y. Kamuro and S. Takatsuto, this volume). Together with brassinolide, these steroids have been the subject of greatest synthetic effort. Their synthesis is less complicated than that of brassinolide because they can be derived more directly from stigmasterol and ergosterol, two very abundant naturally occurring sterols. This is an important consideration for the potential application of brassinosteroids in agriculture.

Chemical synthesis of brassinolide and other brassinosteroids has been accomplished in many laboratories. Except for one, all syntheses employ a C-22 aldehyde or pregnenolone, both derived from stigmasterol, or else a bile acid. The brassinolide side chain with its vicinal diol function and four contiguous chiral centers at C-20, C-22, C-23, and C-24 is built onto the tetracyclic nucleus by a variety of stereoselective reactions. Functionalization of the A and B rings is carried out before or in some cases after construction of the side chain. The various syntheses are described in this chapter more or less in chronological order.

2. Synthesis of Brassinolide

2.1 First Syntheses

The first two syntheses of brassinolide were reported almost simultaneously in 1980. In the synthesis of Fung and Siddall (1980) (Scheme 1), stigmasterol (**4**) was readily converted to the 3,5-cyclosteroidal aldehyde (**5**). Stereoselective alkylation of this aldehyde with lithium butyldimethyl (E)-2,3-dimethylbutenylalanate (**6**) gave the major 22 (S)-allylic alcohol (**7**). Hydroxyl-directed epoxidation of (**7**) afforded threo epoxide (**8**) which on reduction (with inversion at C-24) showed 3:1 regioselectivity for formation of the vicinal diol (**9**). The nucleus was then developed from (**9**) by acid-catalyzed regeneration of the 3β-hydroxy-5-ene as in (**10**). The latter was converted to the acetonide which was tosylated to form (**11**). Hydroboration-oxidation followed by elimination of the tosyl group and then Jones oxidation yielded ketone (**12**). Stereoselective hydroxylation of (**12**) gave the 2α,3α-diol (**13**), which on Baeyer-Villiger oxidation with trifluoroperoxyacetic acid in moist trifluoreacetic acid and methylene chloride led to brassinolide (**1**) mp 273-274°C. This synthesis required 11 steps from aldehyde (**5**), but yields for all steps were not reported.

The synthesis of brassinolide by Ikekawa and co-workers (Ishiguro et al 1980) (Scheme 2) employed a strategy very similar to that of Fung and Siddall. Thus C-22 aldehyde (**14**) on reaction with 3-methylbut-1-ynyl lithium afforded a 1:1 epimeric mixture of the C-22 alcohols. The more polar 22R isomer (**15**) was reduced with Lindlar catalyst yielding the *cis* allylic alcohol (**16**), which on stereospecific Sharpless epoxidation gave the 23R,24R epoxide (**17**). Attempts to directly introduce a methyl group at C-24 of (**17**) failed. Therefore the acetate of (**17**) was subjected to hydrocyanation followed by saponification, acetonide formation, and deprotection

Scheme 1

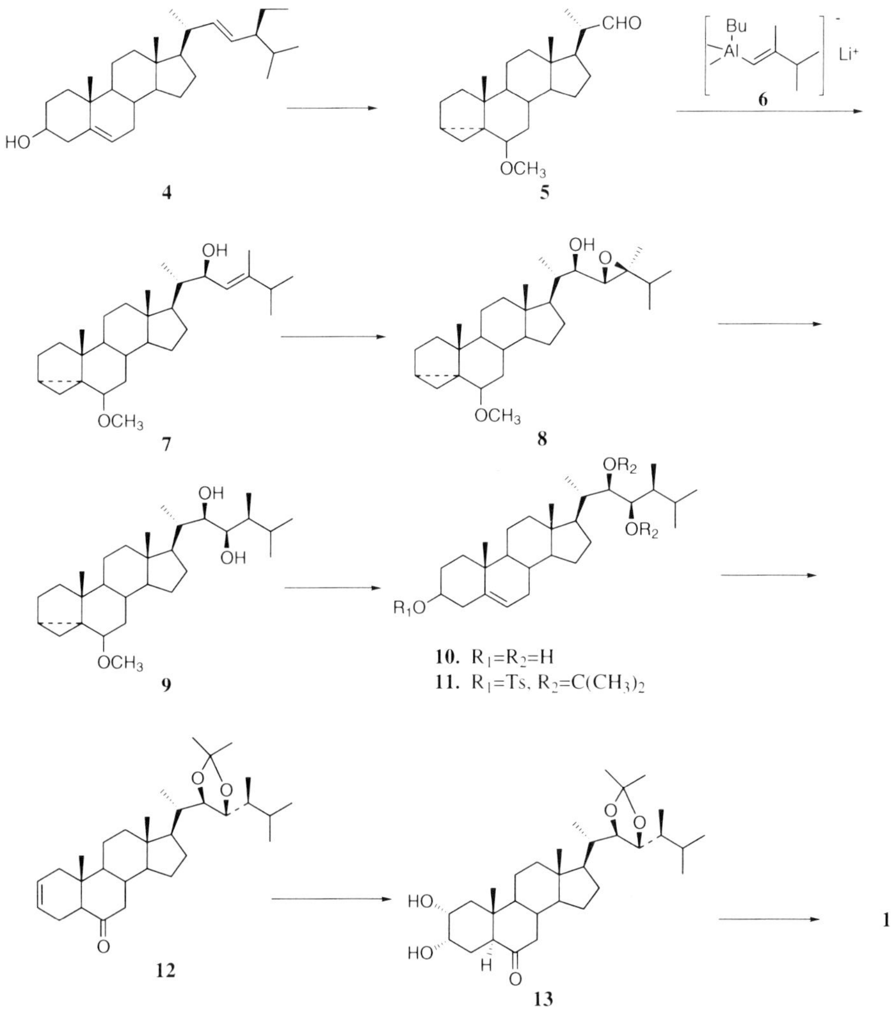

10. $R_1=R_2=H$
11. $R_1=Ts$, $R_2=C(CH_3)_2$

to give the nitrile (**18**). Reduction of (**18**) gave the corresponding aldehyde, which was converted to a methyl group by the sequence: acetylation, $NaBH_4$ reduction, methane sulfonation, iodide substitution and tributyltin hydride reduction to give the acetate (**19**). Modification of the A and B rings was done in a way almost identical to the previous synthesis except that ketone formation in ring B preceded introduction of the double bond in ring A.

The synthesis of brassinolide by the Mori group (Sakakibara et al 1982) is noteworthy because A and B ring functionalization was carried out before construction of the side chain (Scheme 3). Stigmasterol (**4**) was converted to the dienone (**20**) in a manner similar to that described above. Stereoselective dihydroxylation, protec-

Scheme 2

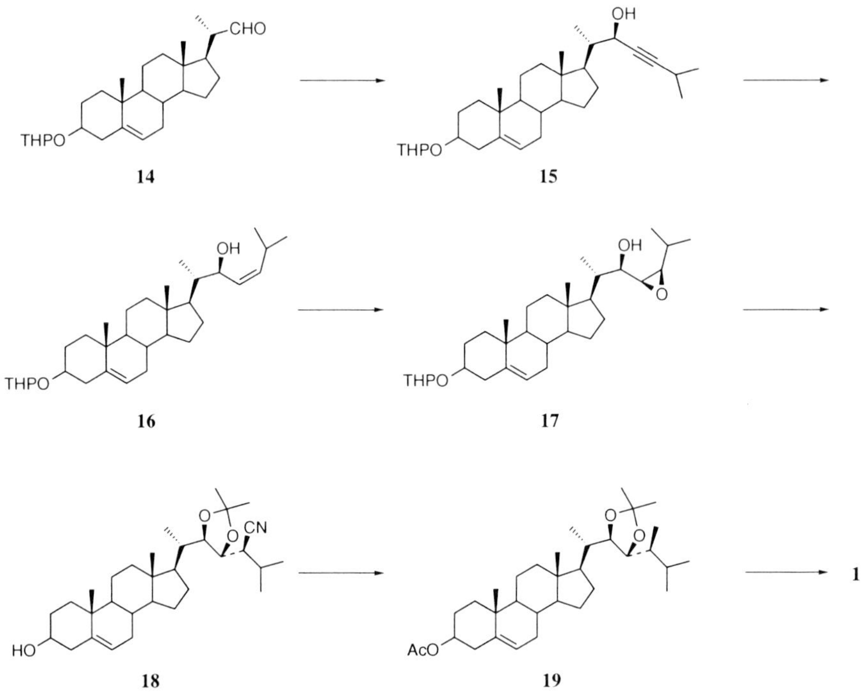

tion of the 2α,3α-diol as the acetonide and the carbonyl group as an ethylene acetal, gave alkene (**21**), which on ozonolysis yielded aldehyde (**22**). Formation of the olefinic side chain with the (24S)-methyl group was accomplished by the Kocienski olefinic synthesis with (**22**) and phenylsulfone (**23**). The latter was prepared from optically pure (R)-(+)-citronellic acid. Reaction of the lithium compound (**23**) with aldehyde (**22**) followed by acetylation gave (**24**), which upon reduction with sodium amalgam gave the olefinic product (**25**). Deprotection, acetylation, and epoxide formation furnished (**26**) as a stereoisomeric mixture. The epoxy ring in (**26**) was cleaved with 30% hydrobromic acid in acetic acid to give the *trans* bromoacetate, which on heating with acetic acid underwent displacement of bromine yielding (**27**). Baeyer-Villiger oxidation of (**27**) followed by hydrolysis of the acetates and acidification gave brassinolide.

Construction of the side chain required many steps with an overall yield of only 5%. In a later synthesis the Mori group followed the method of Ikekewa to construct the side chain (Mori et al 1984). Direct methylation of (**28**) to (**29**) was achieved by use of trimethylaluminum-*n*-butyllithium, which shortened the pathway to brassinolide.

A stereoselective route to the brassinolide side chain (Scheme 4) developed by Hayami et al (1983) involved reaction of C-22 aldehyde (**5**) with the lithio derivative

Scheme 3

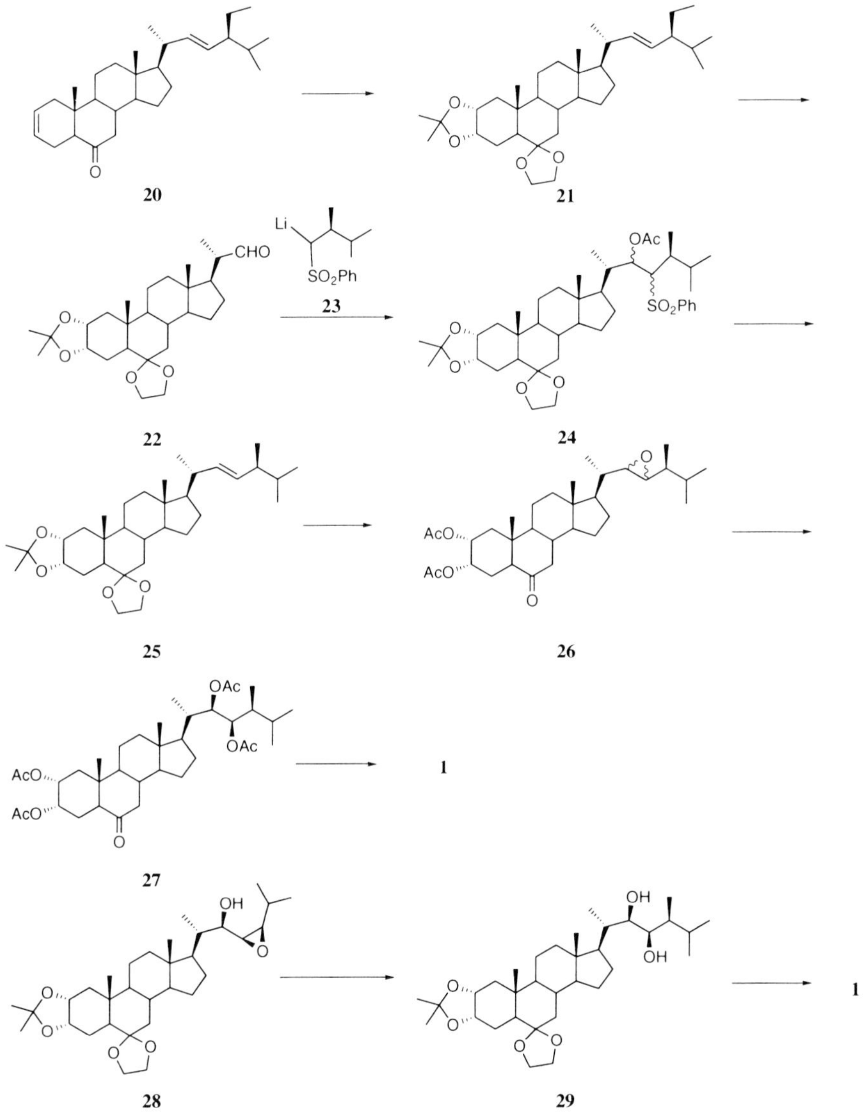

from 2-(dimethylphenylsilyl)-1-iodo-1-propene (**30**) to give the major 22S-allylic alcohol (**31**). Silyl group-assisted Sharpless epoxidation, then elimination of the dimethylphenylsilyl function with tetrabutylammonium fluoride, led to an epoxy alcohol that was transformed to the benzyl ether (**32**). Completion of the side chain construction was achieved by alkylation of (**32**) with lithium diisopropyl cyanocuprate, followed by deprotection with lithium in liquid ammonia leading to intermediate (**9**)

Scheme 4

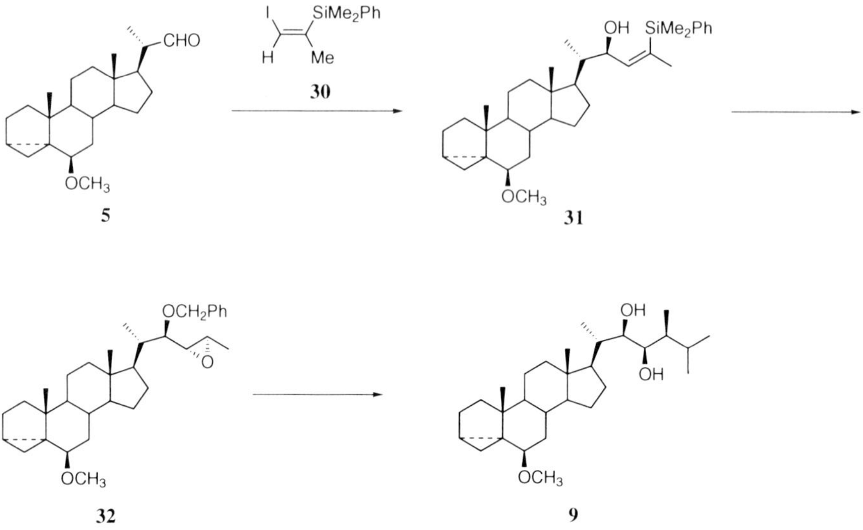

of the Fung and Siddall synthesis.

A route to castasterone (**33**) and its 22S,23S-isomer was reported by Anastasia et al (1983b) (Scheme 5). The C-22 aldehyde (**34**) on Grignard reaction with 3-methyl-but-1-ynyl magnesium bromide gave a 3:2 mixture of alcohols. The 22S-alcohol (**35**) was partially hydrogenated over the Lindlar catalyst to give the Z-allylic alcohol (**36**) which was subjected to Claisen rearrangement with triethyl orthoacetate yielding the Δ^{22}-unsaturated ester (**37**). Diisobutylaluminum hydride reduction of (**37**) furnished the aldehyde (**38**), which was decarbonylated by treatment with tris(triphenylphosphine)rhodium (I) chloride. Appropriate functional groups were introduced in rings A and B by the sequence: oxidation to the ketone with Jones reagent, formation of the 3β-chloro derivative with hydrochloric acid-acetic acid, dehydrohalogenation with dimethyl formamide and lithium bromide. The product dienone (**39**) on treatment with osmium tetraoxide and pyridine gave mainly the 22S,23S isomer of castasterone and about 15-20% of castasterone (**33**).

Takahashi et al (1985) have employed a pregnenolone-derived (20R)-tosyloxy steroid (**40**) to build the brassinolide side chain (Scheme 6). Thus alkylation of (**40**) with a protected eight-carbon cyanohydrin building block (**41**) followed by acid and base treatment gave the α,β-unsaturated ketone (**42**). Stereoselective reduction of (**42**) with DIBAH gave the known allylic alcohol (**7**).

The same intermediate (**42**) has been prepared recently by Hazra et al (1996) starting from aldehyde (**5**) (Scheme 7). The Horner-Wadsworth-Emmons reaction gave enone (**43**), which on reaction with methyllithuim gave a mixture of C-24 epimeric alcohols (**44**). Pyridinium chlorochromate oxidation of the tertiary allylic alcohols furnished the α,β-unsaturated ketone (**42**).

Scheme 5

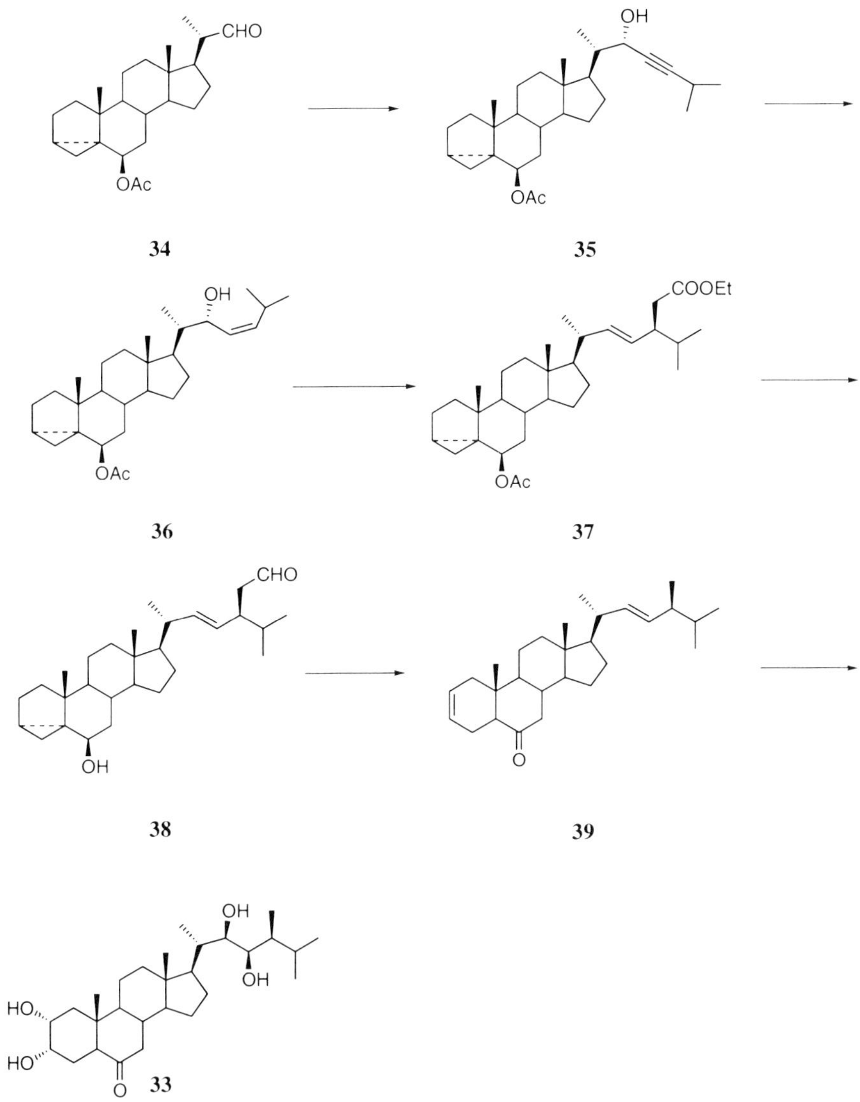

McMorris and co-workers have reported two syntheses of brassinolide. In the first synthesis (Donaubauer et al 1984), stigmasterol was converted to the C-22 aldehyde (**5**), which was used in an aldol reaction with the anion of 3-isopropyl but-2-enolide, or of 2,3-dimethyl but-2-enolide, to construct the side chain. Modification of functional groups in rings A and B was then carried out. The synthesis was later improved (McMorris et al 1996) (Scheme 8) by employing the C-22 aldehyde (**45**)

Scheme 6

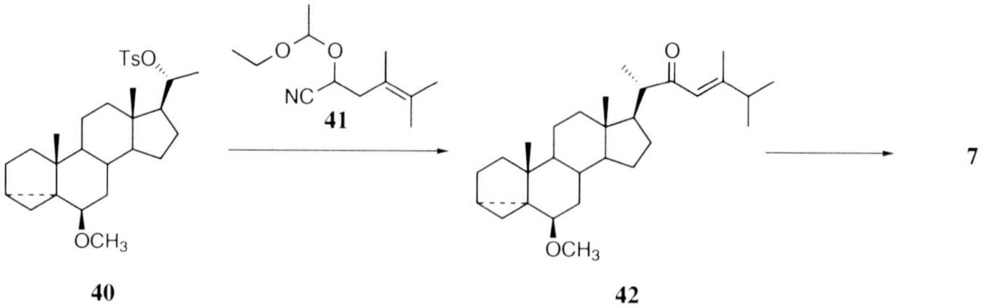

40 → **42** → **7**

(reagent **41**)

Scheme 7

5 → **43** →

44 → **42**

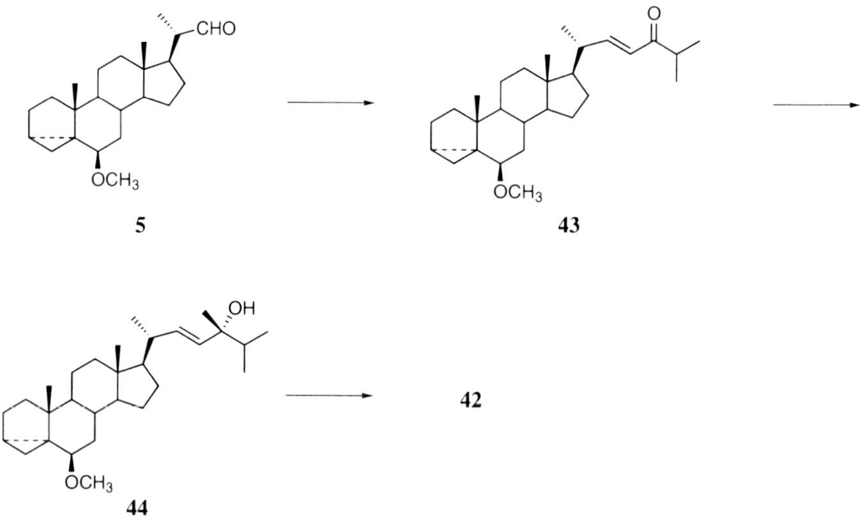

possessing 6-keto and 2α,3α-dihydroxy groups (protected as the acetonide). Condensation of this aldehyde and the lithium enolate (**46**), from 3-isopropyl but-2-enolide, at -70°C gave with high stereoselectivity the 22R,23R product (**47**), in high yield. Catalytic hydrogenation of (**47**) was also highly stereoselective giving the 24S product (**48**). Reduction of the lactone afforded the tetraol (**49**), which, after protection of the vicinal hydroxyl groups, was oxidized to the ketoaldehyde (**50**). Decarbonylation of (**50**) with tris(triphenylphosphine) rhodium (I) chloride (Wilkinson's catalyst) yielded castasterone bisacetonide (**51**). This compound could be converted directly to brassinolide with trifluoroacetic acid and trifluoroperoxyacetic acid. The overall yield of brassinolide from stigmasterol was 7%, which is one of the best yields among the many published syntheses.

A butenolide synthon has also been used by Kametani et al (1988) for the synthe-

Scheme 8

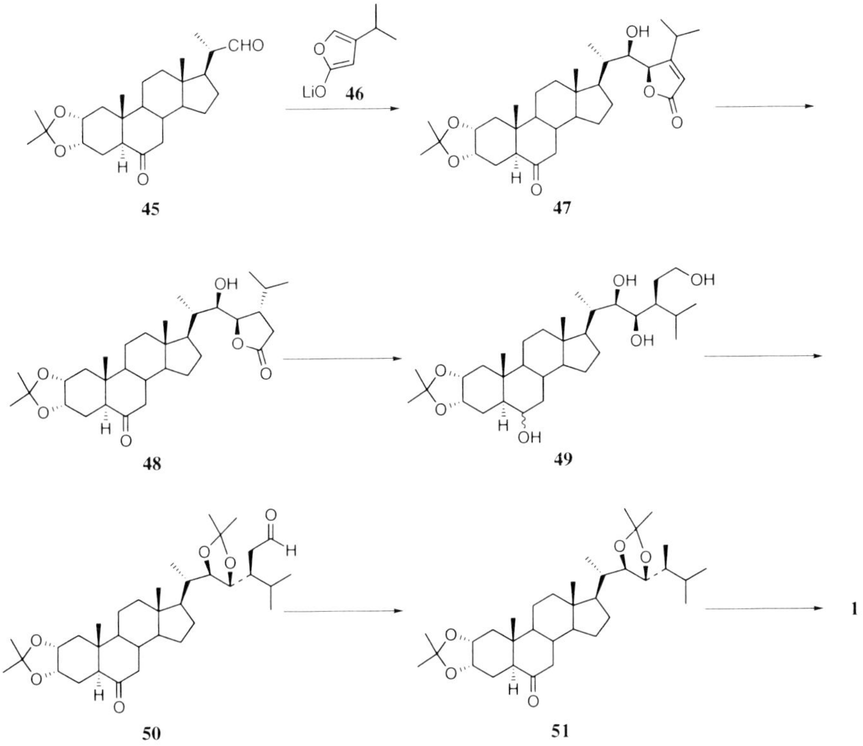

sis of brassinolide from pregnenolone (Scheme 9). The latter was converted to intermediate (**52**), which was reacted with the dilithio derivative of 3-isopropyltetronic acid (**53**) followed by treatment with chloromethyl methyl ether. The major product (**54**) was subjected to syn-dehydration, via the trifluoroacetate, giving a mixture of olefins. Catalytic hydrogenation of the major olefin (**55**) afforded lactone (**56**), which on reduction gave the diol (**57**). Selective mesylation followed by reduction with LiAlH$_4$ yielded the C-24 methyl, and hydrolysis and saponification led to castasterone (**33**), with an overall yield of 34%.

Zhou and Tian (1987) have employed an approach (Scheme 10) similar to that reported by Donaubauer et al (1984). The C-22 aldehyde (**59**) derived from hyodeoxycholic acid was condensed with the lithio derivative (**60**) of 3-methyl butenolide to give a mixture of epimeric alcohols (55:31). Acetylation of the major 22R-alcohol (**61**) followed by catalytic hydrogenation yielded a saturated lactone, which was methylated to give lactone (**62**). Reduction to the lactol, reaction with 1,3-dithiane, acetate hydrolysis, and acetonide formation provided (**63**). Treatment of this intermediate with Raney Ni completed the side chain synthesis as (**64**) in 30% overall yield from (**59**). A more recent synthesis by Zhou and Huang (1992) does not repre-

Scheme 9

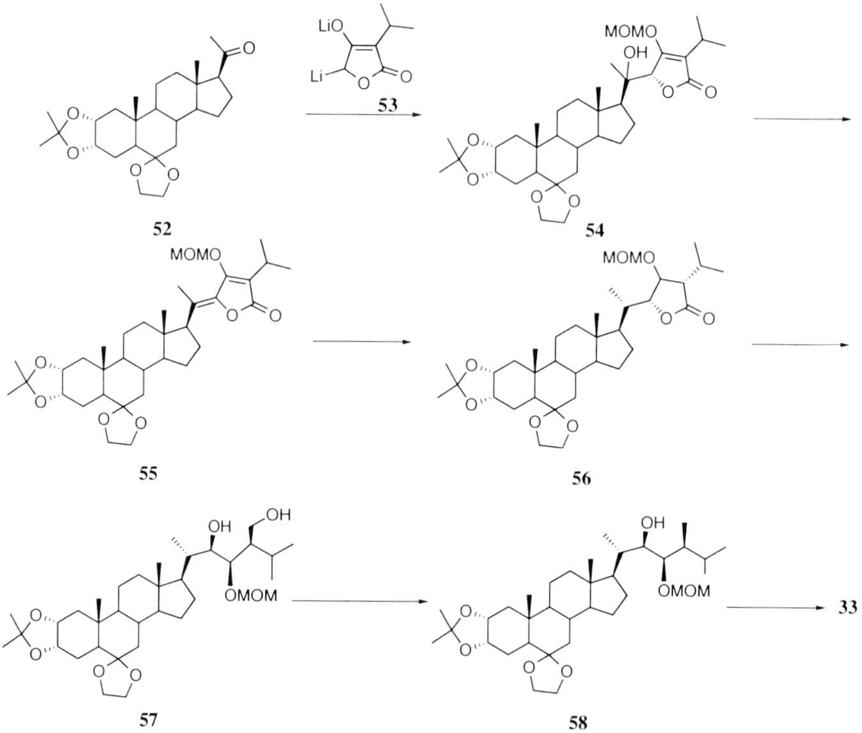

sent an improvement over published methods.

There has been much interest in the chemistry of brassinosteroids in the former Soviet Union. In particular, Khripach and coworkers (1991) have published many papers on the synthesis of brassinosteroids. In one synthesis (Scheme 11) C-22 aldehyde (**65**) was condensed with the anion of the vinyl silane (**66**) giving alcohol (**67**) with high stereoselectivity and yield. Removal of the trimethyl silyl group furnished the Z-allylic alcohol (**68**), which could be converted to brassinolide by a known route (cf Mori et al 1984).

In another approach an isoxazoline side chain was constructed via 1,3-dipolar cycloaddition to alkene (**69**) of isobutyronitrile oxide (Scheme 12). The product (**70**) was converted to the ketol (**71**) and then, after methylation (giving **72**), to the enone (**73**). Stereoselective reduction of the ketone (**73**) gave (**74**) with the same side chain as a known intermediate (**7**) in brassinolide synthesis (Fung and Siddall 1980).

Scheme 10

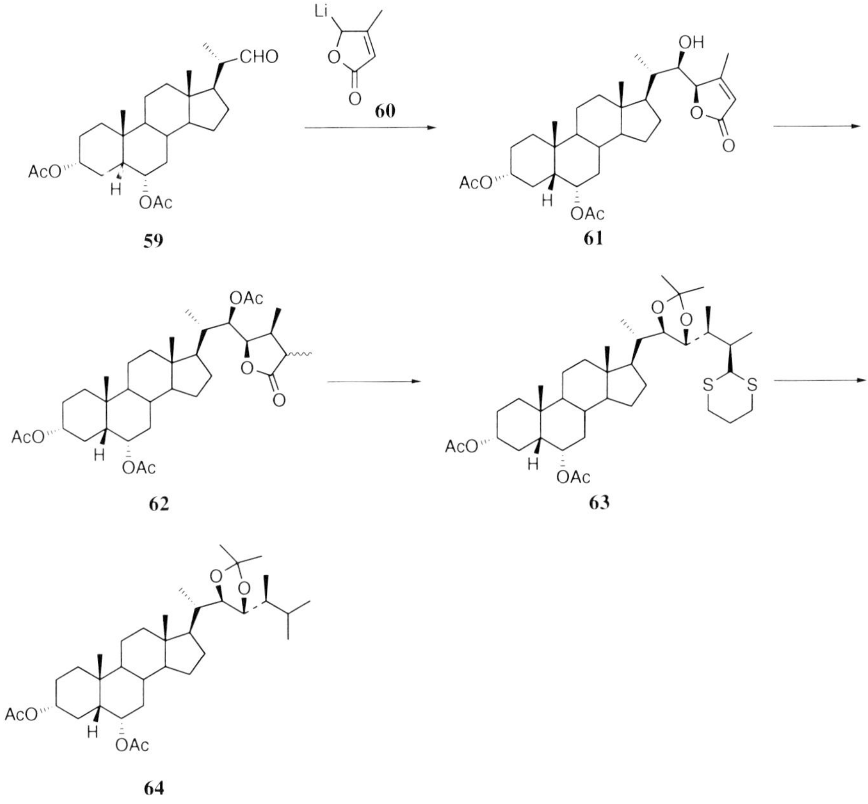

Scheme 11

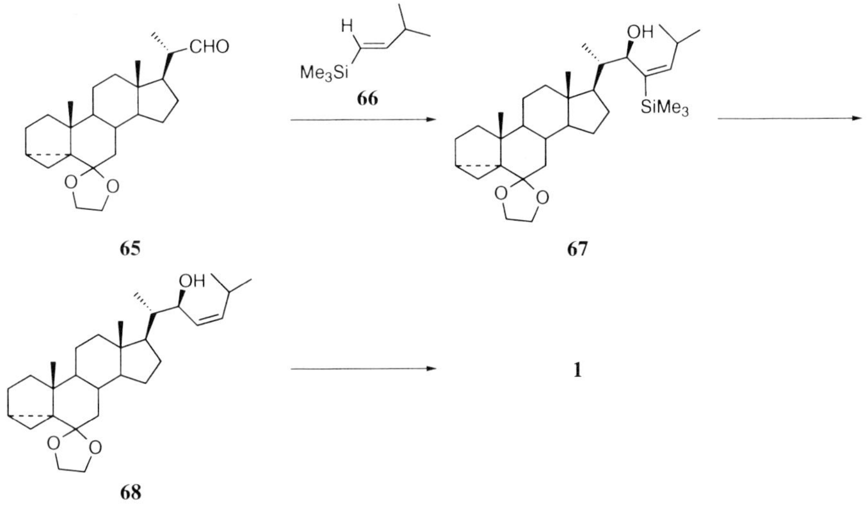

Scheme 12

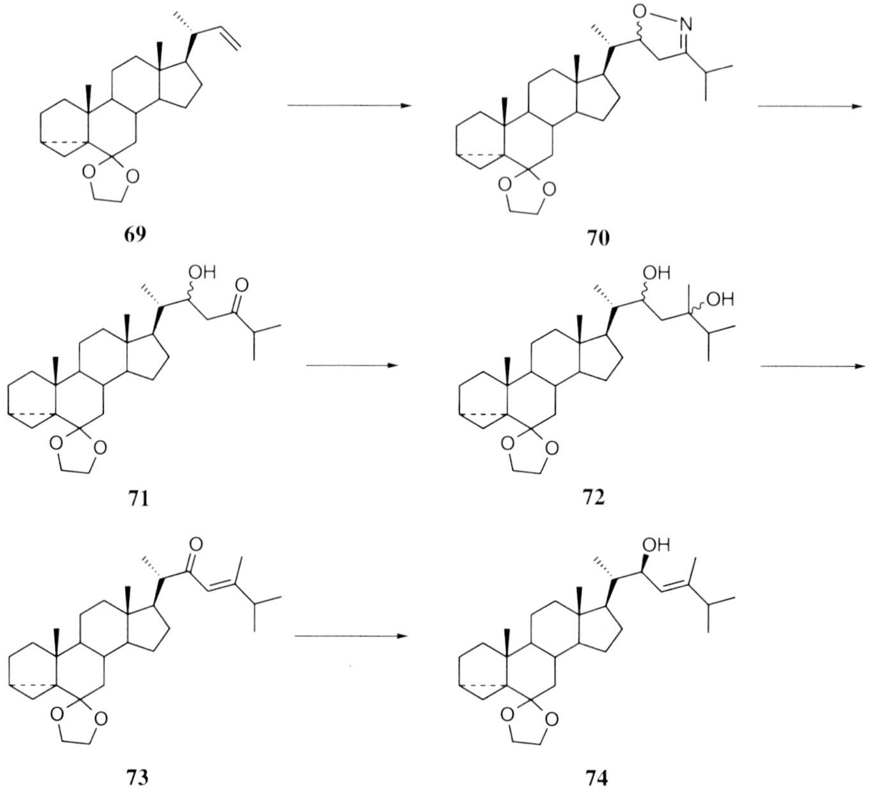

69 → 70

71 → 72

73 → 74

2.2 Recent Syntheses of the Brassinolide Side Chain

Several syntheses of the brassinolide side chain have appeared in the last few years. In one route starting from C-22 aldehyde (**52**) (Scheme 13) Tsubuki et al (1992) built on the side chain by condensation with 2-lithio-4-methyl furan (**75**) yielding an epimeric mixture (3:1) of products. Oxidation of the major product (**76**) with N-bromosuccinimide gave lactol (**77**). Oxidation of (**77**) with pyridinium chlorochromate yielded a keto-lactone, which on reduction with $NaBH_4$-$CeCl_3$ furnished alcohol (**78**) as the sole product. Protection of the alcohol as the ethoxyethyl ether and subsequent methylation with lithium dimethylcuprate afforded lactone (**79**). Reduction of the lactone and selective removal of the resulting primary hydroxyl in (**80**) yielded the brassinolide side chain in 27% overall yield from aldehyde (**52**). Analogues substituted at C-24 and C-25 were also prepared by this method.

The route described by Koreeda and Wu (1995) (Scheme 14) involved condensation of C-22 aldehyde (**5**) with lithio-1,3-dithiane followed by protection of the re-

Scheme 13

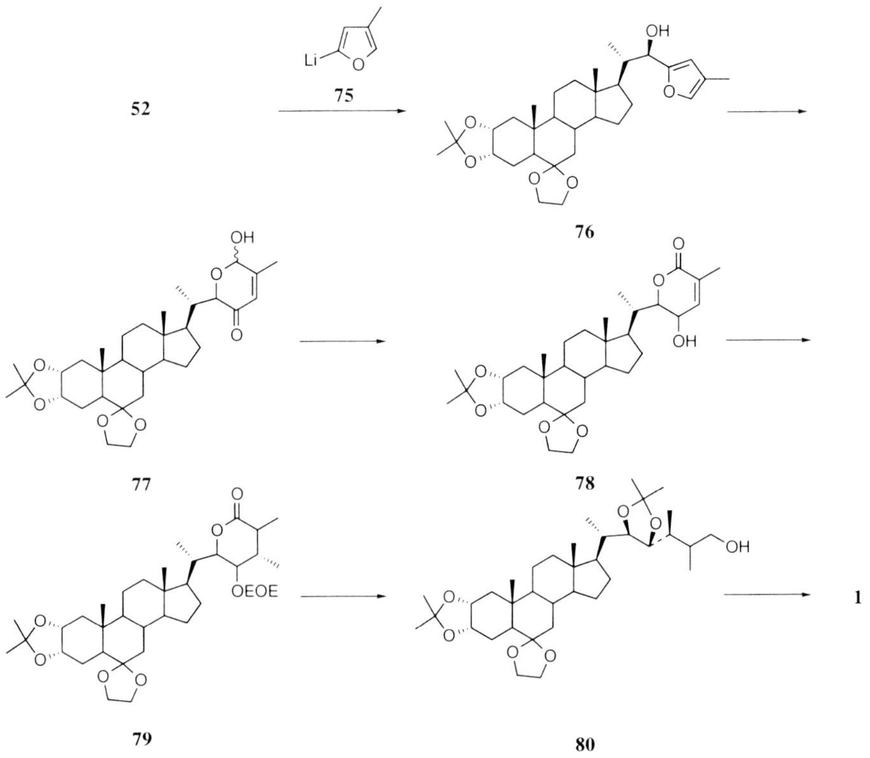

sulting alcohol as the MOM ether and removal of dithiane to give the aldehyde (**81**). A similar aldehyde had been used earlier by Takatsuto and Ikekewa (1983) in their synthesis of dolicholide. This was reacted with 1-lithio-2-methyl propene, which yielded the allylic alcohol (**82**). Both chiral centers (C22 and C23) were obtained with high stereoselectivity. Reaction of (**82**) with bromomethyl-dimethylsilyl chloride gave (**83**), which on treatment with tributyltin chloride, sodium cyanoborohydride, and AIBN in *t*-butanol, gave silane (**84**). Reaction of (**84**) with tetrabutyl ammonium fluoride yielded 23*R*-alcohol (**85**) which on acid treatment gave the brassinolide side chain product (**9**) in a 31% yield from aldehyde (**5**).

Another new high yielding route to the side chain has been described by Marino and co-workers (1996) (Scheme 15). Condensation of C-22 aldehyde (**5**) with the lithio derivative of vinyl sulfoxide (**86**) gave an 80:20 mixture of diastereomeric allylic alcohols. Mesylation of the Cram product (**87**) followed by S_N2' displacement of the mesylate when treated with magnesium methyl cyanocuprate gave (**88**). Treatment of this product with *t*-butyllithium yielded the crinosterol side chain product (**89**).

Scheme 14

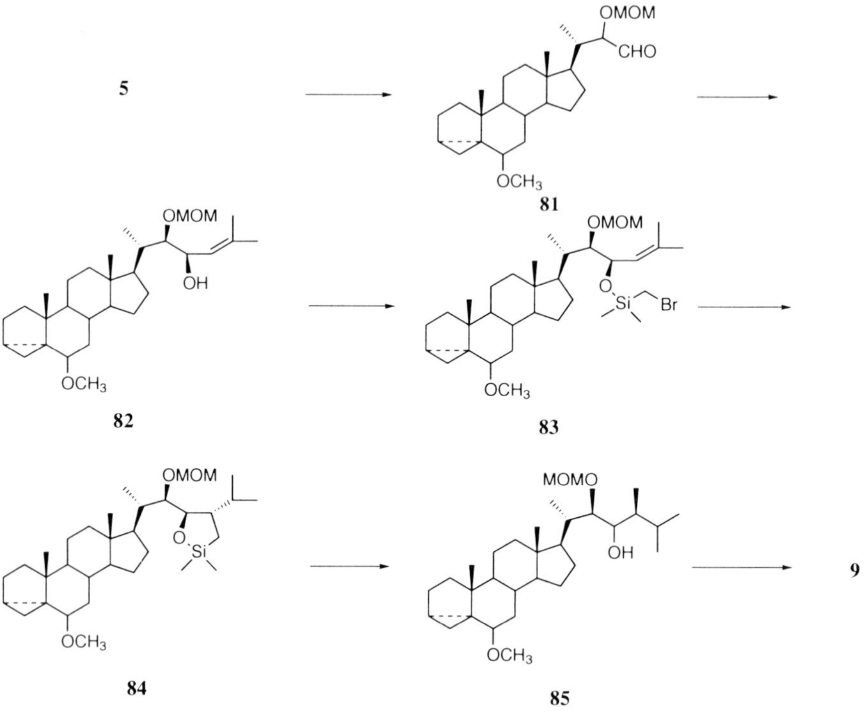

Scheme 15

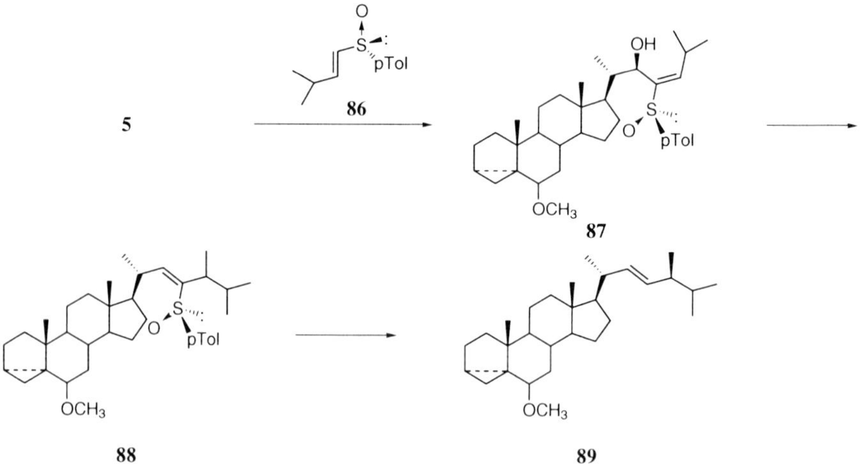

Scheme 16

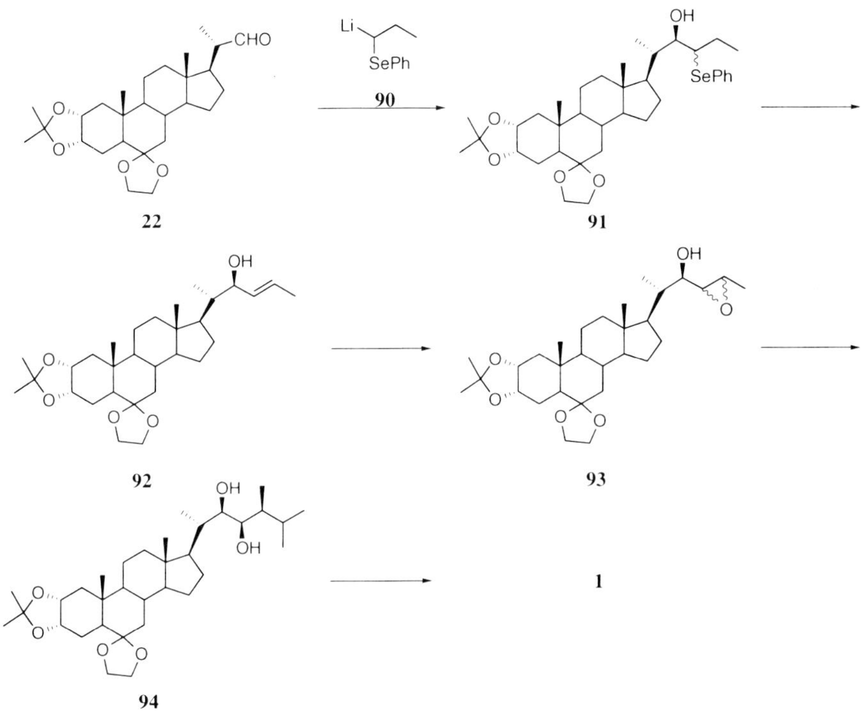

Improved selectivity in condensations of aldehyde (**5**) were obtained with the analogous lithio derivative of an α-iodovinylsulfide, which gave the Cram/anti-Cram products in a ratio of 92:8 and excellent overall yield. Oxidation of the sulfide with Sharpless reagent, cumene hydroperoxide and 1-(+)-diethyl tartrate at low temperature, led to the (Ss)-sulfoxide (**87**) as the only product in 90% yield. By employing the method of asymmetric dihydroxylation of Sharpless et al (1992) alkene (**89**) was converted to diol (**9**) with total stereoselectivity.

Back and co-workers (1993, 1997) have published two methods for constructing the brassinolide side chain. In their newer improved procedure (Scheme 16) the C-22 aldehyde (**22**) was reacted with the selenium-stabilized lithium reactant (**90**) to give a mixture of selenide stereoisomers at C-23 (**91**) but no detectable amount of the corresponding C-22 epimer. Oxidative workup involving selenoxide syn-elimination gave the *trans*-allylic alcohol (**92**). Sharpless oxidation gave a mixture of threo and erythro epoxides (**93**) which on reaction with isopropyl magnesium bromide (with catalytic amount of cuprous cyanide) yielded the 22R, 23R, 24S product (**94**). This was converted to brassinolide in 8% overall yield from stigmasterol.

Schmittberger and Uguen (1997) have recently described the use of the Ramberg-Backlund rearrangement for synthesis of the brassinolide side chain (Scheme 17).

Scheme 17

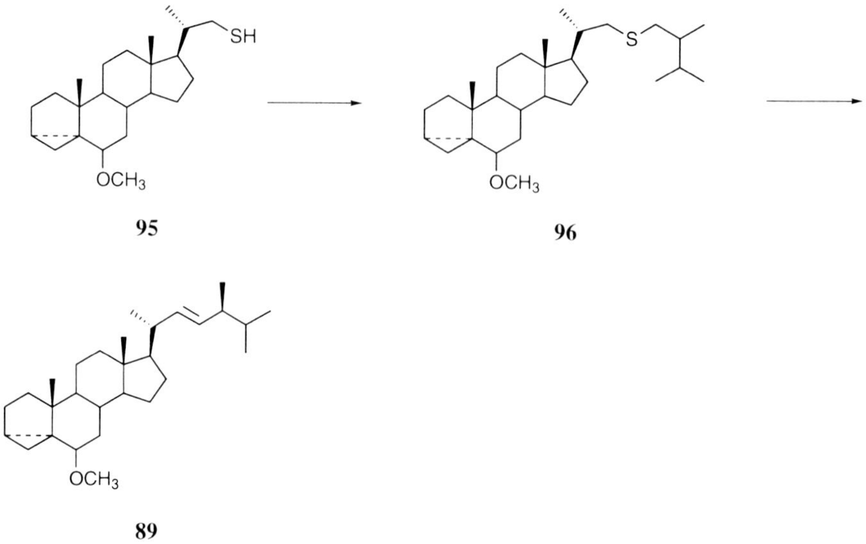

95

96

89

Scheme 18

65 →

97

98

99

39

→ **33**

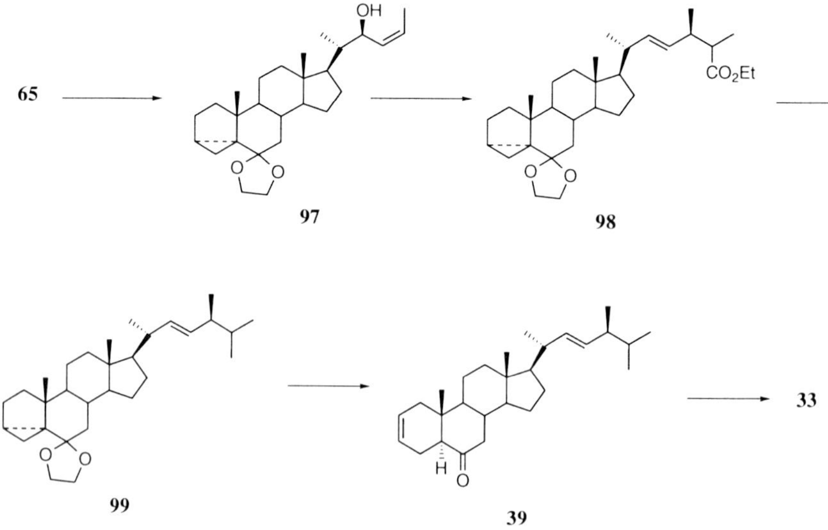

Thus alkylation of thiol (**95**), readily obtained from stigmasterol, with (S)-2,3-dimethyl-1-iodobutane, gave sulfide (**96**), which by a short sequence involving chlorination, oxidation and treatment of the resulting chlorosulfone with excess t-BuOK, gave the unsaturated compound **89**.

Another recent synthesis that takes advantage of the method of asymmetric dihydroxylation of Sharpless et al (1992) has been reported by Watanabe et al (1997) (Scheme 18). Aldehyde (**65**) was reacted with (Z)-prop-1-enylmagnesium bromide to give the (22S, 23Z)-23-en-22-ol (**97**). The latter compound was subjected to orthoester Claisen rearrangement, which furnished the ester (**98**). The ethoxycarbonyl group of (**98**) was transformed to methyl thus giving the crinosterol side chain compound (**99**). It was converted to the unsaturated ketone (**39**), which on asymmetric dihydroxylation afforded castasterone (**33**) in overall yield of 22% from stigmasterol. Takatsuto et al (1997a,b) have employed a similar approach for the synthesis of naturally occurring 6-deoxy brassinosteroids.

3. Synthesis of 28-Homobrassinolide and 24-Epibrassinolide

Synthesis of 28-homobrassinolide (**2**) and 24-epibrassinolide (**3**) have proved to be less challenging than that of brassinolide. Stigmasterol (**4**) and ergosterol (**100**) are abundant sterols and possess the required stereochemical configuration at C-24. The first synthesis of a brassinosteroid in fact employed ergosterol as starting material, reported by Thompson et al (1979) (Scheme 19). Ergosterol was first converted to the iso-sterol (**101**). Oxidation of the alcohol with chromic acid in pyridine followed by reduction of the conjugated double bond with lithium in liquid ammonia gave (**102**). Regeneration of the 3β-hydroxy group followed by elimination in the usual way led to dienone (**103**). Hydroxylation of the dienone with osmium tetroxide and N-methylmorpholine N-oxide yielded a 1:1 mixture of tetraols (22R, 23R, **104**, and 22S, 23S). The isomers were separated and converted to (22R, 23R)-epibrassinolide and (22S, 23S)-epibrassinolide, respectively. The yield of the natural 22R, 23R isomer can be greatly improved by the method of asymmetric dihydroxylation developed by Sharpless et al (1992). This involves hydroxylation with osmium tetroxide, $K_3Fe(CN)_6$ and the chiral ligand dihydroquinidine (DHQD)4-chlorobenzoate in isopropanol-water (1:1). The yield of the desired 22R, 23R isomer increased to 80%. Many groups have reported synthesis of 24-epibrassinolide (Anastasia et al 1983a, Sun et al 1991, Ferrer et al 1990, Traven et al 1991, McMorris and Patil 1993, Brosa et al 1996) including one by Takatsuto and Ikekawa (1984), in which the starting compound was brassicasterol, derived from rape seed oil. An early synthesis of brassinolide by Thompson et al (1981) employed a C-24 epimeric mixture of crinosterol and brassicasterol (60:40) isolated from oysters.

Stigmasterol (**4**) is the sterol of choice for synthesis of 28-homobrassinolide (**2**). Conversion of (**4**) to dienone (**20**) was readily accomplished in good yield by several groups. Early attempts at forming the tetrahydroxyketone (homocastasterone) with osmium tetroxide and N-methylmorpholine N-oxide gave almost exclusively the isomeric 22S, 23S product. Less direct ways involving epoxidation of the Δ^{22} double bond followed by *trans* ring opening of the resulting epoxides by HBr-acetic acid and then an inversion reaction at the carbon-bearing bromine, by acetoxy ion, gave a

Scheme 19

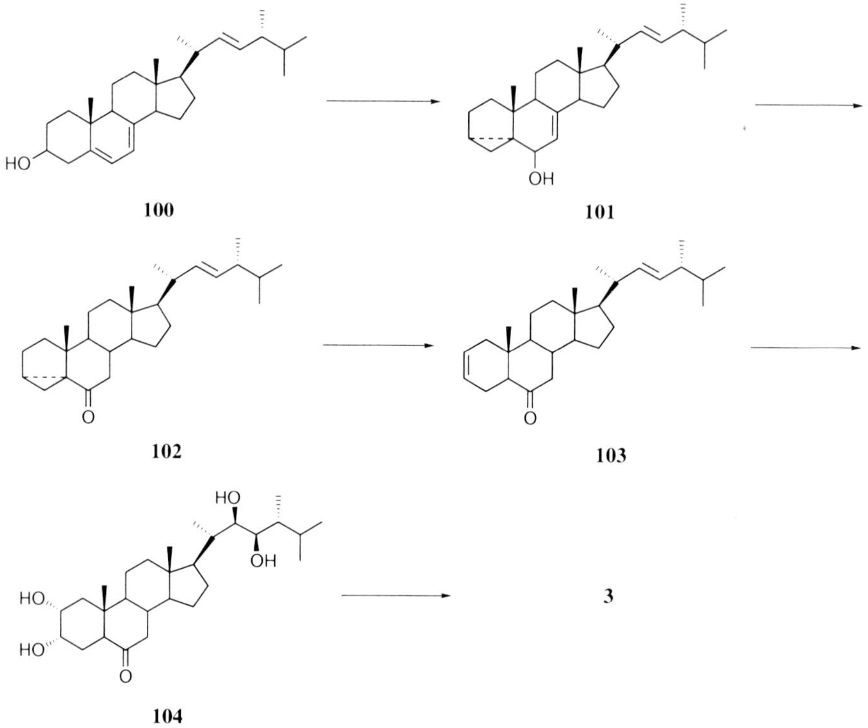

100 → **101** →

102 → **103** →

104 → **3**

Scheme 20

20 → **105** → **2**

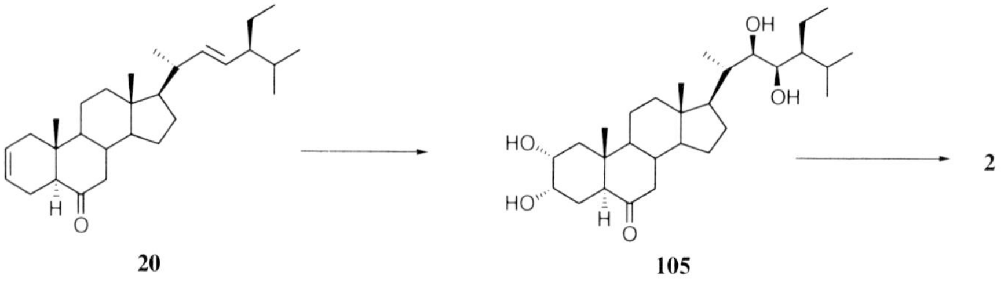

mixture of isomers in low yield (Takatsuto and Ikekawa 1982). The recent application of asymmetric dihydroxylation, as described for 24-epibrassinolide synthesis, gave a greatly improved yield of $2\alpha,3\alpha,22R,23R$-tetrahydroxy-24S-ethyl-5α-cholestan-6-one (**105**) (Scheme 20). This compound was then converted to 28-homobrassinolide (**2**) in high yield (Brosa et al 1992, McMorris et al 1994).

4. Synthesis of ^{14}C and ^{3}H-brassinosteroids

The naturally occurring brassinosteroid dolicholide (**106**) (Scheme 21) has been synthesized by Takatsuto and Ikekawa (1983) from the aldehyde (**107**). The latter was converted via dithiane (**108**) to homologous aldehyde (**109**) as described in the synthesis of Koreeda and Wu (1995). Chelation-controlled coupling of (**109**) with the Grignard reagent derived from 2-bromo-3-methyl-but-1-ene gave exclusively the 22R,23R diol 22-methoxymethyl ether (**110**). Acetylation of the 23-alcohol, followed by acid hydrolysis and subsequent acetylation, provided dolicholide tetraacetate (**111**). Saponification followed by acidification yielded dolicholide (**106**). Dolicholide has been converted by catalytic tritiation to [24,28-^{3}H$_2$]-brassinolide and [24,28-^{3}H$_2$]-epibrassinolide as reported by Yokoto et al (1990) (see the chapter by S. Takatsuto and T. Yokota, this volume).

A ^{14}C-label has been incorporated into 24-epibrassinolide at the C-4 position by a lengthy procedure (Seo et al 1989) that starts from brassicasterol and involves opening and reclosure of ring A. The specific radioactivity of the 24-epibrassinolide was 56.78 mCi/mmol (2130 mBq). More recently, tritium 5,7,7-tris-labeled 24-

Scheme 21

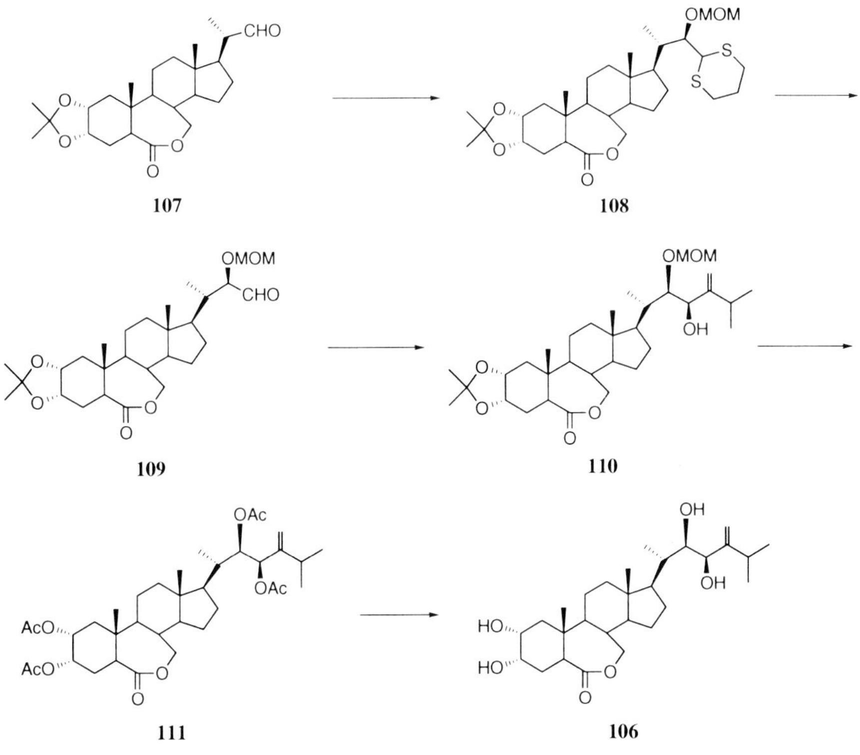

epibrassinolide was prepared by a base-catalyzed exchange reaction using bis-isopropylidene-dioxy-24-epicastasterone and labeled water. Baeyer-Villiger oxidation of the labeled product gave [5,7,7-^{3}H$_3$]-24-epibrassinolide 222 MBq/mmol (Kolbe et al 1993) (see the chapter by S. Tukatsuto and T. Yokota, this volume).

5. Conclusions

This chapter has focused on the many syntheses of brassinolide as well as a few other naturally occurring brassinosteroids. It is not intended to be an exhaustive coverage of all synthetic activity these steroidal plant hormones have generated. Many more related steroids have been synthesized for studies on structure-activity relationships and delineation of biosynthetic pathways. These studies have yielded much information about structural requirements for biological activity and are important when designing new brassinosteroids for practical applications. From a wealth of synthetic investigations in many laboratories, several practical syntheses of brassinolide, 24-epibrassinolide, and 28-homobrassinolide have been developed; and these steroids can now be made stereochemically pure on a large scale.

References

Anastasia M, Ciuffreda P, Fiecchi A (1983a) A new synthesis of brassinosteroids: plant growth promoting steroids. J Chem Soc Perkin Trans 1 1983:379-382

Anastasia M, Ciuffreda P, Del Puppo M et al (1983b) Synthesis of castasterone and its 22S, 23S-isomer: two plant growth promoting ketones. J Chem Soc Perkin Trans 1 1983:383-386

Back TG, Blazecka PG, Krishna MV (1993) A new synthesis of castasterone and brassinolide from strigmasterol: a concise and stereoselective elaboration of the side chain from a C-22 aldehyde. Can J Chem 71:156-163

Back TG, Baron DL, Luo W et al (1997) Concise, improved procedure for the synthesis of brassinolide and some novel side-chain analogues. J Org Chem 62:1179-1182

Brosa C, Peraceula R, Puig R et al (1992) Use of dihydroquinidine 9-0-(9'-phenanthryl) ether in osmium-catalyzed asymmetric dihydroxylation in the synthesis of brassinosteroids. Tetrahedron Lett 33:7057-7060

Brosa C, Puig R, Comas X et al (1996) New synthetic strategy for the synthesis of 24-epibrassinolide. Steroids 61:540-543

Donaubauer JR, Greaves AM, McMorris TC (1984) A novel synthesis of brassinolide. J Org Chem 49:2833-2834

Ferrer JC, Lalueza R, Saavedra O et al (1990) Short step synthesis of (22E, 24R)-5α-ergosta-2,22-dien-6-one, a key intermediate for the preparation of 24-epibrassinolide. Tetrahedron Lett 31:3941-3942

Fung S, Siddall JB (1980) Stereoselective synthesis of brassinolide: a plant growth promoting steroidal lactone. J Am Chem Soc 102:6580-6581

Grove MD, Spencer GF, Rohwedder WK et al (1979) Brassinolide, a plant growth-promoting steroid isolated from *Brassica napus* pollen. Nature 281:216-217

Hayami H, Sato M, Kanemoto S et al (1983) Transition-metal-catalyzed silylmetalation of acetylenes and its application to the stereoselective synthesis of steroidal side chain. J Am Chem Soc 105:4491-4492

Hazra BG, Kumar TP, Pore VS (1996) Synthesis of 3β-hydroxy-24-methylcholesta-5,23-dien-22-one: a brassinolide intermediate. J Chem Res (S) 1996:536-537

Ishiguro M, Takatsuto S, Morisaki M et al (1980) Synthesis of brassinolide, a steroidal lactone with plant-growth promoting activity. JCS Chem Comm 1980:962-964

Kametani T, Katoh T, Fujio J et al (1988) An improved synthesis of plant growth regulating steroid brassinolide and its congeners. J Org Chem 53:1982-1991

Khripach VA, Litvinovskaya RP, Baranovskii AV et al (1991) A new approach to the synthesis of brassinolide. Doklady Akademii Nauk SSSR 318:597-600

Kolbe A, Marquardt V, Adam G (1993) Synthesis of tritium labeled 24-epibrassinolide. J Label Compound Radiopharmaceut 31:801-805

Koreeda M, Wu J (1995) Stereoselective synthesis of the brassinolide side chain by the use of a 5-exo-α-silyl radical cyclization-protiodesilylation sequence. Synlett 1995:850-852

Marino JP, de Dios A, Anna LJ et al (1996) Highly stereocontrolled formal synthesis of brassinolide via chiral sulfoxide-directed S_N2' reactions. J Org Chem 61:109-117

McMorris TC, Patil PA (1993) Improved synthesis of 24-epibrassinolide from ergosterol. J Org Chem 58:2338-2339

McMorris TC, Chavez RG, Patil PA (1996) Improved synthesis of brassinolide. J Chem Soc Perkin Trans 1 1996:295-302

McMorris TC, Patil PA, Chavez RG et al (1994) Synthesis and biological activity of 28-homobrassinolide and analogues. Phytochemistry 36:585-589

Mori K, Sakakibara M, Okada K (1984) Synthesis of naturally occurring brassinosteroids employing cleavage of 23,24-epoxides as key reactions: Synthesis of brassinolide, castasterone, dolicholide, dolichosterone, homodolicholide, homodolichosterone, 6-deoxycastasterone and 6-deoxydolichosterone. Tetrahedron 40:1767-1781

Sakakibara M, Okada K, Ichikawa Y et al (1982) Synthesis of brassinolide, a plant growth promoting steroidal lactone. Heterocycles 17:301-304

Schmittberger T, Uguen D (1997) A formal synthesis of brassinolide. Tetrahedron Lett 38:2837-2840

Seo S, Nagasaki T, Katsuyama Y et al (1989) Synthesis of (22R,23R)- and (22S,23S)-[4-^{14}C]-24-epibrassinolide. J Label Compound Radiopharmaceut 27:1383-1393

Sharpless KB, Amberg W, Bennani YL et al (1992) The osmium-catalyzed asymmetric dihydroxylation: a new ligand class and a process improvement. J Org Chem 57:2768

Sun LQ, Zhou WS, Pan XF (1991) Studies on steroidal plant growth regulators 22. Osmium tetroxide catalyzed asymmetric dihydroxylation of the (22E, 24R) and the (22E, 24S)-24-alkyl steroidal unsaturated side chain. Tetrahedron Asymmetry 2:973-976

Takahashi T, Ootake A, Yamada H et al (1985) Stereoselective reduction of the steroidal 23-en-22-one: a route to the side chain of the plant growth promoter brassinolide. Tetrahedron Lett 26:69-72

Takatsuto S, Ikekawa N (1982) Synthesis of (22R,23R)-28-homobrassinolide. Chem Pharm Bull 30:4181-4185

Takatsuto S, Ikekawa N (1983) Stereoselective synthesis of the plant-growth-promoting steroids dolicholide and dolichosterone. J Chem Soc Perkin Trans 1 1983:2133-2137

Takatsuto S, Ikekawa N (1984) Short-step synthesis of plant growth-promoting brassinosteroids. Chem Pharm Bull 32:2001-2004

Takatsuto S, Watanabe, T, Fujioka S et al (1997a) Synthesis of new naturally occurring 6-

deoxo brassinosteroids. J Chem Research (S) 1997:134-135

Takatsuto S, Kuriyama H, Noguchi T et al (1997b) Synthesis of cathasterone and its related putative intermediates in brassinolide biosynthesis. J Chem Research (S) 1997:418-419

Thompson MJ, Mandava NB, Flippen-Anderson JL et al (1979) Synthesis of brassinosteroids: new plant-growth-promoting steroids. J Org Chem 44:5002-5004

Thompson MJ, Mandava NB, Meudt WJ et al (1981) Synthesis and biological activity of brassinolide and its 22β, 23β-isomer: novel plant growth-promoting steroids. Steroids 38:567-572

Traven, VF Kuznetsova NA, Levinson EE et al (1991) The shortest synthesis of 24-epibrassinolide and its 22S,23S-isomer from ergesterol. Doklady Akademii Nauk SSR 31:901-904

Tsubuki M, Keino K, Honda T (1992) Stereoselective synthesis of plant-growth-regulating steroids: brassinolide, castasterone, and their 24,25-substituted analogues. J Chem Soc Perkin Trans 1 1992:2643-2649

Watanabe, T Takatsuto S, Fujioka S et al (1997) Improved synthesis of castasterone and brassinolide. J Chem Research (S): 360-361

Yokota T, Watanabe S, Ogino Y et al (1990) Radioimmuno assay for brassinosteroids and its use for comparative analysis of brassinosteroids in stems and seeds of *Phaseolus vulgaris*. J Plant Growth Regul 9:151-159

Zhou WS, Tian WS (1987) Study on the synthesis of brassinolide and related compounds III stereoselective synthesis of typhasterol from hyodeoxycholic acid. Tetrahedron 43:3705-3712

Zhou WS, Huang LF (1992) Studies on steroidal plant-growth regulator 25: concise stereoselective construction of side chain of brassinosteroid from the intact side chain of hyodeoxycholic acid: formal syntheses of brassinolide, 25-methylbrassinolide 26,27-bisnorbrassinolide and their related compounds. Tetrahedron 48:1837-1852

5
Biosynthesis

AKIRA SAKURAI

Honorary Scientist of RIKEN (The Institute of Physical and Chemical Research), 1-33-12 Minami-Ogikubo, Suginami-ku, Tokyo 167-0052, Japan

1. Introduction

Naturally occurring brassinosteroids (BRs) so far identified are C_{27}-, C_{28}-, and C_{29}-steroids, whose carbon skeletons of the side chain are common in plant sterols, suggesting that BRs are biosynthesized from the corresponding plant sterols (Fig. 1). Among BRs, brassinolide, belonging to C_{28}-steroids of the same carbon structure as that of campesterol, is the most biologically active and is widely distributed in the plant kingdom. The C_{28}-BRs of castasterone, typhasterol, and teasterone could be regarded as biosynthetic precursors of brassinolide derived from campesterol. Thus, a hypothetical biosynthetic pathway for brassinolide has been proposed (Yokota et al 1991): starting from campesterol to give teasterone, which is converted to typhasterol, followed by hydroxylation at C2 to give castasterone. Brassinolide would then be formed by Baeyer-Villiger type oxidation at C6 of castasterone, as shown in Fig. 1. Experimental evidence for these metabolic conversions was first obtained using the plant cell culture system of *Catharanthus roseus*.

The plant cell cultures derived from *C. roseus* have been found to produce brassinolide and castasterone at levels comparable to those of pollen or immature seeds, which are the most BR-rich plant tissues (Sakurai et al 1991). The cell culture system had a major advantage for biosynthetic studies at the biochemical level, as the effective assimilation of the substrates without the problem of translocation was expected. Biochemical studies using the cultured cells of *C. roseus* elucidated the biosynthetic pathways of brassinolide from campesterol, and the models of two par-

---

Key words: early C6-oxidation pathway, late C6-oxidation pathway, *Catharanthus roseus*, cultured cells of *C. roseus*, cathasterone, *Arabidopsis thaliana*, *det2* mutant, *cpd* mutant, *dwf4* mutant, *diminuto* mutant, *Pisum sativum*, *lkb* mutant, *lk* mutant

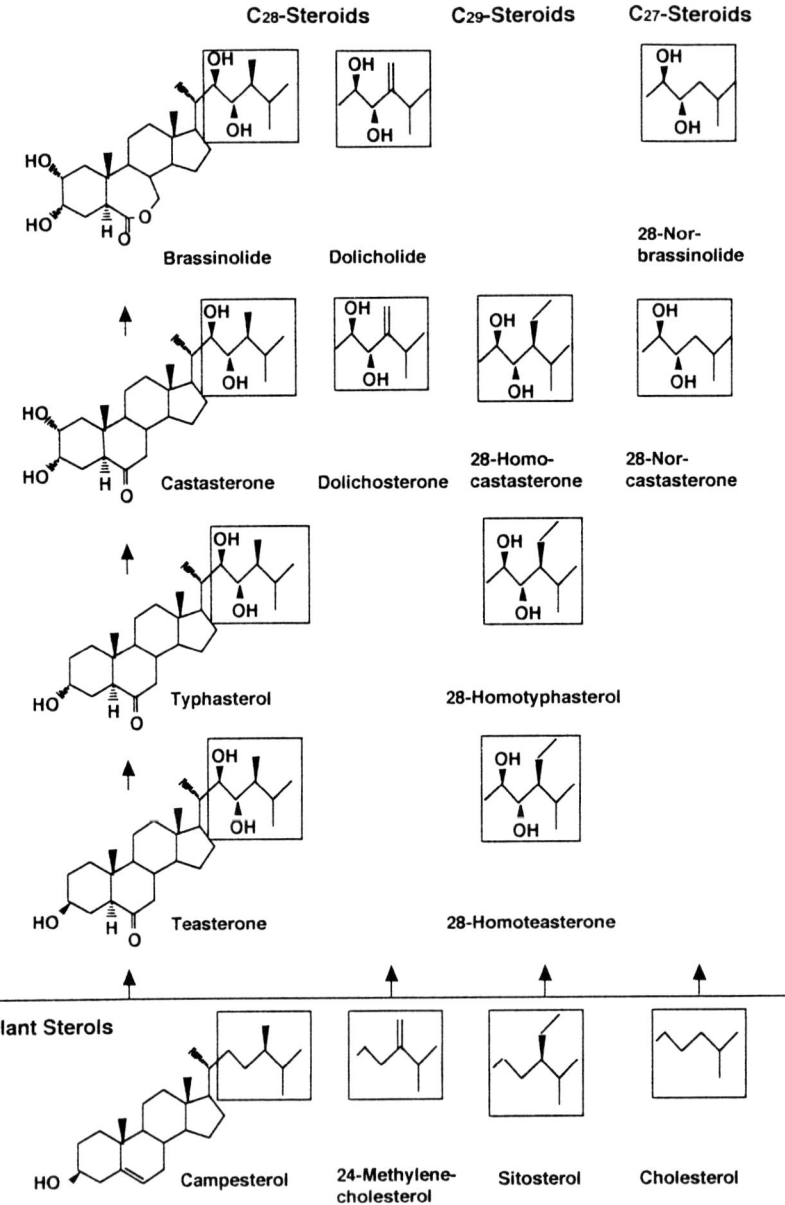

Fig. 1. Structural relationship of the side chain between natural BRs and plant sterols.

allel branched pathways, early C6-oxidation and late C6-oxidation, were established (Sakurai and Fujioka 1997, Fujioka and Sakurai 1997).

The discoveries of BR-deficient mutants among the de-etiolated dwarfs of *Arabidopsis thaliana* and the dwarf mutants of *Pisum sativum* (garden pea) have provided a new methodology in the biosynthetic studies of BRs (see the chapter by S. D. Clouse and K. A. Feldmann, this volume). Genetic, molecular, and biochemical studies on these mutants not only confirm the biosynthetic pathways proposed by the biochemical studies but also give the molecular basis of the biosynthesis.

In this chapter, biosynthetic studies of BRs using plant cell cultures and mutants are described, and the ascertained biosynthetic pathways are introduced.

2. Methods of Metabolic Conversions Using the Cell Culture System

2.1 Cell Cultures of *Catharanthus roseus*

The crown gall cells of V208 (nopaline type) and V277 (octopine type) derived from *Catharanthus roseus* (L.) G. Don (Madagascar periwinkle) were found to produce brassinolide and castasterone (Park et al 1989). Subsequent studies revealed that the non-transformed cells (Vn) of *C. roseus* produced the same BRs (Suzuki et al 1993b), indicating that BR production in the cell cultures does not depend on incorporated T-DNA in the crown gall cells but is one of the indigenous characteristics of the cells derived from *C. roseus*. In any case, the V208 cell line showed the most abundant production of BRs and hence was used for biosynthetic studies.

The V208 cells were grown in Murashige-Skoog medium containing 3% sucrose as suspension cultures at 27°C by reciprocal shaking. The maximum production of brassinolide was observed at the early stationary stage of the growth of the cultures (Sakurai and Fujioka 1996). The endogenous BRs so far identified in the cells of *C. roseus* are brassinolide and castasterone together with their biosynthetically related C_{28}-BRs as shown in Fig. 2.

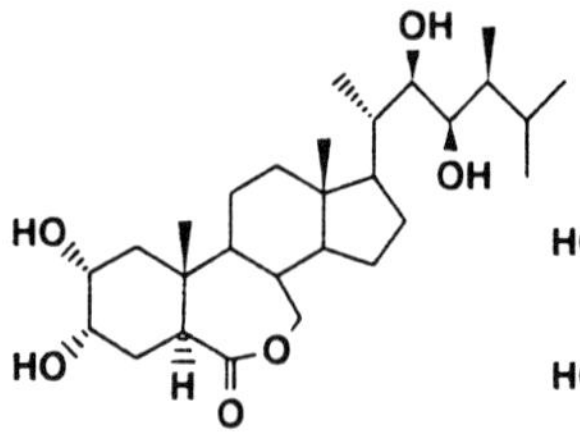

Brassinolide
[8.7 ng/g fr wt]

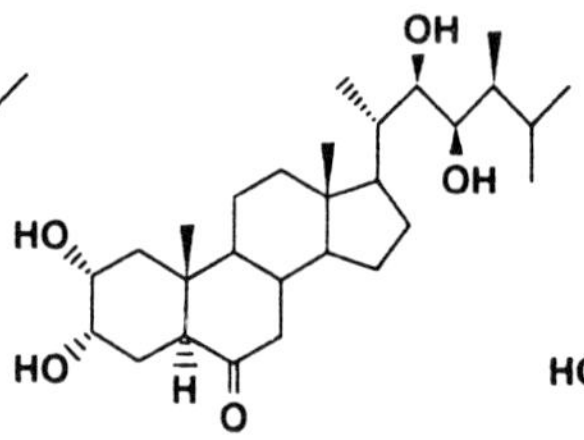

Castasterone
[4.5 ng/g fr wt]

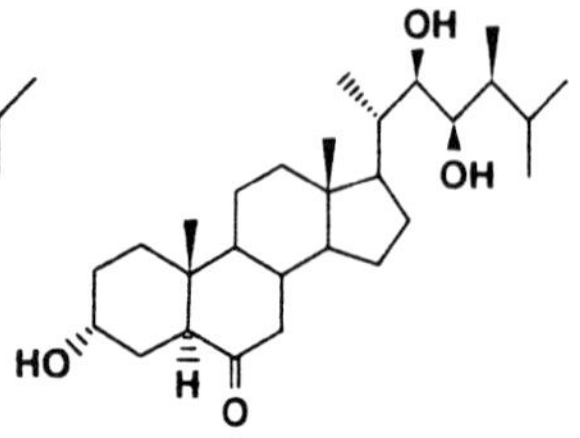

Typhasterol
[0.05 ng/g fr wt]

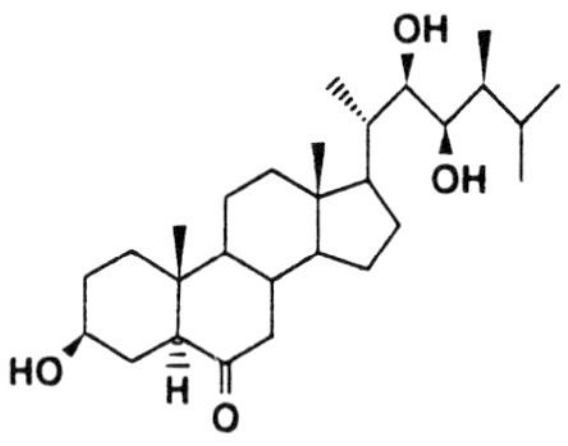

Teasterone
[0.05 ng/g fr wt]

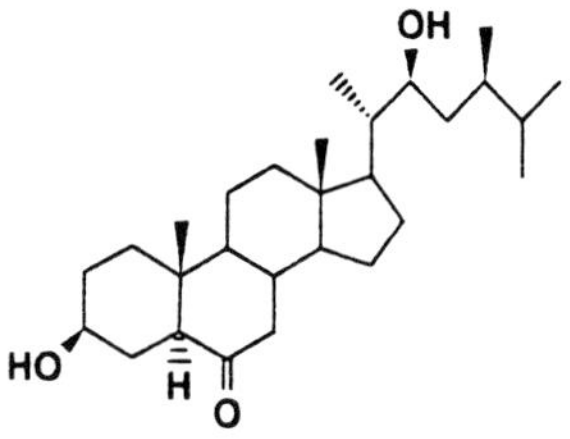

Cathasterone
[3.9 ng/g fr wt]

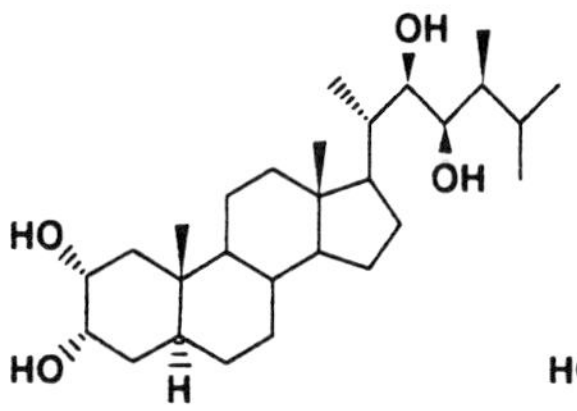

6-Deoxocastasterone
[18.9 ng/g fr wt]

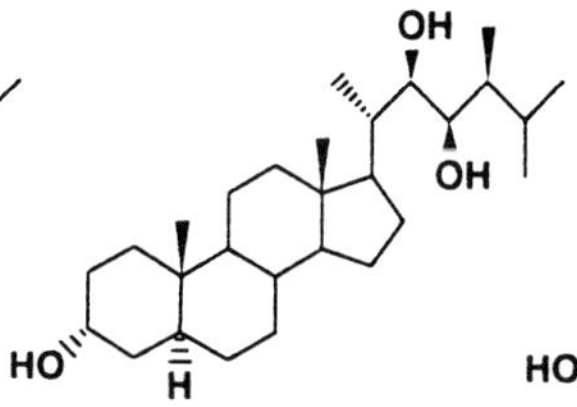

6-Deoxotyphasterol
[0.76 ng/g fr wt]

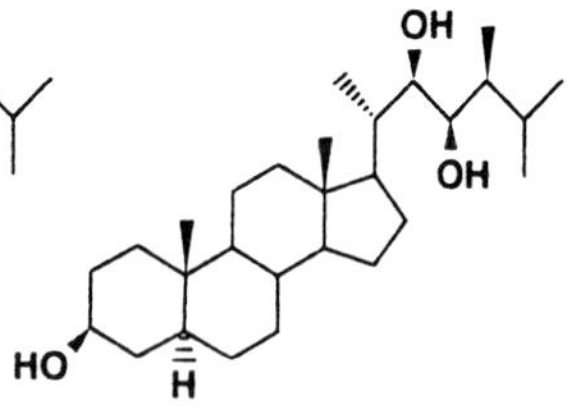

6-Deoxoteasterone
[0.05 ng/g fr wt]

Fig. 2. BRs in the cultured cells of *Catharanthus roseus*. Content of each BR per gram fresh weight of V208 cells is shown in brackets.

2.2 Preparation of Labeled Campesterol and Its Use for Early BR Biosynthesis

Campesterol, the putative precursor of brassinolide, is the major steroid of the cultured cells of *C. roseus* (Suzuki et al 1995b). To elucidate the metabolism of campesterol, labeled campesterol was prepared by the cultured cells. Compactin, an inhibitor of mevalonic acid biosynthesis, was used to minimize the dilution of labeled steroids by endogenous mevalonic acid. Compactin inhibits 3-hydroxyl-3-methylglutaryl (HMG) CoA reductase, which catalyzes the biosynthesis of mevalonic acid (Endo et al 1976). By feeding a mixture of ^{13}C- and ^{14}C-labeled mevalonic acid to the cultured cells starved by use of compactin, ^{13}C- /^{14}C-labeled campesterol was obtained effectively and used for the investigation of early BR biosynthesis (Suzuki et at 1995b). The high-performance liquid chromatography (HPLC) profile of the metabolites obtained by feeding the labeled campesterol to the cells is shown in Fig. 3. Identification of the metabolites that could be the precursors of BR biosynthesis are discussed in sections 4 and 5, this chapter. Actually, the endogenous campesterol in cells of *C. roseus* was found to be a mixture of campesterol and 24-epicampesterol (4:1), which should be called 24-methylcholesterol (Suzuki et at 1995b). Although the metabolites of the steroid were not confirmed to be $(24R)$-24-methyl steroids, the endogenous BRs in the cultured cells were unambiguously identified to have the same configuration as campesterol.

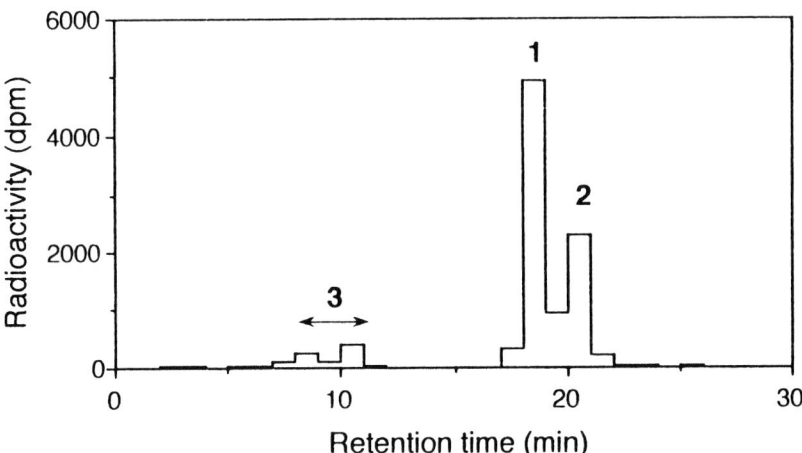

Fig. 3. ODS-HPLC analysis of the metabolites of ^{14}C-labeled campesterol by the cultured cells of *C. roseus* (V208). Radioactive sterol fractions of the cultured cells otained by feeding ^{14}C-labeled campesterol were subjected to HPLC on a Senshu-Pak ODS column with methanol at a flow rate of 0.8 ml/min. Radio-activity of each fraction was measured by a liquid scintillation counter. Peaks are 1: the substrate; 2: campestanol; 3: oxidized metabolites.

2.3 Feeding Experiments and Analysis of Metabolites

Investigations of the pathways followed by the metabolism of campesterol were undertaken using 26,28-^{2}H$_6$-labeled steroids (see the chapter by S. Takatsuto and T. Yokota, this volume). The putative precursors were usually fed to the cells at the early growth stage of the cultures. A solution of the substrate (1-5 ng) was added to a 200 ml conical flask containing the cells grown 8-10 days in 60 ml of the medium, and the culture was continued for 24 to 48 hours. The procedure for the extraction, purification, and analysis of the metabolites in the cells is shown in Fig. 4. The analysis by gas chromatography-mass spectrometry (GC/MS) was carried out according to the method described in the chapter by S. Takatsuto and T. Yokota, this volume. Figure 5 shows a typical result identifying deuterio-labeled castasterone as the metabolite of deuterio-labeled typhasterol in the cells. Repetitive investigation on the metabolites of labeled substrates led to demonstrating the two pathways as described in section 5, this chapter. Steps of conversion were confirmed by identification of the endogenous key intermediates in the cultured cells as shown in Fig. 2.

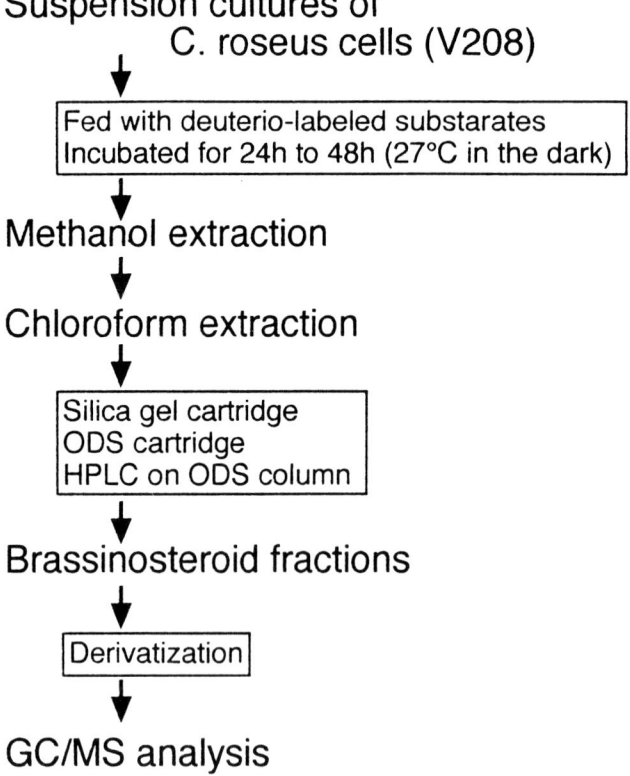

Fig. 4. Procedure for the analysis of metabolites of labeled substrates by the cultured cells of *C. roseus*.

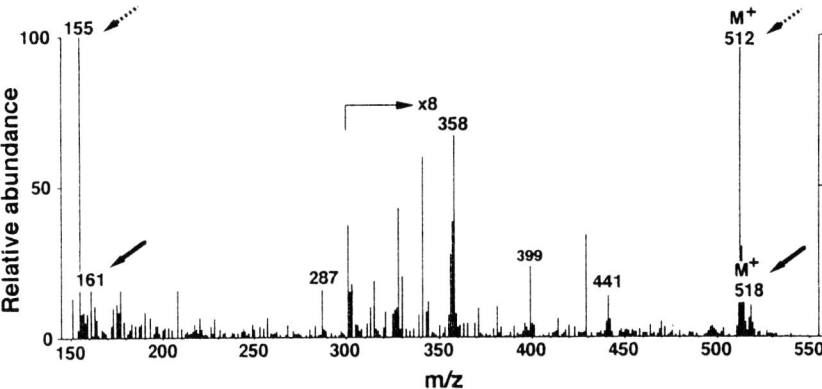

Fig. 5. Mass spectrum of castasterone bismethaneboronate in GC/MS analysis of the metabolites obtained by feeding [^{2}H$_6$] typhasterol to the cultured cells of *C. roseus* (V208). Solid arrows: ions of [^{2}H$_6$] castasterone converted from [^{2}H$_6$] typhasterol; dotted arrows: ions of endogenous castasterone.

3. Biochemical Analysis of Brassinosteroid Biosynthesis Mutants

3.1 Mutants of *Arabidopsis thaliana*

The BR biosynthesis mutants *det2*, *cpd* (=*cbb3*), *diminuto* (=*cbb1*), and *dwf4* have been characterized among the de-etiolated dwarfs of *Arabidopsis thaliana* (see the chapter by S. D. Clouse and K. A. Feldmann, this volume). The growth of the mutants was rescued by exogenous application of brassinolide and its biosynthetic congeners and not by the other hormones gibberellins or indoleacetic acid (IAA), suggesting that the lack of the normal growth in the mutants was caused by the deficiency of endogenous BRs.

Characterization of endogenous BRs in the shoots of Arabidopsis is essential for studies of the mutants. Castasterone, 6-deoxocastasterone, typhasterol, and 6-deoxotyphasterol were identified in the shoots of wild-type Arabidopsis (ecotype Columbia) (Fujioka et al 1996), and their levels were estimated as shown in Table 1 (Fujioka et al 1997). Brassinolide would be the final biologically active BRs in the plant, but its level may be too low to be detected in the seedlings. From seeds and siliques of the same ecotype, brassinolide and 6-deoxoteasterone were identified in addition to the above BRs (Fujioka et al 1998). Schmidt et al (1997) reported the identification of 24-epibrassinolide and castasterone in seeds of wild-type Arabidopsis (ecotype 24).

Brassinosteroid-deficiency in the mutant was shown by the biochemical studies on the *det2* mutant (Fujioka et al 1997). The *DET2* gene has sequence identity with that of mammalian steroid 5α-reductases, which catalyze the reduction of testoster-

Table 1. Brassinosteroid (BR) content in wildtype and *det2* shoots of *A. thaliana* [a] (reproduced from Fujioka et al 1997)

BRs	Wildtype	*det2*
Castasterone	0.75	0.08
6-Deoxocastasterone	0.71	0.07
Typhasterol	0.11	ND [b]
6-Deoxotyphasterol	0.95	ND

[a] BR content is expressed as nanograms per gram fresh weight of tissues.
[b] ND, not detected.

one to dihydrotestosterone, suggesting that the *det2* mutant is blocked in the conversion of campesterol to campestanol (Li et al 1996). Endogenous campestanol in the shoots was analyzed to show that the content in the *det2* mutant was 8-15% that of the wildtype. Deuterio-labeled campesterol, when fed to the shoots, was not converted to campestanol in the *det2*, while deuterio-labeled campestanol was detected in the wildtype. The analysis of the endogenous BRs in *det2* by GC/MS confirmed the deficiency of BRs in the mutant with detectable levels of castasterone and 6-deoxocastasterone showing less than 10% of wildtype levels (Table 1). Thus, the *det2* mutant was shown to be blocked in the formation of campestanol, resulting in BR deficiency.

The dwarf mutants *cpd* (Szekeres et al 1996) and *dwf4* (Choe et al 1998) were predicted to be blocked at hydroxylation steps in BR biosynthesis, since both *CPD* and *DWF4* genes were shown to encode microsomal cytochrome P450 monooxygenases with similarity to mammalian steroid hydroxylases. Although the endogenous BRs in the mutants remained for biochemical analysis, feeding the intermediates of BR biosynthesis to the mutants disclosed the blocked steps of the *cpd* at C_{23}-hydroxylation and of the *dwf4* at C_{22}- hydroxylation (see section 5, this chapter)

Another mutant, *diminuto* (Takahashi et al 1995), was previously thought to be blocked at the step of conversion of teasterone to typhasterol based on feeding experiments of the intermediates (Sakurai and Fujioka 1997). However, the following studies showed that the blocked step of *diminuto* was different from the conversion of teasterone. Biochemical analysis of endogenous sterols of the mutant revealed that the biosynthesis of campesterol was blocked in the mutant (Klahre et al 1997). The level of campesterol reduced to 10% of that of the wildtype and the precursor of campesterol, 24-methylenecholesterol, was accumulated in the mutant.

3.2 Mutants of *Pisum sativum*

A dwarf mutant, *lkb*, of *Pisum sativum* (garden pea) was shown to be BR-deficient (Nomura et al 1997). The mutant is characterized by reduced stem elongation and peduncle length, which is a different phenotype from that of gibberellin-deficient dwarf mutants of pea. The growth of *lkb* was restored by application of brassinolide, castasterone, and teasterone. The analysis of endogenous BRs in the young shoots of *lkb* revealed that the levels of brassinolide, castasterone, and 6-deoxocasatasterone in the *lkb* shoots were approximately 10% of those in wild-type shoots (Table 2), showing that the *lkb* is a BR-deficient dwarf mutant. The blocked step of the mutant has been estimated to be prior to the production of teasterone, and the following studies on the endogenous steroids disclosed that biosynthesis of campesterol was blocked in the mutant (Yokota et al 1997). The sterol composition of the *lkb* showed the decreased levels of campesterol, sitosterol, and stigmasterol compared to those of wildtype, while their precursors, 24-methylenecholesterol and isofucosterol, were accumulated in the mutant, indicating that the reduction of $\Delta^{24(28)}$-steroids is blocked in the *lkb* mutant.

Recently, another dwarf mutant of pea, *lk*, was found to be a BR biosynthesis mutant (Yokota et al 1997). The endogenous levels of castasterone, 6-deoxocastasterone, and 6-deoxotyphasterol in the seedlings of the mutant were decreased to less than 25% of those of wildtype. Moreover, the levels of campestanol and sitostanol were drastically lowered in the mutant, indicating that the *lk* is blocked the formation of campestanol, similar to the *det*2 of Arabidopsis.

Table 2. Brassinosteroid (BR) content in wildtype and *lkb* shoots of *P. sativum* [a] (reproduced from Nomura et al 1997)

BRs	Wildtype	*lkb*
Brassinolide	0.164	0.007
Castasterone	0.355	0.016
Typhasterol	- [b]	-
Teasterone	ND [c]	ND
6-Deoxocastasterone	3.133	0.355

[a] BR content in 3-days-old plants is expressed as nanograms per gram fresh weight of tissues.
[b] No reliable data were obtained.
[c] Not detected.

4. Early Stage of Biosynthesis

4.1 Biosynthesis of Campesterol

Campesterol, the initial precursor of brassinolide, is one of the major plant sterols together with sitosterol, stigmasterol, and cholesterol. Biosynthesis of plant sterols has been extensively studied, and a matrix of the pathways has been proposed (Benveniste 1986). The outline of the biosynthesis of campesterol is shown in Fig. 6 as a generally accepted pathway in vascular plants. Cycloartenol (**1**) is the first

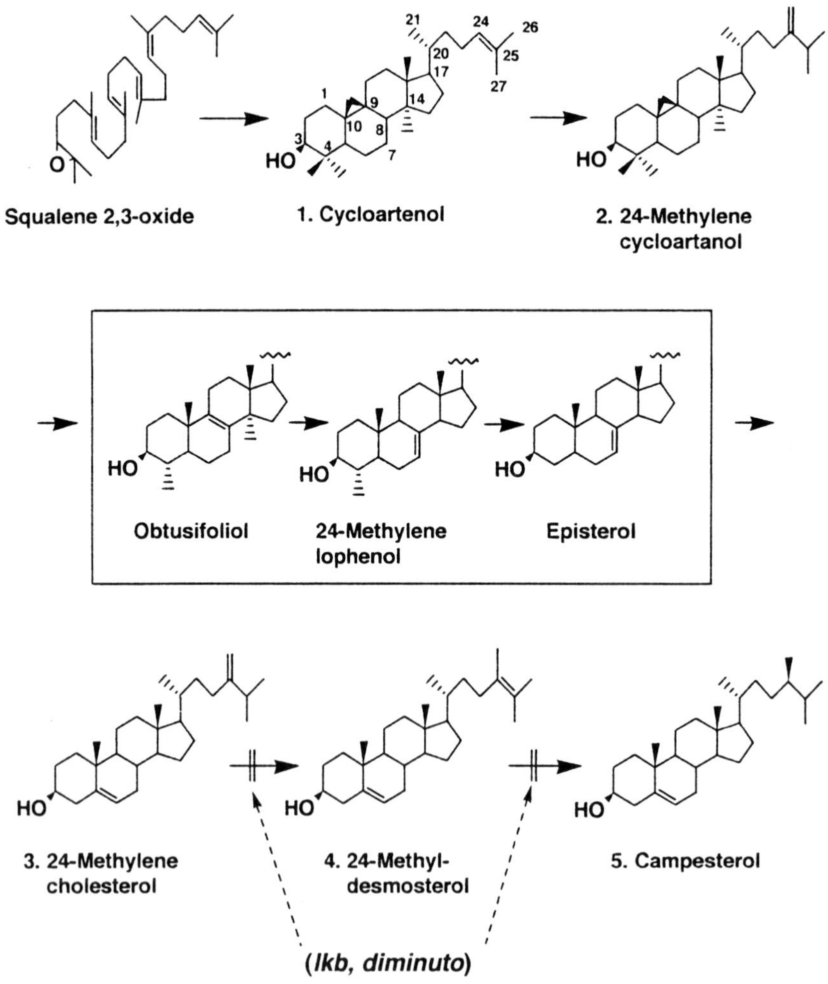

Fig. 6. Biosynthetic pathway of campesterol and blocked steps of the mutants.

cyclic triterpenoid precursor derived from squalene-2,3-oxide in plant-sterol biosynthesis. Methylation of **1** by *S*-adenosylmethionine gives 24-methylenecycloartanol (**2**), which undergoes subsequently 4β-demethylation, opening of the cyclopropane ring, 14-demethylation, 4α-demethylation, and migration of the double bond at C7 to C5 to give 24-methylenecholesterol (**3**). Reduction of a double bond in the side chain of **3** gives campesterol (**5**) via isomerization to 24-methyldesmosterol (**4**) (Yamada et al 1997).

The dwarf mutant *lkb* of garden pea and the *diminuto* mutant of Arabidopsis were shown to be blocked in the conversion of 24-methylenecholesterol (**3**) to campesterol (**5**) by the analysis of endogenous sterols of the mutant (see section 3, this chapter). The conversion is a two-step reaction, namely isomerization of **3** to **4** and reduction of **4** to **5**. Both steps may be affected in the mutants, since labeled 24-methyldesmosterol (**4**) was not converted to **5** in the *diminuto* seedlings (Klahre et al 1997). In the membrane fraction of seedlings of the *lkb* mutant, the shortage of sitosterol, stigmasterol, as well as **5**, and the accumulation of **3** and isofucosterol, were observed indicating that the *LKB* gene is also responsible for the conversion of isofucosterol to sitosterol (Yokota et al 1997). It is noteworthy that a blockage at this step causes not only the deficiency of BR biosynthesis but also abnormality in sterol composition of the membrane in the mutant.

4.2 Conversion of Campesterol to Campestanol

Conversion of campesterol to putative intermediates of BR biosynthesis was investigated using cultured cells of *C. roseus* (Suzuki et al 1995b). A major metabolite of $^{13}C/^{14}C$-labeled campesterol (see section 2.2, this chapter) was campestanol (**6**), which could be the first intermediate of BR biosynthesis.

Biochemical studies on the *det*2 mutant of Arabidopsis confirmed that the conversion of campesterol to campestanol was one of the key steps in BR biosynthesis and disclosed a detour pathway in the conversion of campesterol to campestanol (Fig. 7) (Fujioka et al 1997). The analysis of endogenous levels of campestanol and the feeding experiments of deuterio-labeled campesterol showed the *det*2 mutant to be blocked in the formation of campestanol, resulting in the decrease of BRs in the mutant (see section 3.1, this chapter). However, when the *DET*2 gene was expressed in a human kidney cell culture, DET2 protein catalyzed the reduction of $\Delta^{4(5)}$-steroid to give 5α-steroid but did not reduce the $\Delta^{5(6)}$ bond of campesterol (Li et al 1997). Moreover, endogenous campesterol did not accumulate in the mutant, even though the reduction of campesterol was blocked. Instead, the accumulation of 24-methylcholest-4-en-3-one (**7**) was observed in *det*2 compared to the wildtype, indicating that the reduction of **7** to 24-methylcholestan-3-one (**8**) was blocked in the *det*2 mutant. Therefore, **5** is converted to **7** through dehydration of the 3β-hydroxyl followed by isomerization of the $\Delta^{5(6)}$-bond, and, the resulting **7** is reduced to **8** by the DET2 protein. The additional two enzymes analogous to the mammalian 3β-hydroxy dehydrogenase/$\Delta^{5(6)}$-$\Delta^{4(5)}$ isomerase of steroids (Lachance et al 1990) must be involved in the conversion of **5** to **7**. Reduction of the C_3-ketone in **8** to a 3β-

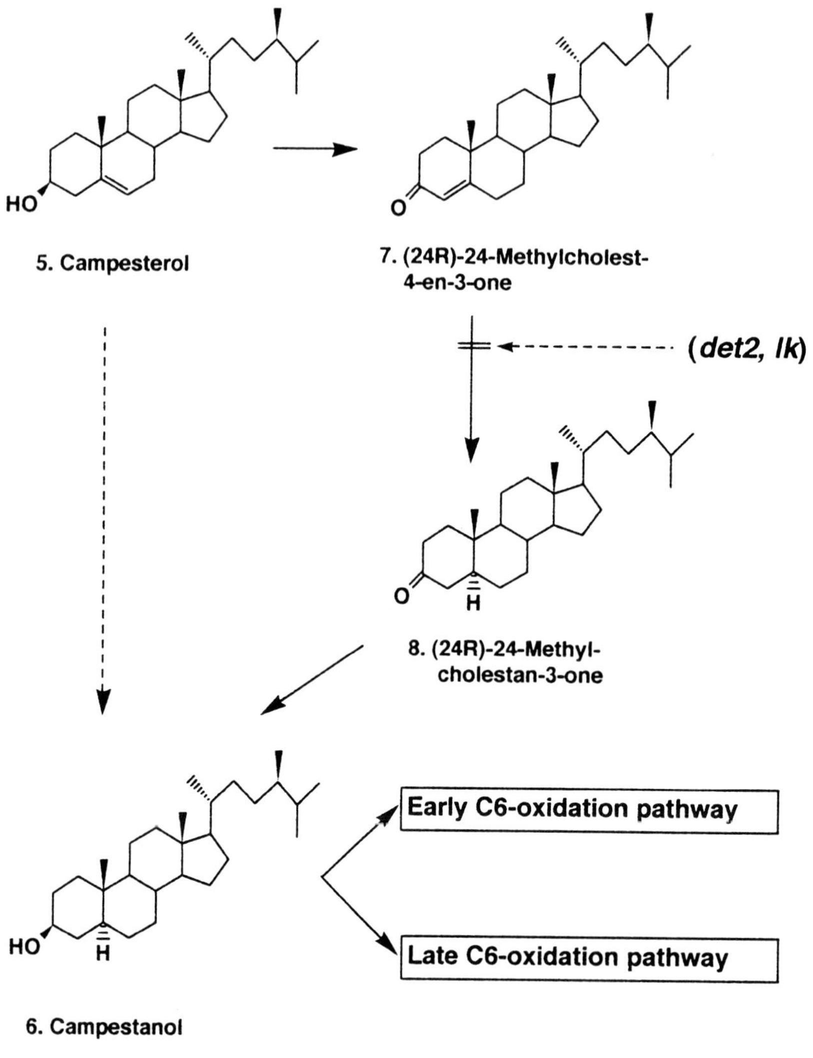

Fig. 7. Early steps in BR biosynthetic pathway and blocked steps of the mutants.

hydroxyl gives campestanol (**6**). Thus, the detour pathway in the conversion of **5** to **6** via **7** and **8** must operate in the plant as shown in Fig. 7, and the *det2* mutant is blocked in the reduction of 24-methylcholest-4-en-3-one (**7**).

The dwarf *lk* mutant of garden pea was found to be a BR-deficient mutant by blocking the formation of campestanol (Yokota et al 1997). The blocked step has recently been shown to be the reduction of **7** to **8** similar to the *det2* mutant of

Arabidopsis (T. Yokota, personal communication, 1998), suggesting that the detour pathway in Fig 7 is ubiquitous in the plants as a process to produce campestanol. The level of sitostanol was also decreased in the mutant, indicating that the process also operates in the conversion of sitosterol to sitostanol.

5. Downstream Pathways to Brassinolide

5.1 Early C6-oxidation Pathway

The wide distribution of 6-oxo type BRs, such as teasterone, typhasterol, and castasterone, which could be biosynthetic precursors of brassinolide, suggested a biosynthetic pathway starting from the oxidation at C6 of campestanol (**6**). The pathway, named early C6-oxidation pathway, was demonstrated as shown in Fig. 8 by the studies using cultured cells of *C. roseus*.

In the oxidized metabolites of $^{13}C/^{14}C$-labeled campesterol in the cells (see section 2.2, this chapter), 6α-hydroxycampestanol (**9**) and 6-oxocampestanol (**10**) were identified. These were found as endogenous steroids in the cultured cells and each step of the conversion of campestanol (**6**) to **9** and **9** to **10** was confirmed by the feeding of chemically prepared deuterio-labeled **6** and **9** to the cells, respectively. An isomer of **9**, 6β-hydroxycampestanol was not detected as an endogenous steroid in the cells nor was it converted to **10** by the cells (Suzuki et al 1995b).

As an intermediate between 6-oxocampestanol (**10**) and teasterone (**12**), 22α-hydroxy-6-oxocampestanol, named cathasterone (**11**), was identified as an endogenous steroid of the cultured cells of *C. roseus* (Fujioka et al 1995a). Feeding of deuterio-labeled cathasterone to the cultured cells gave deuterio-labeled teasterone (**12**) and typhasterol (**14**). Three other possible intermediates, 23α-hydroxy-6-oxocampestanol, $\Delta^{22(23)}$-6-oxocampestanol, and 22,23-epoxy-6-oxocampestanol, did not occur in the cells and did not give **12** in feeding experiments. Consequently, 6-oxocampestanol (**10**) must be hydroxylated at C22 to give cathasterone (**11**), which is converted to teasterone (**12**) by hydroxylation at C23. However, the conversion of **10** to **11** has not been detected by feeding experiments using labeled 6-oxocampestanol (**10**), because the pool size of **11** in the cells was smaller than 1/500 of **10**. This step in the pathway must be rate-limiting in the biosynthesis of brassinolide.

The dwarf *cpd* mutant of Arabidopsis was predicted to be blocked in the hydroxylation at C23, because teasterone (**12**) and the downstream intermediates **13** to **15** were able to rescue the growth of the mutant, while cathasterone (**11**) was inactive (Szekeres et al 1996). On the other hand, the *dwf4* mutant phenotype was rescued by feeding C_{22}-hydroxylated BRs such as cathasterone (**11**), indicating that the *dwf4* mutant was blocked in the hydroxylation at C22 (Choe et al 1998). Both *CPD* and *DWF4* genes were shown to encode cytochrome P450 monooxygenases (see the chapter by S. D. Clouse and K. A. Feldmann, this volume). Thus, the sequential hydroxylations at C22 and C23 in BR biosynthesis was revealed to be catalyzed by cytochrome P450 enzymes with similarity to mammalian steroid hydroxylases.

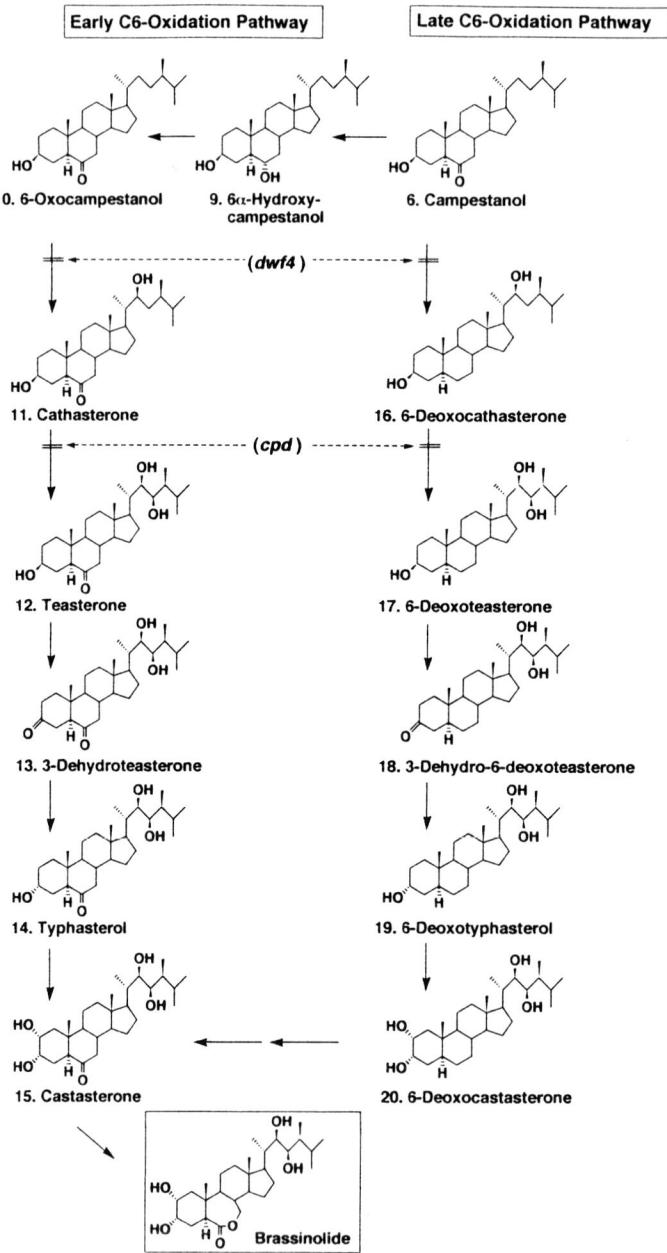

Fig. 8. Downstream pathways of early C6-oxidation and late C6-oxidation in BR biosynthesis and blocked steps of the mutants.

Teasterone (**12**) was converted to typhasterol (**14**) by feeding experiments in cultured cells of *C. roseus* (Suzuki et al 1994a) and in seedlings of *C. roseus, Oryza sativa*, and *Nicotiana tabacum* (Suzuki et al 1995a). The conversion, being an inversion of 3β-hydroxyl to 3α-hydroxyl, was reversible with predominant formation of **14**. In this conversion, the involvement of a 3-oxo form, 3-dehydroteasterone (**13**), was confirmed (Suzuki et al 1994b) similar to the 3-epimerization process of ecdysone (Milner and Rees 1985). In cultured cells of *C. roseus*, deuterio-labeled **13** gave typhasterol (**14**) as the major metabolite accompanied by minor amounts of teasterone (**12**). The failure to detect endogenous **13** in the cultured cells may indicate a facile metabolism of **12** to **14**. Nevertheless, 3-dehydroteasterone (**13**) was identified as a natural BR in wheat grain (Yokota et al 1994), anthers of *Lilium longiflorum*, and leaves of *Distylium racemosum* (Abe et al 1994). Moreover, feeding of **13** to the cultured cells of *L. longiflorum* was reported to give **14**, though the cells did not produce any endogenous BRs.

Typhasterol (**14**) was hydroxylated at C2 to give castasterone (**15**) by feeding labeled **14** to cultured cells of *C. roseus* (Suzuki et al 1994a) and to seedlings of *C. roseus, O. sativa*, and *N. tabacum* (Suzuki et al 1995a). The minor conversion of **14** to **12** was observed in every experiment as the reversible inversion of the 3-hydroxyl via **13**.

The intermediates of the early C6-oxidation pathway, teasterone, typhasterol, and/or castasterone, have been found together with brassinolide in a wide variety of plants. Biological activities of the intermediates in the rice lamina inclination test and the wheat leaf unrolling test increased with their order in the pathway (Fujioka et al 1995b). The same tendency was observed in the activity to rescue the growth of Arabidopsis mutants *cpd* (Szekeres et al 1996) and *det*2 (Fujioka et al 1997). Such observations indicates that the pathway is operating to convert the intermediates to brassinolide, which must be a biologically active end-product of BRs in the plants.

5.2 Late C6-oxidation Pathway

6-Deoxocastasterone (**20**) that has no ketone function at C6 has been identified as a naturally occurring BR from varieties of plants (see the chapter by S. Fujioka, this volume). Previously, **20** has been considered to be a dead-end product not to be converted to biologically active forms in the plants, because it showed little biological activity in BR bioassays, such as the rice lamina inclination test (Yokota et al 1983). However, the subsequent identification of 6-deoxotyphasterol (**19**) and 3-dehydro-6-deoxoteasterone (**17**) as putative precursors of **20** in the pollen of *Cupresse arizonica* (Griffiths et al 1995) suggested the involvement of these BRs in the biosynthetic pathways 6-Deoxocastasterone (**20**) was also identified in cultured cells of *C. roseus* in addition to 6-oxo type BRs and shown to be converted to castasterone (**15**) and brassinolide by feeding deuterio-labeled substrate (Choi et al 1996). These indicate the existence of another pathway to produce castasterone, named the late C6-oxidation pathway (Fig. 8).

Endogenous occurrence of 6-deoxoteasterone (**17**) and 6-deoxotyphasterol (**19**) was also shown in cultured cells of *C. roseus*. Feeding experiments with deuterio-labeled substrates disclosed the conversion of **17** to **19** (Choi et al 1997). When the putative intermediate of deuterio-labeled 3-dehydro-6-deoxoteasterone (**18**) was fed to the cells, it gave 6-deoxotyphasterol (**19**). 3-Epimerization of **17** must occur via a 3-oxo form (**18**) similar to the early C6-oxidation pathway. 6-Deoxotyphasterol (**19**) was converted to 6-deoxocastasterone (**20**), which was oxidized at C6 to give castasterone (**15**) (Choi et al 1996). It is most likely that only 6-deoxocastasterone (**20**) undergoes C6-oxidation to give castasterone, since branched pathways such as C6-oxidation of **17** or **19** were not detected in the cells.

Consequently, the late C6-oxidation pathway constituting the reactions parallel to those of early C6-oxidation was demonstrated as shown in Fig. 8 using cultured cells of *C. roseus*. 6-Deoxocathasterone (**16**) could be the intermediate of the late C6-oxidation pathway from campestanol, though its occurrence and conversion require further investigation.

Conversion of 6-deoxocastasterone (**20**) to castasterone (**15**) was also observed in the seedlings of *C. roseus*, *O. sativa*, and *N. tabacum* (Choi et al 1996), suggesting that the late C6-oxidation pathway is operating in intact plants. Indeed, as the endogenous BRs in the seedlings of Arabidopsis, 6-deoxotyphasterol (**19**) and 6-deoxocastasterone (**20**) were identified together with typhasterol (**14**) and castasterone (**15**) (Fujioka et al, 1996). For the *det2* mutant of Arabidopsis, the intermediates in the both pathways (**12** to **15** and **17** to **20**) showed the ability to rescue the growth of the mutant (Fujioka et al 1997), confirming that both the early C6-oxidation and late C6-oxidation pathways are playing roles in the growth of plants. The gene products of *CPD* and *DWF4* may also hydroxylate 6-deoxo steroids at C23 and C22 in the late C6-oxidation pathway, since *dwf4* mutants showed response to 6-deoxo BRs hydroxylated at C_{22} and C_{23} (Choe et al 1998).

5.3 Conversion of Castasterone to Brassinolide

Transformation of castasterone (**15**) to brassinolide through Baeyer-Villiger type oxidation of C_6-ketone was shown first by feeding tritium-labeled castasterone to cultured cells of *C. roseus* (Yokota et al 1990) and later confirmed by feeding the deuterio-labeled substrate to the cultured cells (Suzuki et al 1993b) and seedlings of *C. roseus* (Suzuki et al 1995a). In the case of feeding **15** to the seedlings, conversion to 3-epicastasterone together with brassinolide was observed in *C. roseus*, while only 3-epicastasterone was formed in *N. tabacum* and *O. sativa*. The conversion of **15** to 3-epicastasterone may be an inactivation process, since 3-epicastasterone has been identified as a weak biologically active BR in immature seeds of *Phaseolus vulgaris* (Kim 1991). In the cultured cells of lily, teasterone-3-myristate was shown to be converted to free teasterone, castasterone, and brassinolide using ^{14}C-labeled substrates (Abe et al 1996). Teasterone-3-myristate, identified as a conjugate form of BR from lily anthers (Asakawa et al 1994), must act as a reversible storage form during BR biosynthesis.

Brassinolide is the most biologically active among the naturally occurring BRs, suggesting that brassinolide, with a 6-oxo-7-oxa lactone in the B ring, is the end-product of biosynthesis. However, castasterone, the direct precursor of brassinolide, was found in the widest range of species of plants, where brassinolide was not identified in many cases. This may be explained by the extremely low level of brassinolide as an active form in plants and/or the lower sensitivity of GC/MS to detect brassinolide compared to castasterone (Fujioka et al 1996). On the other hand, metabolic conversion of castasterone to brassinolide has been studied in a variety of plants. Although the conversion was observed in cultured cells and seedlings of *C. roseus* as described above, radiolabeled castasterone was metabolized to water-soluble conjugated forms only in mung bean cuttings (Suzuki et al 1993a) and etiolated rice leaf explants (Yokota et al 1992). These facts suggest the possibility that castasterone is biologically active per se. Nevertheless, isolation of a mutant blocked in the conversion of castasterone to brassinolide will be necessary to define the active form of BRs.

6. Regulation of Biosynthesis

Brassinolide and its biosynthetically related C_{28}-steroids with a 24R-methyl are the most abundant and widely occurring in plants, indicating that campesterol is the most preferred precursor of BRs (Yokota 1997). Although sitosterol is generally the major C_{29}-sterol, accounting for 50-80% of the total sterols in plants, C_{29}-BRs with the same carbon skeleton as sitosterol are rare. Similarly, C_{27}-BRs and C_{28}-BRs with 24S-methyl or 24-methylene are identified in limited species of plants. These BRs are mostly found as minor components together with the major C_{28}-BRs derived from campesterol. These observations suggest that BR biosynthetic enzymes prefer C_{28}-steroids of campesterol as substrates.

The endogenous levels of plant sterols are generally measured in millimoles to micromoles in plant tissues, and campestanol or 6-oxocampestanol in the early steps of the pathway also exists in plants at a similar order of magnitude. On the other hand, BRs with hydroxyls at the side chain occur at less than the nanomole level in plants, even at their most abundant occurrence in pollen or immature seeds, suggesting that hydroxylation of the side chain in the biosynthetic pathways must be strictly regulated. In cultured cells of *C. roseus*, the endogenous levels of cathasterone and of the downstream intermediates are dramatically reduced to 1/500 that of 6-oxocampestanol, which is a direct precursor of cathasterone (Fujioka et al 1995a). This suggests that the hydroxylation at C22 of the side chain is the limiting step in BR biosynthesis.

The expression of the Arabidopsis *CPD* gene, which encodes a cytochrome P450 steroid hydroxylase involved in the hydroxylation at C23 of cathasterone, was investigated by Mathur et al (1998). They constructed a *CPD* promoter-*uidA* reporter gene in the promoter test vecter pPCV812, which was introduced into Arabidopsis. Expression of the *CPD* gene in the transgenic plant was monitored by GUS enzyme activities. GUS activity appeared in cotyledons and leaf primodia in etiolated seedlings and was detectable in adaxial parenchyma of expanding leaves in light-grown plants but not in the lower section of hypocotyls or in roots. *CPD* expression was not detectable in actively elongating cells throughout the hypocotyls and roots, suggesting active transport of BRs from source to target tissues and/or repression of the *CPD* gene at the target tissues by a negative regulatory factor. Transcription of the *CPD* gene was not affected by auxins, gibberellins, cytokinins, ethylene, or jasmonic acid but was specifically down-regulated by brassinolide in both dark and light. Intermediates in the BR biosynthetic pathways and biologically active synthetic analogs carrying C22 and/or C23 hydroxyls efficiently inhibit the activity of the *CPD* promoter (Mathur et al 1998). Repression of *CPD* transcription by BRs attests to the functioning of a negative feedback regulation of BR biosynthesis at this enzyme step. Feedback control represents a mode of self-regulation of plant hormone levels, since gibberellin biosynthesis was shown to be modified by the action of gibberellin itself in a type of feedback regulation (Hedden and Kamiya 1997).

Both pathways of early C6-oxidation and late C6-oxidation appeared to operate equally in cultured cells of *C. roseus*, since no significant difference in the metabolic conversion and endogenous levels of the intermediates of the two pathways was observed (Sakurai and Fujioka 1997). Differential growth effects on light- and dark-grown Arabidopsis were observed between 6-oxo type and 6-deoxo type BRs when fed to the *det*2 mutants (Fujioka et al 1997) and the *dwf*4 mutants (Choe et al 1998). 6-Oxo BRs were more active in the dark and less active in the light than 6-deoxo BRs with respect to rescue of the mutants. The observation suggests that the early C6-oxidation pathway predominantly operates in the dark, while the late C6-oxidation pathway is dominant in the light. Although the difference in endogenous levels of BRs between dark-grown and light-grown plants has not yet been investigated, BR biosynthesis could be regulated by the light as has been shown for gibberellin biosynthesis (Hedden and Kamiya 1997).

7. Conclusion

The first step in the study of biosynthesis was to elucidate the pathways leading to brassinolide, which have now been demonstrated by biochemical studies using the cultured system of *C. roseus*. Brassinolide is biosynthesized from campesterol in two parallel pathways, namely the early C6-oxidation and the late C6-oxidation pathways, that branch after the formation of campestanol. Studies on dwarf mutants deficient in BR biosynthesis found in Arabidopsis and garden pea have provided genetic and biochemical evidence confirming that the above biosynthetic pathways operate in diverse species of plants. However, multiple pathways including branched and/or modified steps can be expected due to the complex pattern of natural BRs.

Genetic studies of BR-deficient mutants of Arabidopsis have disclosed so far three genes encoding the enzymes involved in the biosynthesis of brassinosteroids. One is a steroid 5α-reductase catalyzing the formation of campestanol, and the other two are cytochrome P450 monooxygenases involved in the hydroxylation steps of the side chain. As expected, these genes have close similarity with those of corresponding mammalian steroid enzymes. Further characterization of the enzymes involved in the biosynthesis and development of specific enzyme inhibitors will contribute to understanding the whole picture of BR biosynthesis.

For all that, the findings so far obtained provide useful probes for molecular approaches, such as cloning genes encoding biosynthetic enzymes from various plant species, regulation of the expression of these genes by environmental conditions, localization/compartmentation of the biosynthetic enzymes, and so on. These lines of research are now progressing as the next step of biosynthetic studies of BRs and are essential for understanding the physiological roles of BRs in the growth and development of plants.

References

Abe H, Honjo C, Kyokawa Y, et al (1994) 3-Oxoteasterone and the epimerization of teasterone: Identification in lily anthers and *Distylium racemosum* leaves and its biotransformation into typhasterol. Biosci Biotech Biochem 58: 986-989

Abe H, Asakawa S, Natsume M. (1996) Interconvertible metabolism between teasterone and its conjugate with fatty acid in cultured cells of lily. Proc Plant Growth Reg Soc Amer 23: 9

Asakawa S, Abe H, Kyokawa Y, et al (1994) Teasterone 3-myristate: A new type of brassinosteroid derivative in *Lilium longiflorum* anthers. Biosci Biotech Biochem 58: 219-220

Benveniste P (1986) Sterol biosynthesis. Annu Rev Plant Physiol 37: 275-308.

Choe S, Dilkes BP, Fujioka S, et al (1998) The *DWF4* gene of Arabidopsis encodes a cytochrome P450 that mediates multiple 22α-hydroxylation steps in brassinosteroid biosynthesis. Plant Cell 10: 231-243

Choi Y-H, Fujioka S, Harada A, et al (1996) A brassinolide biosynthetic pathway via 6-deoxocastasterone. Phytochemistry 43: 593-596

Choi Y-H, Fujioka S, Nomura T, et al (1997) An alternative brassinolide biosynthetic pathway via late C-6 oxidation. Phytochemistry 44: 609-613

Endo A, Kuroda M, Tanzawa K (1976) Competitive inhibition of 3-hydroxy-3-methylglutaryl coenzyme A by ML-236A and ML-236B fungal metabolites, having hypocholesterolemic activity. FEBS Lett 72: 323-326

Fujioka S, Sakurai A (1997) Biosynthesis and metabolism of brassinosteroids. Physiol Plant 100: 710-715

Fujioka S, Inoue T, Takatsuto S, et al (1995a) Identification of a new brassinosteroid, cathasterone, in cultured cells of *Catharanthus roseus* as a biosynthetic precursor of teasterone. Biosci Biotech Biochem 59: 1543-1547

Fujioka S, Inoue T, Takatsuto S, et al (1995b) Biological activities of biosynthetically-related congeners of brassinolide. Biosci Biotech Biochem 59: 1973-1975

Fujioka S, Choi Y-H, Takatsuto S, et al (1996) Identification of castasterone, 6-deoxocastasterone, typhasterol and 6-deoxotyphasterol from the shoots of *Arabidopsis thaliana*. Plant Cell Physiol 37: 1201-1203

Fujioka S, Li J, Choi Y-H, et al (1997) The Arabidopsis *deetiolated2* mutant is blocked early in brassinosteroid biosynthesis. Plant Cell 9: 1951-1962

Fujioka S, Noguchi T, Yokota T, et al (1998) Brassinosteroids in *Arabidopsis thaliana*.

Phytochemistery 48: 595-599

Griffiths PG, Sasse JM, Yokota T, et al (1995) 6-Deoxotyphasterol and 3-dehydro-6-deoxoteasterone, possible precursors to brassinosteroids in the pollen of *Cupressus arizonica*. Biosci Biotech Biochem 59: 956-959

Hedden P, Kamiya Y (1997) Gibberellin biosynthesis: enzymes, genes and their regulation. Annu Rev Plant Physiol Plant Mol Biol 48: 431-460

Kim S-K (1991) Natural occurrences of brassinosteroids. In: Cutler HG, Yokota T, Adam G (Eds) Brassinosteroids: Chemistry, Bioactivity and Applications. ACS Symp Ser 474, Amer Chem Soc, Washington, DC, pp 26-35

Klahre U, Fujioka S, Yokota T, et al (1997) Characterization of the *diminuto* mutant and genes regulated by brassinosteroids. Proc Plant Growth Reg Soc Amer 24: 99-100

Lachance Y, Luu-The V, Labrie C, et al (1990) Characterization of human 3β-hydroxysteroid dehydrogenase/Δ^5-Δ^4-isomerase gene and its expression in mammalian cells. J Biol Chem 265: 20469-20475

Li J, Nagpal P, Vitart V, et al (1996) A role for brassinosteroids in light-dependent development of *Arabidopsis*. Science 272: 398-401

Li J, Biswas M, Chao A, et al (1997) Conservation of function between mammalian and plant steroid 5α-reductases. Proc Natl Acad Sci USA 94: 3554-3559

Mathur J, Molnár, Fujioka S, et al (1998) Transcription of *Arabidopsis CPD* gene, encoding a steroidogenic P450, is negatively controlled by brassinosteroids. Plant J 14: 593-602

Milner NP, Rees HH (1985) Involvement of 3-dehydroecdysone in the 3-epimerization of ecdysone. Biochem J 231: 369-374

Nomura T, Nakayama M, Reid JB, et al (1997) Blockage of brassinosteroid biosynthesis and sensitivity causes dwarfism in garden pea. Plant Physiol 113: 31-37

Park K-H, Saimoto H, Nakagawa S, et al (1989) Occurrence of brassinolide and castasterone in crown gall cells of *Catharanthus roseus*. Agric Biol Chem 53: 805-811

Sakurai A, Fujioka S (1996) *Catharanthus roseus* (*Vinca rosea*): In vitro production of brassinosteroids. In: Bajaj YSP (Ed) Biotechnology in Agriculture and Forestry, Vol 37: Medicinal and Aromatic Plants IX. Springer-Verlag, Berlin Heidelberg, pp 87-96

Sakurai A, Fujioka S (1997) Studies on biosynthesis of brassinosteroids. Biosci Biotech Biochem 61: 757-762

Sakurai A, Fujioka S, Saimoto H (1991) Production of brassinosteroids in plant-cell cultures. In: Cutler HG, Yokota T, Adam G (Eds) Brassinosteroids: Chemistry, Bioactivity and Applications. ACS Symp Ser 474, Amer Chem Soc, Washington, DC, pp 97-106

Schmidt J, Altmann T, Adam G (1997) Brassinosteroids from seeds of *Arabidopsis thaliana*. Phytochemistry 45: 1325-1327

Suzuki H, Kim S-M, Takahashi N, et al (1993a) Metabolism of castasterone and brassinolide in mung bean explant. Phytochemistry 33: 1361-1367

Suzuki H, Fujioka S, Takatsuto S, et al (1993b) Biosynthesis of brassinolide from castasterone in cultured cells of *Catharanthus roseus*. J Plant Growth Regul 12: 101-106

Suzuki H, Fujioka S, Takatsuto S, et al (1994a) Biosynthesis of brassinolide from teasterone via typhasterol and castasterone in cultured cells of *Catharanthus roseus*. J Plant Growth Regul 13: 21-26

Suzuki H, Inoue T, Fujioka S, et al (1994b) Possible involvement of 3-dehydroteasterone in the conversion of teasterone to typhasterol in cultured cells of *Catharanthus roseus*. Biosci Biotech Biochem 58: 1186-1188

Suzuki H, Fujioka S, Takatsuto S, et al (1995a) Biosynthesis of brassinosteroids in seedlings of *Catharanthus roseus, Nicotiana tabacum*, and *Oryza sativa*. Biosci Biotech Biochem

59: 168-172
Suzuki H, Inoue T, Fujioka S, et al (1995b) Conversion of 24-methylcholesterol to 6-oxo-24-methylcholestanol, a putative intermediate of the biosynthesis of brassinosteroids, in cultured cells of *Catharanthus roseus*. Phytochemistry 40: 1391-1397
Szekeres M, Németh K, Koncz-Kálman Z, et al (1996) Brassinosteroids rescue the deficiency of CYP90, a cytochrome P450, controlling cell elongation and de-etiolation in Arabidopsis. Cell 85: 171-182
Takahashi T, Gasch A, Nishizawa N, et al (1995) The *DIMINUTO* gene of *Arabidopsis* is involved in regulating cell elongation. Genes Dev 9: 97-107
Yamada J, Morisaki M, Iwai K, et al (1997) 24-Methyl and 24-ethyl-$\Delta^{24(25)}$-cholesterols as immediate biosynthetic precursors of 24-alkylsterols in higher plants. Tetrahedron 53: 877-884
Yokota T (1997) The structure, biosynthesis and function of brassinosteroids. Trends Plant Sci 2: 137-143
Yokota T, Morita M, Takahashi N (1983) 6-Deoxocastasterone and 6-deoxodolichosterone: putative precursor for brassinolide-related steroids form *Phaseolus vulgaris*. Agric Biol Chem 47: 2149-2151
Yokota T, Ogino Y, Takahashi N, et al (1990) Brassinolide is biosynthesized from castasterone in *Catharanthus roseus* crown gall cells. Agric Biol Chem 54: 1107-1108
Yokota T, Ogino Y, Suzuki H, et al (1991) Metabolism and biosynthesis of brassinosteroids. In: Cutler HG, Yokota T, Adam G (Eds) Brassinosteroids: Chemistry, Bioactivity and Applications. ACS Symp Ser 474, Amer Chem Soc, Washington DC, pp 86-96
Yokota T, Higuchi K, Kosaka Y, et al (1992) Transport and metabolism of brassinosteroids in rice. In: Karssen CM, van Loon LC, Vreugdenhil D (Eds) Progress in Plant Growth Regulation. Kluwer Academic Publishers, Dordrecht, pp 298-305
Yokota T, Nakayama M, Wakisaka T, et al (1994) 3-Dehydroteasterone, a 3,6-diketobrassinosteroid as a possible biosynthetic intermediate of brassinolide from wheat grain. Biosci Biotech Biochem 58: 1183-1185.
Yokota T, Nomura T, Kitasaka Y, et al (1997) Biosynthetic lesions in brassinosteroid-deficient pea mutants. Proc Plant Growth Reg Soc Amer 24: 94

6
Uptake, Transport and Metabolism

GÜNTER ADAM[1] AND BERND SCHNEIDER[2]

[1] Institut für Pflanzenbiochemie, Weinberg 3, D-06120 Halle, Germany
[2] Max-Planck-Institut für Chemische Ökologie, Tatzendpromenade 1a, D-07745 Jena, Germany

1. Introduction

Knowledge on uptake, transport, and metabolism of brassinosteroids (BRs) is an important prerequisite for a deeper understanding of the physiological effects, mode of action, and the practical application of these plant hormones. Whereas at the present stage, our knowledge on uptake and translocation is still rather limited, in the field of metabolism of BRs remarkable progress has been made within the last three years. This field has been part of recent reviews on BR research (Yokota et al 1991; Adam et al 1996; Fujioka and Sakurai 1997; Sasse 1997).

2. Uptake and Transport in Plants

Indirect evidence for uptake and transport of exogenously applied BRs (Fig. 1) was obtained by several groups describing effects of BRs on various plant species (see the chapter by J. Sasse, this volume). Thus, for example, treatment of the bases of mung bean hypocotyl cuttings with brassinolide (BL, **1**) caused elongation of the epicotyls (Gregory and Mandava 1982) whereas treatment of young tomato and radish plants with **1** via the roots promoted the elongation of petioles and hypocotyls (Takatsuto et al 1983). The normalization of Arabidopsis and other mutants by BRs (see the chapter by S. D. Clouse and K. A. Feldmann, this volume) also requires preceding uptake and translocation to the receptor site. However, studies immedi-

Key words: brassinosteroids, plant hormones, uptake, transport, metabolism, epimerisation, hydroxylation, conjugation, glycosidation, side-chain degradation, cell suspension cultures, explants, fatty acid esters, pregnanes.

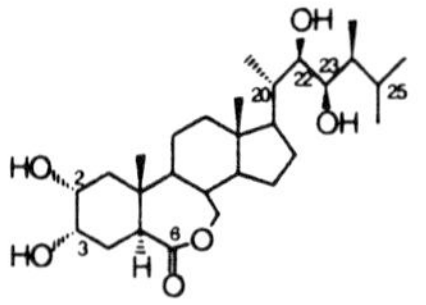

Brassinolide (1)

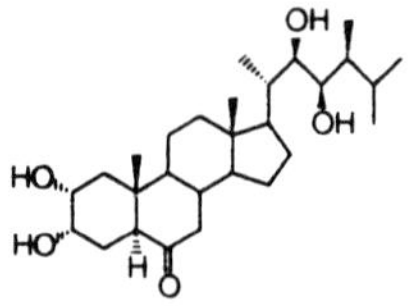

Castasterone (2)

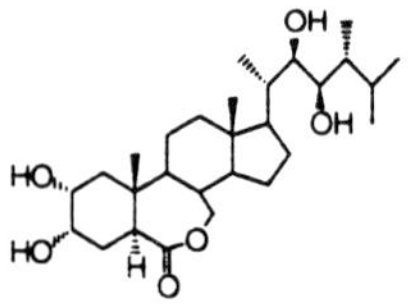

24-Epibrassinolide (3)

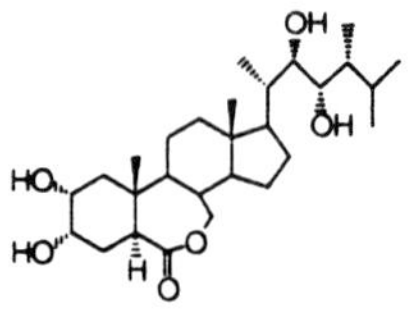

22,23,24-Triepibrassinolide (4)

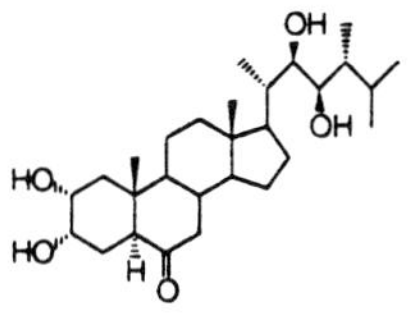

24-Epicastasterone (5)

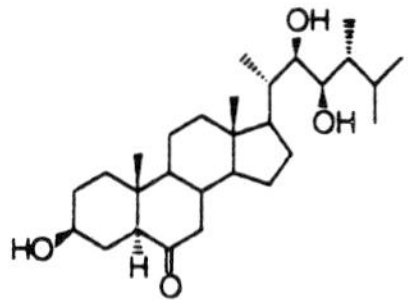

24-Epiteasterone (6)

Fig. 1. Structures of BRs used in metabolic studies.

ately directed to investigate absorption and translocation processes are rather rare and can only be efficiently carried out using labeled BRs.

Experiments on the uptake of 24-epiBL (**3**) in growing maize root segments using multiple-selected ion monitoring with [5,7,7-^{2}H]-**3** as standard indicated an accumulation of BR independent of the energy supply, whereas studies with frozen-thawed roots showed a large adsorption to cell structures. In the case of fresh roots 30% of BR (**1**) was bound irreversibly while in frozen-thawed roots the adsorption was freely reversible (Allevi et al 1988).

Based on their earlier findings that in tomato application of BRs via the roots resulted in stimulation of ethylene biosynthesis and epinasty, Schlagnhaufer and Arteca (1991) investigated the uptake of the synthetic analogue 22,23,24-triepiBL (**4**) in this plant. Thus, upon treatment of hydroponically grown tomato plants with tritiated **4** for 24 h followed by transfer to a BR-free culture medium, the BR-induced ethylene production decreased under BR-free conditions and polar metabolites (see section 3.5, this chapter) were formed. These results demonstrated that uptake, transport, and metabolism are closely related processes.

Earlier, Yokota et al (1987) had shown preliminary evidence for the formation of obviously polar metabolites upon root application of [^{3}H]BL (**1**) and [^{3}H] castasterone

(CS, **2**) to seedlings of rice, followed by more detailed studies (Yokota et al 1991). After feeding **1** and **2** as well as [^{3}H]epiBL (**3**) to the roots of rice seedlings, the labeled BRs were incorporated in roots and shoots. In all cases the shoots contained about one-tenth of the radioactivity found in the roots, clearly indicating uptake via roots followed by transport to the shoots. Upon application to the leaf surface of rice, after 24 h the incorporated radioactivity was found in the treated leaves. Only after 72 h of application did some radioactivity appear also in the roots, which showed a slight but again basipetal transport of the BRs or their metabolites after leaf application (Yokota et al 1992).

Upon feeding mung bean explants with [^{3}H]CS (**2**) and [^{3}H]BL (**1**), both compounds were incorporated rapidly into hypocotyls with faster incorporation of **2** than that of **1** throughout the incubation. On the other hand, movement of **1** into epicotyls was faster than that of **2**, reaching about 8% of the total applied radioactivity after 72 h (Suzuki et al 1993).

The uptake and transport of exogenously applied [^{14}C]epiBR (**3**) in intact seedlings of cucumber and wheat has been studied using autoradiography. Upon application to the roots, ready uptake and a quick distribution (probably via the xylem) was observed in both plant species. When labeled **3** was applied to the adaxial surface of young cucumber leaves a ready uptake but very slow transport was observed. Whereas the transport throughout the leaves occurred within 3 days in these experiments, the transport to the upper leaves from the treated leaf took place within 7 days after treatment. Also [^{3}H]epiBL (**3**) was taken up easily from the young leaves of wheat but was transported only in the apical direction within 7 days. Altogether, these results indicate an acropetal transport for the exogenously applied BR (Nishikawa et al 1994).

3. Metabolic Studies with Explants

3.1 Prerequisites

To study the metabolism of low-concentration natural products such as plant hormones in the plant cell, several prerequisites are required. Usually substrates labeled with radioactive or heavy isotopes must be used to detect small amounts of metabolites using very sensitive radiodetection or gas chromatography-mass spectroscopy (GC-MS). Most of the studies on the metabolism of BRs hitherto described used ^{3}H or ^{14}C, while in biosynthetic studies deuterium labeling very often is also applied. The analytical methods employed to study metabolism are MS, GC-MS, and nuclear magnetic resonance (NMR). The latter method is available only for metabolites that can be obtained in amounts of at least several micrograms.

The choice of the plant material for metabolic studies is another important prerequisite. Intact plants, plant parts, and cell cultures have been employed to study the metabolism of BRs. When intact plants or explants are used, the amount of formed metabolites in general is relatively low due to uptake barriers and other limiting factors. The advantage of studies with intact plants compared to explants or cell cultures is, however, that they reflect natural metabolic reactions occurring in the entire plant system.

3.2 Metabolism in Bean Plants

The structures of the compounds hitherto applied in metabolic studies are shown in Fig. 1. Preliminary investigations on BR metabolism in mung bean explants were reported by Yokota et al (1991) and published later in more detail (Yokota et al 1992). In the mung bean tissues, [^{3}H]CS (**2**) was readily converted to unknown water-soluble metabolites. Enzymatic hydrolysis of this hydrophilic fraction by an enzyme preparation containing high activity afforded small amounts of a CS-like compound, indicating the glycosidic character of these components. However, the major part of the hydrophilic fraction could not be enzymatically hydrolyzed and therefore was considered to be nonglycosidic. Although CS (**2**) is generally accepted as a biogenetic precursor of BL (**1**) and was shown to be converted to **1** in *Catharanthus roseus*, this conversion could not be observed in mung bean tissues.

The behavior of [^{3}H]BL (**1**) was shown to be different from that of [^{3}H]CS (**2**). Again, conversion to both glycosidic and nonglycosidic compounds was observed. However, a glycosidic metabolite was the major component and the nonglycosidic metabolites were shown to occur only in minor amounts. The structure of the BR glucoside was elucidated by its FAB-MS and by GC-MS of the methanolysis products after derivatization, trimethylsilylation in the case of the resulting anomeric methyl glycosides, and bismethanoboronation in the case of the aglycone. The position of the glucose attachment to the BL skeleton was established by ^{1}H-NMR, ^{1}H^{1}H-COSY, and selected relayed COSY of the metabolite and its peracetyl derivative as 23-O-β-D-glucopyranosyl-BL (**7**) (Yokota et al 1991) (Fig. 3). 23-O-Glucosides of BRs, namely 23-O-β-D-glucopyranosyl-25-methyldolichosterone and 23-O-β-D-glucopyranosyl-2-epi-25-methyldolichosterone, have been found as naturally occurring glucosides in seeds of *Phaseolus vulgaris* (Kim 1991). This type of 23-O-glucosides might be characteristic for the Leguminosae because hitherto there is no indication for the occurrence of such conjugates in other species.

3.3 Metabolism in Plants of Rice, Tobacco, and Periwinkle

The *Oryza sativa* cultivar "Koshihikari," which has been used for the rice lamina inclination test (RLIT) (Maeda 1965), was employed in metabolic studies (Yokota et al 1992). Several ^{3}H-labeled BRs were applied to roots and leaves of light-grown rice seedlings. Furthermore, the metabolism of [^{3}H]CS (**2**) was investigated in etiolated leaves during the RLIT. Water-soluble metabolites obtained after feeding [^{3}H]BL (**1**) and [^{3}H]CS (**2**), respectively, to roots of green rice seedlings behaved like

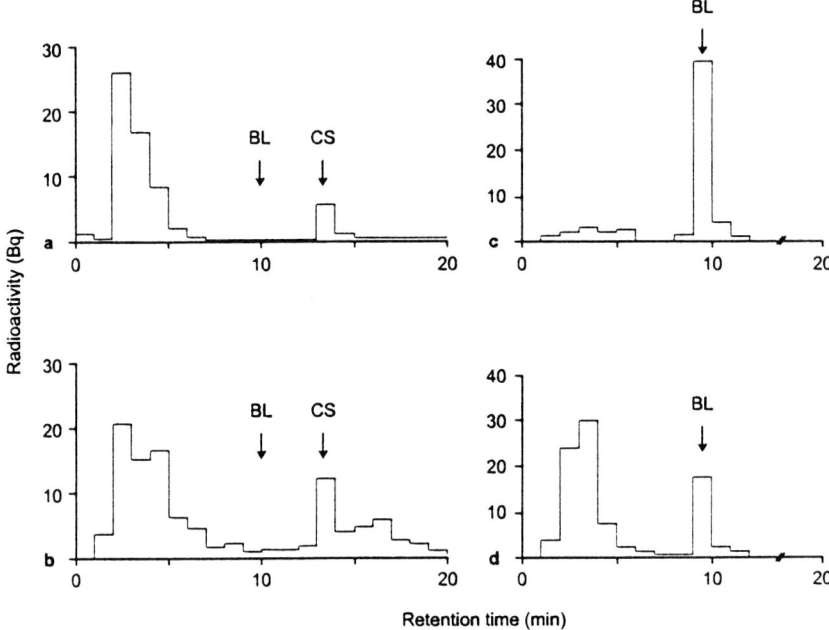

Fig. 2. Reversed-phase HPLC of chloroform-soluble fractions of shoots (**a**) and roots (**b**) obtained from rice seedlings fed with [³H]CS (**2**) via roots for 6 h and grown for 7 days thereafter, and of roots obtained from seedlings fed with [³H]BL (**1**) via roots for 6 h (**c**) and grown for 7 days thereafter (**d**). (From Yokota et al. 1992 with permission).

sulfate esters of animal steroid hormones. Upon solvolysis, the metabolites of **1** and **2** released chloroform-soluble substances with polarities similar to those of free BRs as examined by thin-layer chromatography (TLC) and reversed phase high performance liquid chromatography (HPLC). A similar result was obtained by alkaline hydrolysis of the metabolites of [³H]CS (**2**) (Fig. 2).

Most of the radioactivity found in the roots after application to the leaf surface of light-grown rice seedlings was also water-soluble. During the incubation time of 72 h in the RLIT, the portion of water-soluble metabolites increased continuously as indicated by radiodetection of HPLC fractions. Characterization of these hydrophilic metabolites by means of basic and acid hydrolysis or treatment with alkaline phosphatase followed by methylation with diazomethane did not show changed behavior as examined by reversed phase HPLC. Thus, it was concluded that etiolated rice leaf explants metabolize [³H]CS (**2**) differently from intact rice roots (Yokoto et al 1992).

Different rates or types of metabolization of BRs in the rice lamina explant would imply different degrees of deactivation. Consequently, the metabolism as well as transport and uptake would have to be taken into consideration on evaluation of the results of the RLIT. For this purpose metabolism of a series of relevant BRs should be studied.

In many biosynthetic studies CS (**2**) was used as a substrate to establish its transformation to BL (**1**), which is the most bioactive BR. However, this transformation hitherto could be demontrated only in *C. roseus* but not in other species. After application of a mixture of [^{3}H]-**2** and [^{2}H$_6$]-**2** to seedlings of *O. sativa* and *Nicotiana tabacum*, for example, epimerization of the 3α-hydroxyl group of CS was observed to afford 3-epiCS (**9**) (Suzuki et al 1995), which was already known from immature seeds of *P. vulgaris* (Kim 1991). Because this inversion of configuration is accompanied by loss of bioactivity, **9** must be considered a product of deactivation rather than a biosynthetic product. In analogous experiments with seedlings of *C. roseus*, **9** was found along with BL (**1**).

3.4 Metabolism in Cucumber and Wheat

The metabolism of [^{14}C]24-epiBL (**3**) was studied in intact seedlings of a monocot species, wheat (*Triticum aestivum*), and a dicot species, cucumber (*Cucumis sativus*) (Nishikawa et al 1995b). The substrate was applied to seedlings of both species via the roots in a pulse labeling procedure and further incubated for up to 7 days.

Radio-TLC and reversed phase HPLC analysis of extracts obtained from second leaves of wheat after 7 days' incubation indicated conversion of **3** to several components having higher polarity than the parent compound. The metabolic pattern changed over the incubation time, but the metabolites were not further characterized (Nishikawa et al 1995a).

In cucumber seedlings, exogenously applied [^{14}C]24-epiBL (**3**) was differentially metabolized in leaves and epicotyls. The major radioactivity was found to be soluble in chloroform. More polar and less polar metabolites in comparison with **3** were detected in leaves, and only less polar metabolites were found in epicotyls. Most of them remained unidentified. Based on hydrolyses and MS data of the aglycone, it was suggested that the polar metabolite is a glucoside and the less polar metabolites are acyl conjugates of a BR epimeric to **3** only at one position in ring A. However, the assignment of the stereochemistry of the ring A of this aglycone was based only on derivatization experiments using methylboronate and TMS. The metabolite did not form a bismethanoboronate at ring A and thus was referred to as a 2,24-diepiBL (**8**). The authors also discussed 3,24-diepiBL (**12**) as a candidate for this metabolite (Nishikawa et al 1995a). In our opinion, this suggestion should be seriously taken into consideration because it would nicely confirm our later finding that only BRs having a 3β, but not a 3α, hydroxyl group are able to be conjugated at the 3-position (Kolbe et al 1998b).

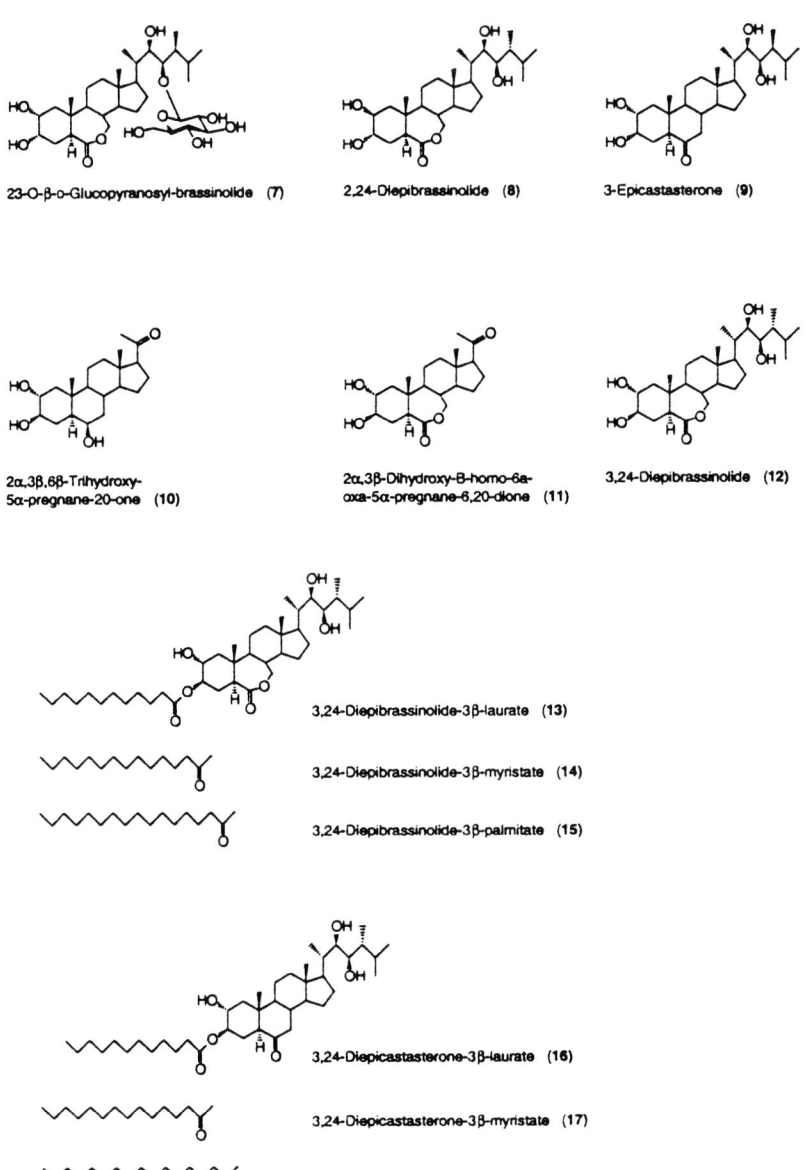

Fig. 3. Structures of BR metabolites found in plants and plant cell cultures.

3.5 Metabolism in Tomato Plants

In studies of the influence of BRs on the accumulation of ethylene and its biosynthetic precursor 1-aminocyclopropane-1-carboxylic acid (ACC), a synthetic analogue of BRs, 22,23,24-triepiBL (**4**) was labeled with ^{3}H and applied to hydroponically grown tomato plants. After increasing incubation times between 4 and 24 h, up to

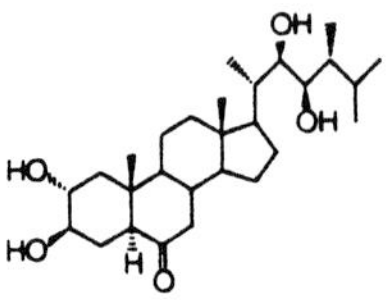

3,24-Diepicastasterone (**19**)

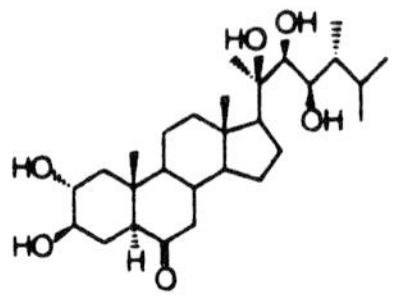

20R-Hydroxy-3,24-diepicastasterone (**20**)

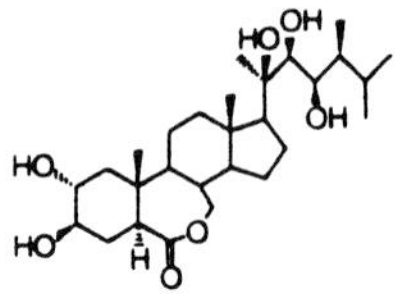

20R-Hydroxy-3,24-diepibrassinolide (**21**)

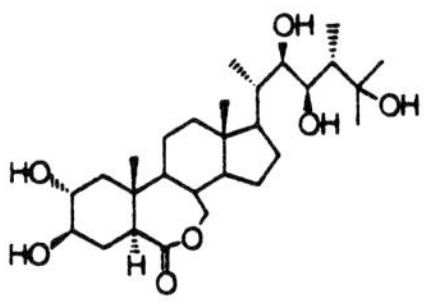

25-Hydroxy-3,24-diepi-brassinolide (**22**)

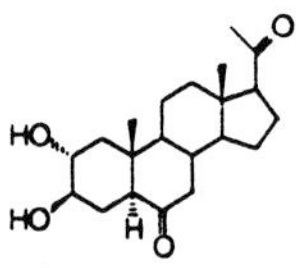

2α,3β-Dihydroxy-5α-pregnane-6,20-dione (**23**)

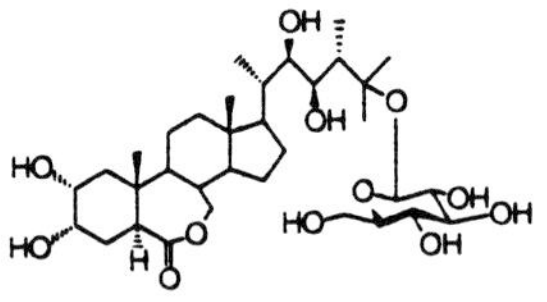

25-β-D-Glucopyranosyloxy-24-epibrassinolide (**24**)

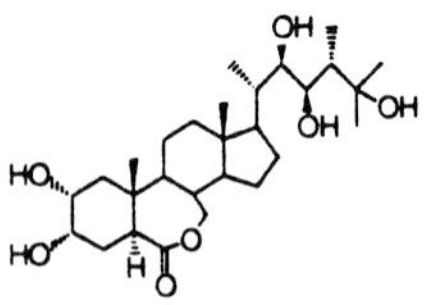

25-Hydroxy-24-epibrassinolide (**25**)

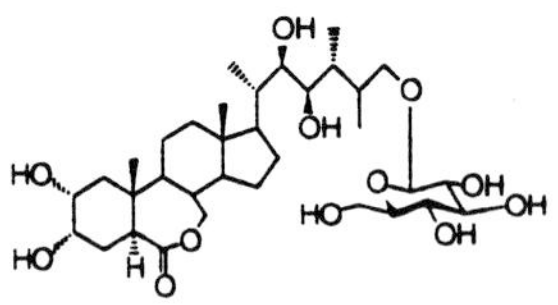

26-β-D-Glucopyranosyloxy-24-epibrassinolide (**26**)

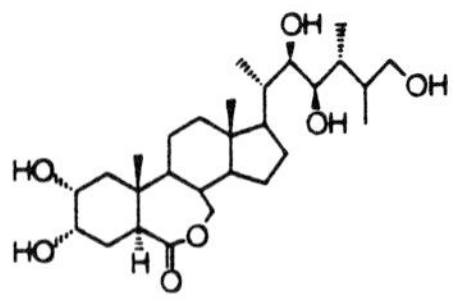

26-Hydroxy-24-epibrassinolide (**27**)

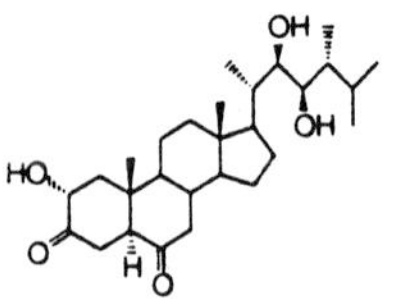

3-Dehydro-24-epicastasterone (**28**)

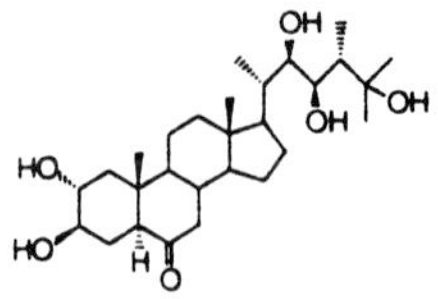

25-Hydroxy-3,24-diepicastasterone (**29**)

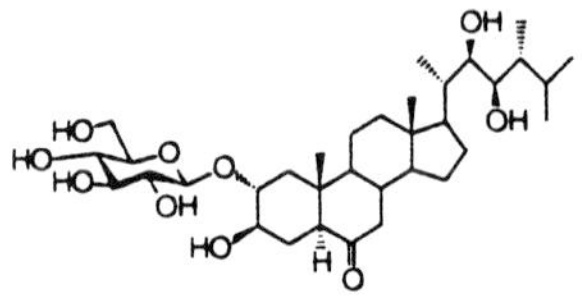

2-O-β-D-Glucopyranosyl-3,24-diepicastasterone (**30**)

Fig. 3. *Continued.*

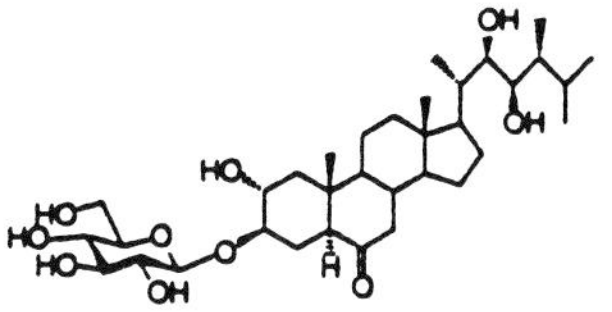

3-O-β-D-Glucopyranosyl-3,24-diepicastasterone (31)

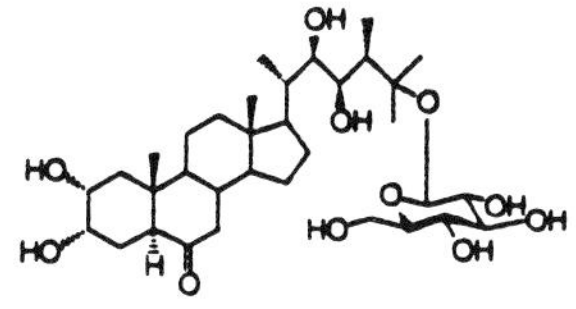

25-β-D-Glucopyranosyloxy-24-epicastasterone (32)

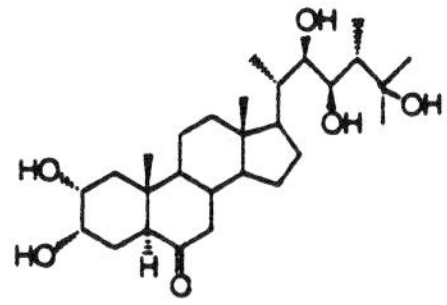

25-Hydroxy-24-epicastasterone (33)

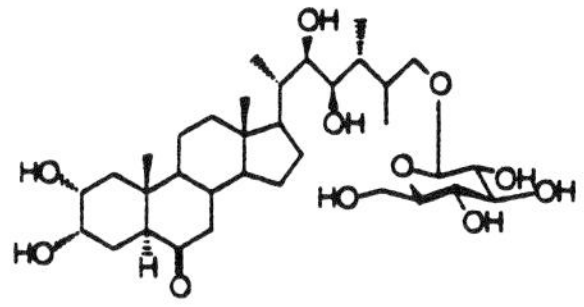

26-β-D-Glucopyranosyloxy-24-epicastasterone (34)

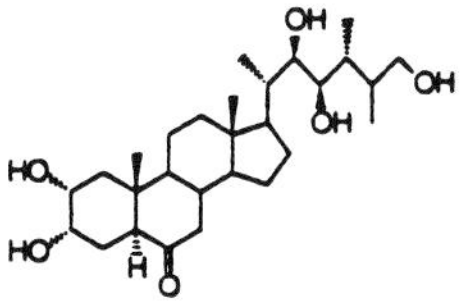

26-Hydroxy-24-epicastasterone (35)

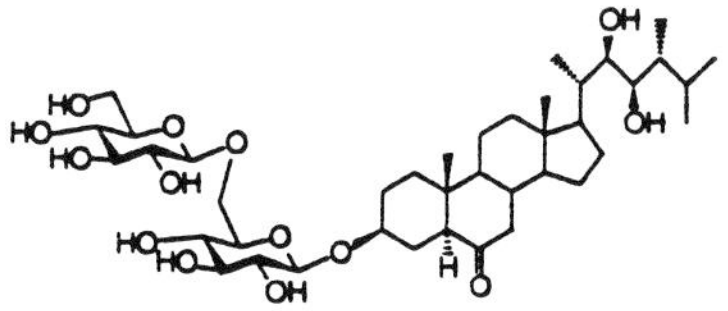

3-O-(β-D-Glucopyranosyl-(1→6))-β-D-glucopyranosyl-24-epiteasterone (36)

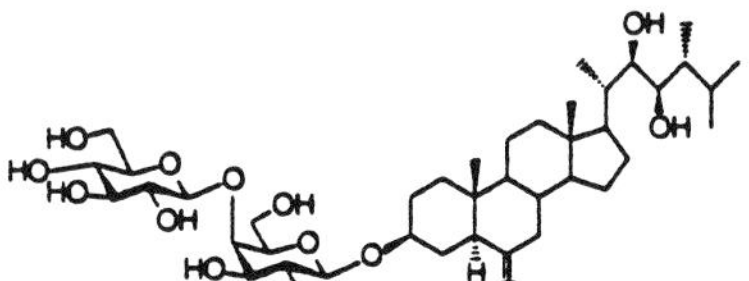

3-O-(β-D-Glucopyranosyl-(1→4))-β-D-galactopyranosyl-24-epiteasterone (37)

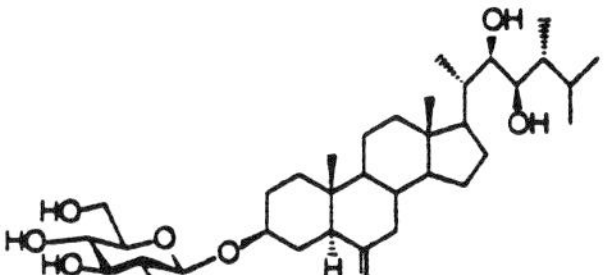

3-O-β-D-Glucopyranosyl-24-epiteasterone (38)

Fig. 3. *Continued.*

three metabolites were found that were more polar than the parent compound. The structures of these metabolic products remained unknown (Schlagnhaufer and Arteca 1991).

4. Metabolism in Cell Cultures

4.1 Cell Growth and Uptake

Cell suspension cultures have been extensively employed to study biogenetic and metabolic pathways of natural products including plant hormones. In comparison with the intact plants, cell cultures provide a number of advantages. For example, they are cultivated under aseptic conditions and allow therefore investigation of metabolism without microbial influence. Cell cultures are independent from seasonal and daily growth cycles. Another important advantage of suspension cultured cells is the opportunity to expose them immediately to exogenous substrates, avoiding complex translocation processes.

Before using cell cultures in metabolic studies on compounds with an impact on plant growth and development, the effect on the cell growth and differentiation should be studied. BRs are known to influence both cell division and cell elongation of various plant species (Fujioka and Sakurai 1997; see the chapter by J. Sasse, this volume). In carrot cell cultures, for example, BRs induce cell enlargement but not cell division (Sala and Sala 1985), whereas they act as effective inhibitors of the

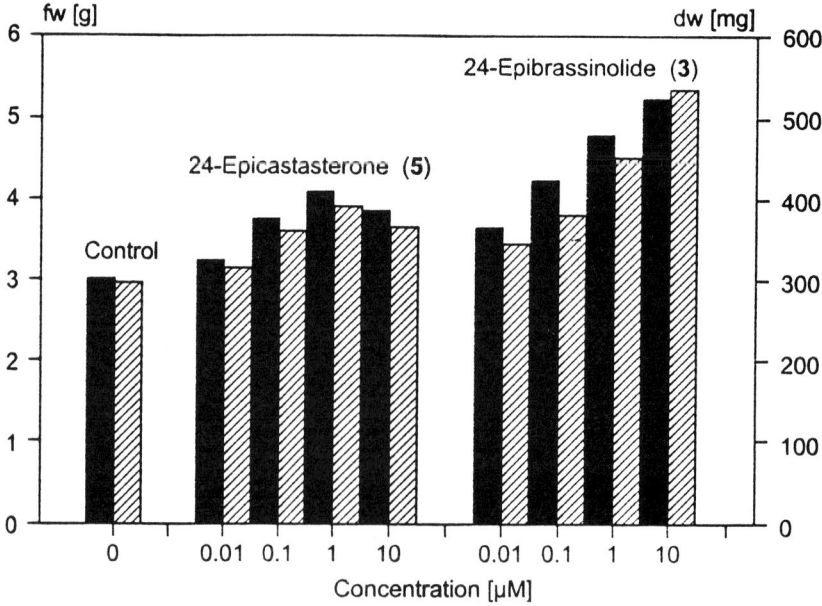

Fig. 4. Effect of 24-epiCS (**5**) and 24-epiBL (**3**) on the growth of cell suspension cultures of serradella (*Ornithopus sativus*). Full columns: fresh weight (fw); hatched columns: dry weight (dw).

growth of tobacco crown gall cells (Roth et al 1989).

The influence of BRs on the growth of suspended cells of tomato (*Lycopersicon esculentum*) and serradella (*Ornithopus sativus*) has been investigated in some detail (B. Schneider, T. Hai, A. Kolbe, and G. Adam, unpublished results). In these experiments the most active compound in the 24*R* series, 24-epiBL (**3**), was applied in different concentrations to study the rate of cell growth under the experimental conditions used. No significant changes of the fresh weight and dry weight in the range of concentrations between 10^{-8} to 10^{-5} M were observed in *L. esculentum* (data not shown). In contrast, in the case of *O. sativus,* enhancement of both the fresh weights and dry weights of suspended cells was observed after application of the same concentrations of 24-epiBL (**3**) (Fig. 4). Also, 24-epiCS (**5**) enhanced both fresh and dry weights of *O. sativus* cells in a similar way except that the highest applied concentration of 10^{-5} M did not further promote cell growth.

After the influence of BRs on the growth of suspended cells had been studied, in further experiments the absorption of exogenously applied radiolabeled 24-epiBL (**3**) and 24-epiCS (**5**), respectively, by the cells was measured (Adam et al 1996). The distribution of radioactivity between the suspended cells of *O. sativus* and the culture medium did not significantly change during the entire growth period of 7 days (Fig. 5a). This unusual phenomenon was observed after application of both

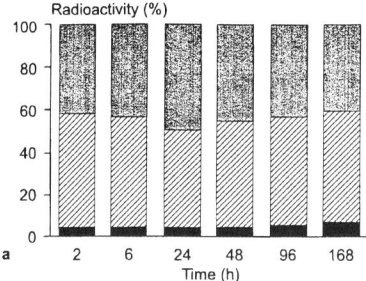

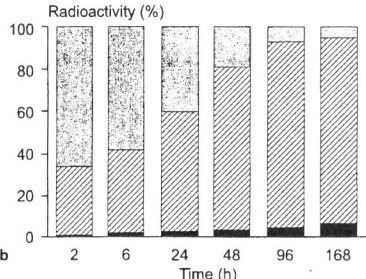

Fig. 5. Distribution of radioactivity in cell suspension cultures of serradella (*Ornithopus sativus*) (**a**) and tomato (*Lycopersicon esculentum*) (**b**) after feeding [^{3}H]24-epiCS (**5**) and [^{3}H]24-epiBL (**3**), respectively. Half-tone columns: medium; hatched columns: extract; full columns: residue.

compounds **3** and **5**. About 40% of the radioactivity after feeding **3** and about 25% after feeding **5** were found in the medium and the remainder was associated with the cells. It is likely that there is rapid uptake of the substrates from the medium into the cells and, at the same time, active reverse transport of the metabolites through the cell wall into the surrounding medium because the various metabolites were present in both medium and cells. In contrast, tomato cell cultures exhibited a totally different absorption profile (Fig. 5b). Very rapid uptake occurred during the initial incubation period (about 46% uptake after 2 h) and most of the substrate and/or metabolites were confined within the cells after longer incubation. The absorption curve was used to define the optimal time of cell harvest, which was 4 days for tomato cells.

4.2 Metabolism in Cell Cultures of Serradella

The first BR metabolites from *O. sativus* were isolated from the chloroform extract of the cell culture medium (Kolbe et al 1994). After application of both [^{3}H]24-epiCS (**5**) and [^{3}H]24-epiBL (**3**), the cells were cultured for 7 days before extraction. In each experiment a major peak was detected by radio-TLC and finally purified by reversed phase HPLC. The structures were established by EI-MS and NMR spectroscopy of non-derivatized metabolites and acetyl derivatives (Figs. 6, 7). In both metabolites of **3** and **5**, loss of the side chain was indicated by the presence of metabolites showing only two methyl signals in the high field region of the ^{1}H-NMR spectra. The broad signal δ 3.88 in the spectrum of 2α,3β,6β-trihydroxy-5α-pregnane-20-one (**10**), isolated upon application of **5**, was assigned to 6α-H by a 2D ^{1}H,^{1}H-COSY experiment. As a metabolite of **3**, 2α,3β-dihydroxy-B-homo-6a-oxa-5α-pregnane-6,20-dione (**11**) was identified as a product of side chain degradation. To get deeper insight into this catabolic side chain degradation, the incubation time was shortened in further experiments to accumulate putative metabolic intermediates. Thus, after only 24 h of application of 24-epiBL (**3**), 3,24-diepiBL (**12**) was the new major metabolite in the cell culture medium. This compound was also isolated and its structure was assigned by NMR and MS. While the relevant ^{1}H-NMR side chain signals of compound **12** were identical with those of **3**, the ring A signals were different. In particular 3-H exhibited coupling constants that indicated an axial position of this proton and hence epimerization at C-3 (Kolbe et al 1995). The presence of the 3β-OH group in **12** and in the pregnane-type **10** and **11** suggested that the epimerization at C-3 might be the primary step of BR metabolism in *O. sativus*. This assumption was further supported by the identification of 3,24-diepiCS (**19**) as a metabolite of exogenously applied **3** (Kolbe et al 1996). The subtance **19** is known as a natural product from seeds of *P. vulgaris* (Kim 1991). The general mechanism of steroid sidechain degradation is assumed to proceed via 20-hydroxylation as proposed for the metabolism of 20-hydroxyecdysone in insects (Galbraith et al 1969) and in crustaceous organisms (Lachaise and Lafont 1984). An alternative pathway for the formation of pregnanes from Δ^{22}-sterols via 22,23-epoxides has recently been established in a marine organism (Kerr et al 1995). Although a number of plants contain pregnanes of various structural types, representing also the key intermedi-

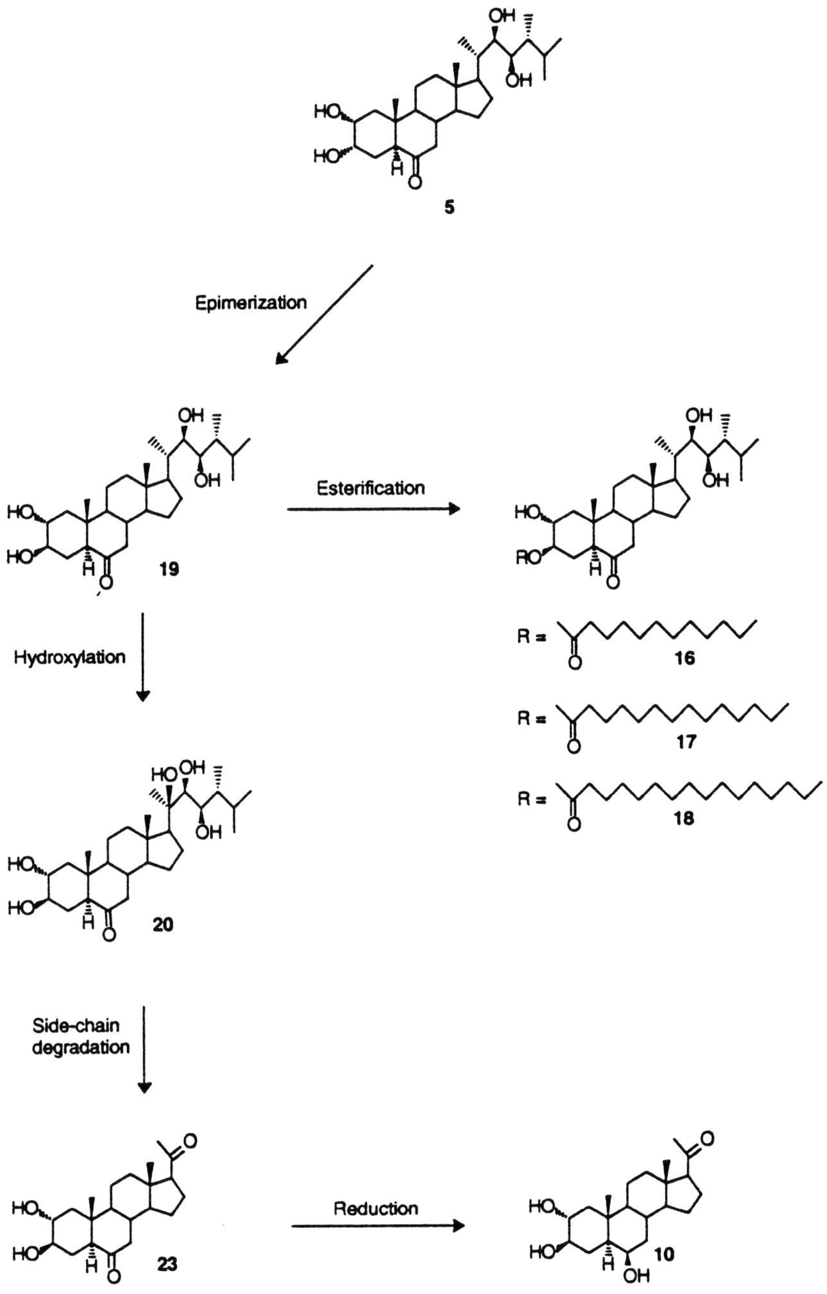

Fig. 6. Metabolism of 24-epiCS (**5**) in cell suspension cultures of serradella (*Ornithopus sativus*).

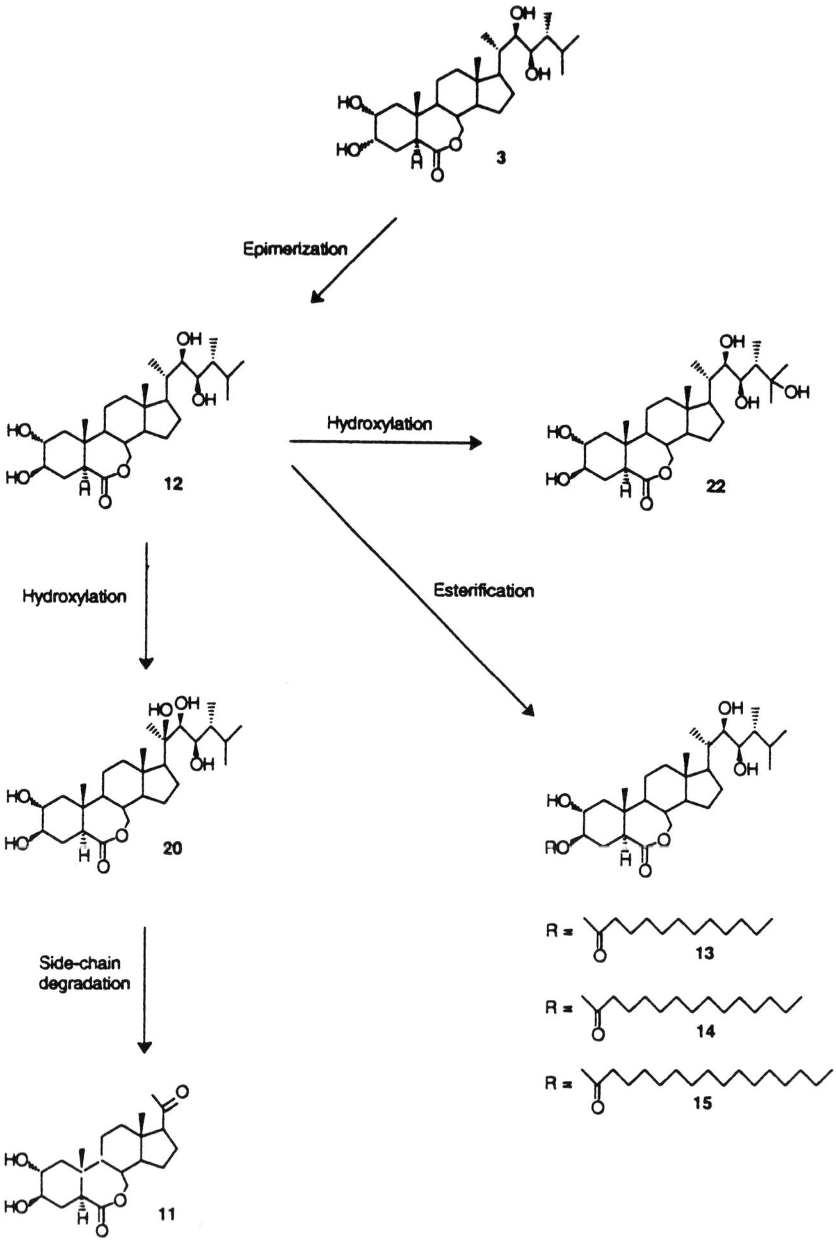

Fig. 7. Metabolism of 24-epiBL (**3**) in cell suspension cultures of serradella (*Ornithopus sativus*).

ates in the biosynthesis of cardenolides and bufadienolides (Deepak et al 1989), their biosynthesis in plants hitherto is not clear.

Cell cultures of *O. sativus* might be a suitable experimental system to overcome this gap and to study further pregnane biosynthesis. In such a context it was quite interesting that after application of 24-epiCS (**5**) and 24-epiBL (**3**) to cell cultures of *O. sativus*, in both cases intermediates bearing a 20R-hydroxy group could be isolated from the cell culture medium. Both metabolites occurred in only very low concentrations, which indicated the transient character of these compounds. Obviously the bond between C-20 and C-22 is destabilized by the vicinal hydroxyl group and, consequently, becomes accessible to enzymatic attack resulting in bond cleavage. This clearly indicates that the formation of pregnane-type compounds in *O. sativus* proceeds via 20-hydroxylation, a result only available by employing BRs labeled in the ring system. Labeling procedures for the preparation of [5,7,7-^{3}H]BRs have been decribed elsewhere (Kolbe et al 1992, 1998a). In contrast, in studies using side chain labeled BRs (Suzuki et al 1993, 1995), e.g., [24,28-^{3}H]BL (**1**) and [24,28-^{3}H]CS (**2**) (Yokota et al 1990), it is not possible to detect metabolites formed by loss of the side chain moiety. Detection of hydroxypentanoic acids as side chain fragments could account for that mechanism. For example, 4-hydroxy-4-methyl-pentanoic acid was formed by side chain cleavage of 20-hydroxyecdysone (Galbraith et al 1969; Lachaise and Lafont 1984). In *O. sativus* a hypothetical pentanoic acid moiety metabolically derived from ring-labeled [5,7,7-^{3}H]24-epiBL (**3**) and [5,7,7-^{3}H]24-epiCS (**5**), respectively, was not detectable due to lack of label within the side chain of the parent compounds (Kolbe et al 1994). Besides hydroxylation at C-20, introduction of an additional hydroxyl group occurred also in the terminal part of the side chain at C-25 of 3,24-diepiBL (**12**). Hydroxylation at C-25 is well known for a variety of other steroid compounds, e.g., ecdysones (Rees 1989), and was observed also for BRs fed to tomato cell cultures (Schneider et al 1994; Hai et al 1996).

In the experiments with *O. sativus*, a fraction of lipophilic metabolites was observed (Kolbe et al 1995). As indicated by chromatographic properties on TLC and reversed phase HPLC, these metabolites showed significantly less polarity than **3** and **5**. Three of these compounds were formed in nearly the same quantity. The ^{1}H-NMR spectra were very similar to those of 3,24-diepiCS (**19**) and 3,24-diepiBL (**12**), respectively. In comparison to the spectra of the latter compounds, the signals of 3α-H exhibited a downfield shift of about 1.2 ppm due to an ester bond at this position. Additionally, a broad singlet at δ 1.25 indicated the presence of fatty acid methylene protons, and a triplet at δ 0.88 owing to terminal methyl protons was also present in the spectra. These NMR data, suggesting fatty acid conjugates of **12** and **19**, respectively, were supported by the EI mass spectra. Lauric acid, myristic acid, and palmitic acid were found as conjugating moieties in **13 – 18** for both 3,24-diepimeric BRs. Acyl conjugates of BRs hitherto have been identified only from lily pollen, exhibiting also acylation of 3β-hydroxyl (Asakawa et al 1996). The function of these fatty acid conjugates is not yet known. As discussed for phytosterol acyl conjugates (Woichiechowski 1991) they might be involved in compartmentation processes.

All the metabolites of 3,24-diepiCS (**19**) and 3,24-diepiBL (**12**), so far isolated and identified after feeding ^{3}H-labeled precursors to *O. sativus*, casually fit well into the metabolic pathways outlined in Figs. 6 and 7. Several further hydrophilic metabolites that occur in minor concentrations inside the cells seem to be glycosides. These sequences represent the first comprehensive metabolic pathways of BRs in a particular plant species.

4.3 Metabolism in Cell Cultures of Tomato

Tomato cell suspension cultures were used to study the metabolism of 24-epiteasterone (24-epiTE, **6**), 24-epi-CS (**5**), and 24-epiBL (**3**). Compound **6**, located in the biosynthetic sequence of BRs before inversion of 3β- to the 3α-configuration, was readily

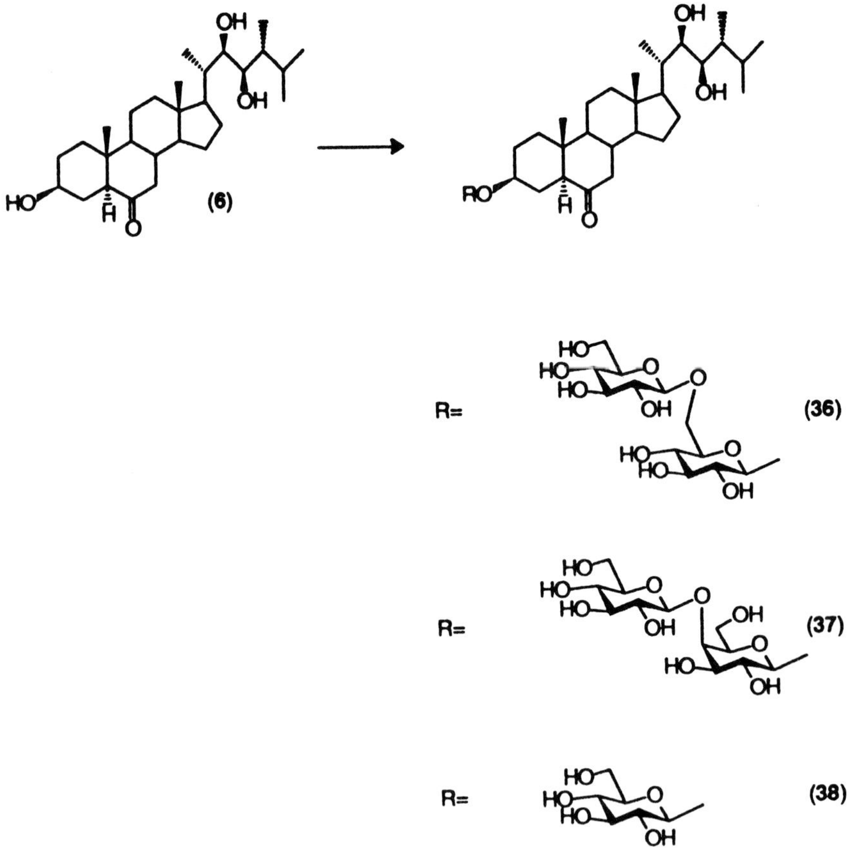

Fig. 8. Metabolism of 24-epiTE (**6**) in cell suspension cultures of tomato (*Lycopersicon esculentum*).

transformed into hydrophilic glycosides. Detailed structural analysis revealed that conjugation occurred exclusively at the 3β-position as indicated by the ^{1}H chemical shift and coupling constants of the ring A protons. The conjugating sugar moieties were identified as β-D-glucopyranosyl-(1→6)-β-D-glucopyranoside (gentiobiose), β-D-glucopyranosyl-(1→4)-β-D-galactopyranoside (lycobiose) (Kolbe et al 1997), and β-D-glucose (Kolbe et al 1998b) (Fig. 8). Since conjugation of plant hormones in general is a reversible process, one might speculate that these 24-epiTE glycosides (**36**, **37**, and **38**) are involved in the regulation of the level of the intermediate 24-epiTE (**6**) in the biosynthesis of 24R-BRs. Otherwise the metabolism through simple deactivation of the exogenously applied precursor cannot be excluded. The observed 3β-conjugation of **6** proceeds very rapidly. After 5 h of feeding, two-thirds (67%) of the applied precursor were found to be conjugated with disaccharides and glucose. At that time only 33% of the parent compound **6** was still present in nonconjugated form (Kolbe et al 1997).

To establish the metabolic transformation of 24-epiCS (**5**) in tomato cells and to demonstrate the assumed Bayer-Villiger oxidation to give 24-epiBL (**3**), radiolabeled **5** was administered to these cell cultures (Hai et al 1996). Chromatography on Amberlite XAD-2 and DEAE A-25 afforded a fraction containing several glycosidic metabolites. The ^{1}H-NMR and MS analysis of the obtained compounds **32** and **34** and their corresponding peracetylated derivatives indicated hydroxylation and glycosidation at the terminal part of the side chain. The position of the glucosyl units at the newly functionalized carbon atoms C-25 and C-26, respectively, was also established by NMR spectroscopic methods. Interestingly, 25-hydroxy- and 26-hydroxy-24-epiCS (**33** and **35**) were not detected in free form. Consequently, the hydroxylation must be the rate-limiting reaction in that two-step metabolic process. There were also some indications of a disaccharide conjugate of a pentahydroxylated BR in tomato cell cultures. However, the structure could not be identified due to the low amount of the sample.

Two further glucosides, 2-O-β-D-glucopyranosyl-3,24-diepiCS (**30**) and 3-O-β-D-glucopyranosyl-3,24-diepiCS (**31**), could not be separated by HPLC, and therefore the structures of these compounds were elucidated by NMR analysis of the mixture. They represent the first BR glucosides bearing the sugar moiety on ring A hydroxyl groups (Hai et al 1996). As in the disaccharide conjugates **36** and **37**, which are derived from 24-epiTE (**6**), the conjugating sugar unit of compound **31** is bound to a hydroxyl group having 3β-configuration. Therefore, inversion of the 3-configuration of the parent compound 24-epiCS (**5**) must have taken place before conjugation. To identify further metabolic intermediates, a part of the cell extract was hydrolyzed by means of a glycosidase preparation. From that aglycon fraction 3,24-diepiCS (**19**), also found as a metabolite in cell cultures of *O. sativus*, was isolated. This compound represents a branching point in the metabolism of 24-epiCS (**5**) and is the intermediate in the formation of both glucosides **30** and **31** as well as 25-hydroxy-3,24-diepiCS (**29**). The occurrence of 3-dehydro-24-epiCS (**28**), a new 3,6-diketo BR, in the cell culture extract of *L. esculentum* indicated that the inversion of configuration at C-3 proceeds via an oxidation/reduction mechanism (Hai et al 1996).

The opposite reaction was found to be a step in the biosynthetic pathway of BRs between teasterone (TE) and typhasterol (TY) (Suzuki et al 1994; Griffith et al 1995) and between 24-epiTE and 24-epiTY (Kolbe et al 1998b). Figure 9 shows the metabolic pathways of 24-epiCS (5) as deduced from detection of the different metabo-

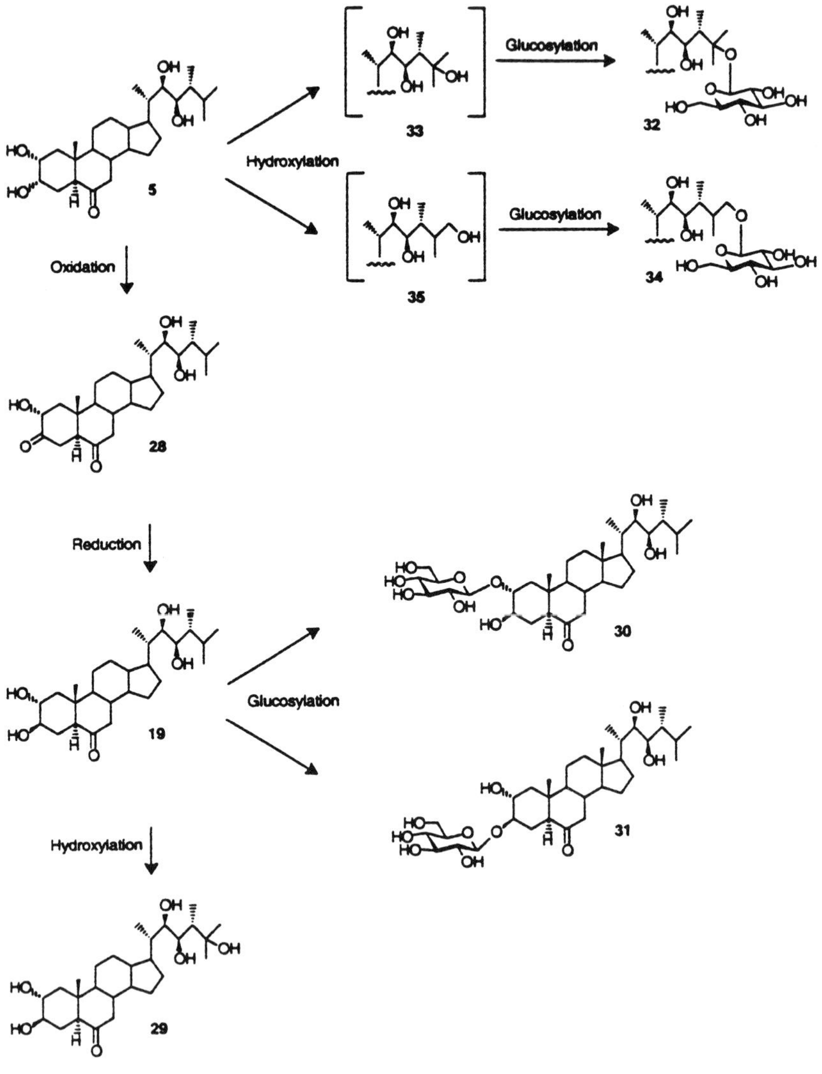

Fig. 9. Metabolism of 24-epiCS (5) in cell suspension cultures of tomato (*Lycopersicon esculentum*).

lites and conjugates. All these metabolites were isolated exclusively from cultured cells. The cell culture medium did not contain significant amounts of any BR metabolite. 24-EpiBL (**3**), hitherto considered to be the final biosynthetic product in the 24R series and therefore expected to be formed from exogenously applied 24-epiCS (**5**), was not found in cell cultures of *L. esculentum*.

The formation of both 25-β-D-glucopyranosyloxy-24-epiBL (**24**) and 26-β-D-glucopyranosyloxy-24-epiBL (**26**) represents a metabolic sequence that involves hydroxylation at C-25 and C-26, respectively, to give **25** and **27**, each of which is subsequently glucosylated at the newly introduced hydroxyl group to afford both regioisomeric glucosides in a ratio of about 1:1 (Fig. 10). Other minor glucosidic metabolites in cell cultures of *L. esculentum* were not further characterized.

The first experiments to characterize both types of enzymes involved in these

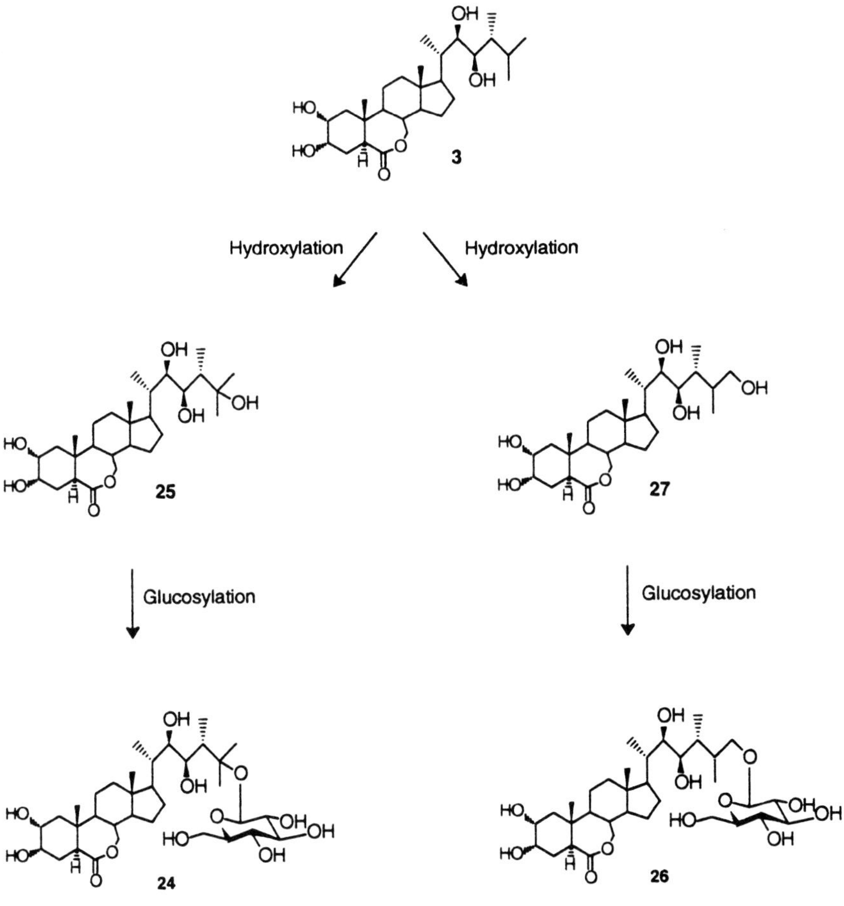

Fig.10. Metabolism of 24-epiBL (**3**) in cell suspension cultures of tomato (*Lycopersicon esculentum*).

reactions, hydroxylases and glucosyltransferases, indicated involvement in biosynthesis rather than nonspecific detoxification. Thus, treatment of tomato cell cultures with the specific cytochrome P450 inhibitors clotrimazole and ketoconazole resulted in a strong decrease of only C-25 hydroxylation; hydroxylation at C-26 was not affected (Winter et al 1997). These studies demonstrated clearly that hydroxylation at C-25 and C-26 is catalyzed by two different enzymes, whereas only the C-25-hydroxylase is a typical cytochrome P450 protein but the C-26 hydroxylase is not. In further experiments, the common cytochrome P450 inducers ethanol, $MnCl_2$, phenobarbital, pregnenolone 16α-carbonitrile or clofibrate with diverse inducing spectra, were applied along with substrate (**3**) but did not induce hydroxylation activity at C-25 or at C-26. In addition, the substrate analogues (22S,23S)-28-homoBL, 24-epiCS (**5**), ecdysone, and 20-hydroxy-ecdysone were not accepted. Only application of 24-epiBL (**3**) and BL (**1**) resulted in increased activity of both the C-25 and C-26 hydroxylases indicating the high specificity of induction of that metabolic process. Further examination of this inducing effect showed that under the influence of cycloheximide (a protein biosynthesis inhibitor) no hydroxylase activity could be detected. Thus, the increase of both hydroxylase activities is obviously based on gene expression by specific action of the substrates 24-epiBL (**3**) and BL (**1**).

The 25-O-β-D-glucosyltransferase is supposed to be highly specific because biosynthetically prepared compound **25** after reapplication to cell cultures of *L. esculentum* was glucosylated in a highly regiospecific manner exlusively at the 25-hydroxyl group while the other positions remain non-glucosylated. In contrast, compound **27** afforded at least four glucosidic products under identical conditions (T. Hai, B. Schneider, and G. Adam, unpublished results).

Despite BL (**1**) and 24-epiBL (**3**) being the most active compounds of the 24S and 24R series, respectively, there is no evidence whether they are the active plant hormones per se or further biosynthetic steps are required. For evaluation of the bioactivity of novel metabolic products the RLIT (Maeda 1965) and several additional bioassay systems (Mandava 1988) are in use. The metabolites hitherto isolated from *O. sativus* and *L. esculentum* did not show significant bioactivity in the RLIT using the Japanese rice variety Koshihikari. An exception was observed when the Vietnamese rice variety Nep IR-415 was applied to test 25-hydroxy-24-epiBL (**25**) (Hai et al 1995). This pentahydroxylated metabolite was prepared by enzymatic hydrolysis of 25-β-D-glucopyranosyloxy-24-epiBL (**24**), which has been isolated from cell cultures of *L. esculentum* after feeding **3** (Schneider et al 1994). In this case the RLIT indicated significantly higher bioactivity of **25** in comparison with **3** (Hai et al 1995). This remarkable bioactivity could be confirmed by strong induction of epinasty in *Vicia faba* plants. However, a synthetic sample of **25** (Voigt et al 1996) caused only moderate lamina bending using the Koshihikari rice cultivar. Furthermore, treatment with **25** did not rescue the BR deficiency of Arabidopsis mutants (T. Altmann, personal communication). Thus, there is no final answer on the question of whether the hydroxylation of BRs at C-25 is a general activation step in the biosynthesis or the first step of oxidative degradation. Otherwise, as was found for other classes of plant hormones it could be speculated that several BRs show spe-

Table 1. Metabolic studies on brassinosteroids

Substrate	Plant species	Plant part	Metabolites	Reference
[³H]BL (**1**) [³H]24-epiBL (**3**) [³H]CS (**2**)	Oryza sativa	Seedlings Explants	Unknown non-glucosidic conjugates	Yokota et al 1992
[³H]22,23,24-triepiBL (**4**)	Lycopersicon esculentum	Plants	Unknown polar metabolites	Schlagnhaufer and Arteca 1991
[¹⁴C]24-epiBL (**3**)	Cucumis sativus, Triticum aestivum	Seedlings, Seedlings	2,24-DiepiBL (**8**), unknown metabolites unknown metabolites	Nishikawa et al 1995a,b
[³H]BL (**1**) [³H]CS (**5**)	Vigna radiata	Explants	23-O-β-D-Glucopyranosyl-BL (**7**) Non-glycosidic (major), glucosidic (minor)	Yokota et al 1991 Suzuki et al 1993
Mixture of [³H]CS and [²H₆]CS (**5**)	Catharanthus roseus, Nicotiana tabacum, Oryza sativa	Seedlings	3-epiCS (**9**)	Suzuki et al 1995
[³H]24-epiBL (**3**) [³H]24-epiCS (**5**)	Ornithopus sativus	Cell suspension	2α,3β-Dihydroxy-B-homo-6a-oxa-5α-pregnane-6,20-one (**11**) 2α,3β,6β-Trihydroxy-5α-pregnane-20-one (**10**)	Kolbe et al 1994
[³H]24-epiBL (**3**) [³H]24-epiCS (**5**)	Ornithopus sativus	Cell suspension	3,24-DiepiBL (**12**) 3,24-DiepiBL-3β-laurate (**13**), -3β-myristate (**14**) and -3β-palmitate (**15**) 3,24-DiepiCS-3β-laurate (**16**), -3β-myristate (**17**) and -3β-palmitate (**18**)	Kolbe et al 1995
[³H]24-epiBL (**3**) [³H]24-epiCS (**5**)	Ornithopus sativus	Cell suspension	3,24-DiepiCS (**19**) 20R-Hydroxy-3,24-diepiCS (**20**) 20R-Hydroxy-3,24-diepiBL (**21**) 25-Hydroxy-3,24-diepiBL (**22**) 2α,3β-Dihydroxy-5α-pregnane-6,20-dione (**23**)	Kolbe et al 1996
[³H]24-epiBL (**3**)	Lycopersicon esculentum	Cell suspension	25-β-D-Glucopyranosyloxy-24-epiBL (**24**) 25-Hydroxy-24-epiBL (**25**) (hydrolysate of **24**)	Schneider et al 1994
[³H]24-epiBL (**3**)	Lycopersicon esculentum	Cell suspension	26-β-D-Glucopyranosyloxy-24-epiBL (**26**) 26-Hydroxy-24-epiBL (**27**) (hydrolysate of **26**) Unknown glycosides	Hai et al 1995
[³H]24-epiCS (**5**)	Lycopersicon esculentum	Cell suspension	3-Dehydro-24-epiCS (**28**), 3,24-DiepiCS (**19**), 25-Hydroxy-3,24-diepiCS (**29**) 2-O-β-D-Glucopyranosyl-3,24-diepiCS (**30**) 3-O-β-D-Glucopyranosyl-3,24-diepiCS (**31**) 25-β-D-Glucopyranosyloxy-24-epiCS (**32**) 25-Hydroxy-24-epiCS (**33**) (hydrolysate of **32**) 26-β-D-Glucopyranosyloxy-24-epiCS (**34**) 26-Hydroxy-24-epiCS (**35**) (hydrolysate of **34**) Unknown diglucosidic conjugate	Hai et al 1996
[³H]24-epiTE (**6**)	Lycopersicon esculentum	Cell suspension	3-O-{β-D-Glucopyranosyl-(1→6)}-β-D-glucopyranosyl-24-epiteasterone (**36**), 3-O-{β-D-glucopyranosyl-(1→4)}-β-D-galactopyranosyl-24-epiteasterone (**37**)	Kolbe et al 1997
[³H]24-epiTE (**6**)	Lycopersicon esculentum	Cell suspension	3-O-β-D-Glucopyranosyl-24-epiteasterone (**38**)	Kolbe et al 1998b

cies-specific activities. This possibility was discussed for CS (**2**), which except in *C. roseus* could not be transformed into BL in various plant systems (Yokota 1997). 26-Hydroxy-24-epiBL (**27**), prepared by enzymatic hydrolysis of 26-β-D-glucopyranosyloxy-24-epiBL (**26**), did not show enhanced bioactivity in the RLIT (Hai et al 1995).

5. Conclusions

At present, knowledge on uptake and transport of BRs is limited. Besides several indirect studies, more recent investigations with rice, cucumber, and wheat indicate a direct uptake of exogenously applied labeled compounds via the roots. Also subsequent transport to the shoots is clearly confirmed, occurring probably via the xylem. Upon application via the leaves, the transport takes place to a much lesser extent, but again predominantly acropetal. Further studies including other important plant species are necessary to get deeper insight into uptake and transport of BRs and especially to clarify the correlation between long-distance transport and the effects of other plant hormones.

Investigations on the metabolism of BRs have made considerable progress during the last few years (Table 1). Until now more than 30 BR metabolites have been identified, most of them new compounds not yet found in plants. Two different metabolic pathways were established in cell cultures of serradella and tomato, respectively. We have no knowledge about whether these pathways are specific to the species in which they were found. Nevertheless, epimerization at C-3 and side chain hydroxylation at different positions as well as conjugation reactions, both at the side chain moiety and at ring A, seem to be essential metabolic reactions of a more general distribution in various species. Side chain cleavage observed in serradella may be also deemed as an important metabolic pathway in other plants. Without doubt most of the compounds clearly have to be considered as deactivation or even degradation products. However, it cannot be excluded that new bioactive compounds will be found via that approach.

References

Adam G, Porzel A, Schmidt J, et al (1996) New developments in brassinosteroid research. In: Atta-ur-Rahman (Ed), Studies in natural products chemistry, Vol 18. Elsevier, Amsterdam, pp 495-549

Allevi P, Anastasia M, Cerana R, et al (1988) 24-Epibrassinolide uptake in growing maize root segments evaluated by multiple-selected ion monitoring. Phytochemistry 27:1309-1313

Asakawa S, Abe H, Nishikawa N, et al (1996) Purification and identification of new acyl-conjugated teasterones in lily pollen. Biosci Biotech Biochem 60:1416-1420

Deepak D, Khare A, Khare M P (1989) Plant pregnanes. Phytochemistry 28:3255-3263

Fujioka S, Sakurai A (1997) Brassinosteroids. Nat Prod Rep 14:1-10
Galbraith MN, Horn DHS, Middleton EJ, et al (1969) The catabolism of crustecdysone in the blowfly *Calliphora stygia*. J Chem Soc Chem Commun 1969: 1134-1135
Gregory LE, Mandava NB (1982) The activity and interaction of brassinolide and gibberellic acid in mung bean epicotyls. Physiol Plant 54:239-243
Griffith PG, Sasse JM, Yokota T, et al (1995) 6-Deoxotyphasterol and 3-dehydro-6-deoxoteasterone, possible precursors to brassinosteroids in the pollen of *Cupressus arizonica*. Biosci Biotech Biochem 59:956-959
Hai T, Schneider B, Adam G (1995) Metabolic conversion of 24-*epi*-brassinolide into pentahydroxylated brassinosteroid glucosides in tomato cell. Phytochemistry 40:443-448
Hai T, Schneider B, Porzel A, et al (1996) Metabolism of 24-*epi*-castasterone in cell suspension cultures of *Lycopersicon esculentum*. Phytochemistry 41:197-201
Kerr RG, Kelly K, Schulman A (1995) A novel biosynthetic route to pregnanes in the marine sponge *Amphimedon compressa*. J Nat Prod 58:1077-1080
Kim S-K (1991) Natural occurrences of brassinosteroids. In: Cutler HG, Yokota Y, Adam G (Eds) Brassinosteroids – chemistry, bioactivity and applications. ACS Symp Ser No 474, Amer Chem Soc, Washington, DC, pp 26-35
Kolbe A, Marquardt V, Adam G (1992) Synthesis of tritium labelled 24-epi-brassinolide. J Lab Comp Radiopharm 31:801-805
Kolbe A, Schneider B, Porzel A, et al (1994) Pregnane-type metabolites of brassinosteroids in cell suspension cultures of *Ornithopus sativus*. Phytochemistry 36:671-673
Kolbe A, Schneider B, Porzel A, et al (1995) Acyl-conjugated metabolites of brassinosteroids in cell suspension cultures of *Ornithopus sativus*. Phytochemistry 38:633-636
Kolbe A, Schneider B, Porzel A, et al (1996) Metabolism of 24-*epi*-castasterone and 24-*epi*-brassinolide in cell suspension cultures of *Ornithopus sativus*. Phytochemistry 41:163-167
Kolbe A, Porzel A, Schneider B, et al (1997) Diglycosidic metabolites of 24-epi-teasterone in cell suspension cultures of *Lycopersicon esculentum* L. Phytochemistry 46:1019-1022
Kolbe A, Schneider B, Voigt B, et al (1998a) Labelling of biogenetic brassinosteroid precursors. J Lab Comp Radiopharm 41: 131-137
Kolbe A, Schneider B, Porzel A, et al (1998b) Metabolic inversion of the 3-hydroxy function of brassinosteroids. Phytochemistry 48: 467-470
Lachaise F, Lafont R (1984) Ecdysteroid metabolism in a crab: *Carcinus maenas*. Steroids 43:243-259
Maeda E (1965) Rate of lamina inclination in excised rice leaves. Plant Physiol 18:813-827
Mandava NB (1988) Plant growth-promoting brassinosteroids. Annu Rev Plant Physiol Plant Mol Biol 39: 23-52
Nishikawa N, Toyama S, Shida A, et al (1994) The uptake and transport of ^{14}C-labeled epibrassinolide in intact seedlings of cucumber and wheat. J Plant Res 107:125-130
Nishikawa N, Abe H, Natsume M, et al (1995a) Epimerization and conjugation of ^{14}C-labeled epibrassinolide in cucumber seedlings. J Plant Physiol 147: 294-300
Nishikawa N, Shida A, Toyama S (1995b) Metabolism of ^{14}C-labeled epibrassinolide in intact seedlings of cucumber and wheat. J Plant Res 108:65-69
Rees HH (1989) Pathways of biosynthesis of ecdysone. In: Koolman J (Ed) Ecdysone - from chemistry to mode of action. Thieme, New York, pp 152-160
Roth PS, Bach TJ, Thompson MJ (1989) Brassinosteroids: potent inhibitors of growth of transformed tobacco callus cultures. Plant Sci 59:63-70

Sala C, Sala F (1985) Effect of brassinosteroid on cell division and enlargement in cultured carrot (*Daucus carota* L.) cells. Plant Cell Rep 18:144-147

Sasse JM (1997) Recent progress in brassinosteroid research. Physiol Plant 100:696-701

Schlagnhaufer CD, Arteca RN (1991) The uptake and metabolism of brassinosteroid by tomato (*Lycopersicon esculentum*) plants. J Plant Physiol 138:191-194

Schneider B, Kolbe A, Porzel A, et al (1994) A metabolite of 24-*epi*-brassinolide in cell suspension cultures of *Lycopersicon esculentum*. Phytochemistry 36:319-321

Suzuki H, Kim S-K, Takahashi N, et al (1993) Metabolism of castasterone and brassinolide in mung bean explant. Phytochemistry 33:1361-1367

Suzuki H, Inoue T, Fujioka S, et al (1994) Possible involvement of 3-dehydroteasterone in the conversion of teasterone to typhasterol in cultured cells of *Catharanthus roseus*. Biosci Biotech Biochem 58:1186-1188

Suzuki H, Fujioka S, Takatsuto S, et al (1995) Biosynthesis of brassinosteroids in seedlings of *Catharanthus roseus, Nicotiana tabacum*, and *Oryza sativa*. Biosci Biotech Biochem 59:168-172

Takatsuto S, Yazawa N, Ikekawa N, et al (1983) Structure-activity relationships of brassinosteroids. Phytochemistry 22:2437-2441

Voigt B, Porzel A, Golsch D, et al (1996) Regioselective oxyfunctionalization of brassinosteroids by methyl(trifluoromethyl)dioxirane: Synthesis of 25-hydroxybrassinolide and 25-hydroxy-24-epi-brassinolide by direct C-H insertion. Tetrahedron 52: 10653-10658

Winter J, Schneider B, Strack D, et al (1997) Role of a cytochrome P450-dependent monooxygenase in the hydroxylation of 24-*epi*-brassinolide. Phytochemistry 45:233-237

Woichiechowski ZA (1991) Biochemistry of phytosterol conjugates. In: Patterson GW, Nes WD (eds) Physiology and biochemistry of sterols. American Oil Chemists' Society, Champaign, Illinois, pp 361-395

Yokota T, Kim S-K, Kosaka Y, et al (1987) Conjugation of brassinosteroids. In: Schreiber K, Schütte HR, Sembdner G (Eds) Conjugated plant hormones – structure, metabolism and function. Deutscher Verlag der Wissenschaften, Berlin, pp 288-296

Yokota T, Watanabe S, Ogino, Y, et al (1990) Radioimmunoassay for brassinosteroids and its use for comparative analysis of brassinosteroids in stems and seeds of *Phaseolus vulgaris*. J Plant Growth Regul 9:151-159

Yokota T, Ogino Y, Suzuki H, et al (1991) Metabolism and biosynthesis of brassinosteroids. In: Cutler, HC, Yokota T, Adam G (Eds) Brassinosteroids: chemistry, bioactivity and applications. ACS Symp Ser 474, Amer Chem Soc, Washinton, DC, pp 86-96

Yokota T, Higuchi K, Kosaka Y, et al (1992) Transport and metabolism of brassinosteroids in rice. In: Karssen CM, van Loon LC, Vreugdenhil D (Eds) Progress in plant growth regulation. Kluwer, Dordrecht, pp 298-305

Yokota T (1997) The structure, biosynthesis and function of brassinosteroids. Trends Plant Sci 2:137-143

7
Physiological Actions of Brassinosteroids

Jenneth Sasse

Forestry School, Institute of Land and Food Resources, University of Melbourne, Parkville, Vic. Australia 3052

1. Introduction

Following the exploratory work of Mitchell (e.g., Mitchell and Whitehead 1941) studies of the physiological effects of brassinosteroids (BRs) began with almost pure "brassins" before the identification of brassinolide (BL) in pollen of *Brassica napus* (Steffens 1991). As well, there had been an intriguing report in 1968 of the effects in bioassays of incompletely identified plant growth regulators from *Distylium* and other plants, and these were shown to be BRs some years later (see the chapter by T. Yokota, this volume). The possibility that the bioactive component(s) of brassins might be a new family of plant hormones was appreciated even earlier than a public demonstration of its effects on beans in 1963 (Cutler 1991), and now, with recent evidence from Arabidopsis mutants (see the chapter by S.D. Clouse and K.A. Feldmann, this volume) adding to the data from physiological studies (Sasse 1991a), there is widespread consensus that BRs are such a family. There are at least 40 members of the family, and more are likely; for the structures of the native BRs, their precursors, and their metabolites, see the chapters by S. Fujioka, A. Sakurai, and G. Adam and B. Schneider, this volume, and for synthetic analogues, the chapter by Y. Kamuro and S. Takatsuto, this volume.

After the application of brassins to young bean seedlings, a marked elongation of the second and third internodes was noted and with higher dosages swelling and a distinctive splitting of the young stem tissue. The response was claimed to be histologically different from the elongation induced by gibberellin (GA). A later paper claimed enhancement and acceleration of overall plant growth, and that the relative

Key words: brassinolide, brassinosteroids, brassins, bioactivity, development, environmental signals, growth, insects, interactions, source/sinks, stress responses, transport

response of the smaller plants in a population was greater than that of the larger ones. Interactions between brassins concentration and intensity and spectral quality of light were also described in bean, and an interaction with phytochrome signalling was postulated. Differential sensitivity to brassins in the stem was also noted, and the influence of brassins was largely confined to immature tissues in woody plants (reviewed in Mandava 1988, Steffens 1991).

Immature vegetative tissues were also studied in the work of Yopp et al (1979), who reported that brassins were active in some auxin, GA, and cytokinin assays but only weakly active or inactive in others. They thought enhanced elongation and hook closure in beans after brassins treatment might be mediated by endogenous auxin, but that it mimicked GA in its site of elongation. Reversal of light-induced retardation of extension was also observed, and increased ethylene production. A student of Yopp investigated the effects of brassins on oat throughout its life cycle and found growth was increased, with the most significant increase in the weight of seeds. The stage of coleoptile growth most sensitive to brassins was similar to that of exogenous auxin, and brassins acted synergistically with auxin (Miller 1972).

Later, most of these responses were confirmed with BRs. BL itself, its 24-epimer, or a mixture of epimers at the 22-, 23-, and 24-positions, or 28-homoBL or its 22S,23S-epimer were the BRs used most frequently. The distinctive elongation, swelling, and splitting of the second internode occurred in bean seedlings treated with BL (Grove et al 1979), and effects in selected bioassays were confirmed, as was synergism between BL and auxin and additivity with GA in elongation (Mandava et al 1981, Yopp et al 1981a, Gregory and Mandava 1982). Barley and lettuce were shown to mature at an accelerated rate, with plants grown in soil treated with lower levels of fertilizer responding more favourably (Meudt et al 1983); Braun and Wild (1984) also noted accelerated growth of whole plants.

Early physiological studies of effects induced by BR treatment, either alone or together with other plant growth regulators, have been reviewed frequently (e.g., Mandava and Thompson 1984; Adam and Marquardt 1986; Yokota and Takahashi 1986; Meudt 1987; Mandava 1988; Katsumi 1991; Marquardt and Adam 1991; Sasse 1991a,b; Takeuchi et al 1992; Khripach et al 1993; Sakurai and Fujioka 1993). As well as early and subsequent reports in Japanese and Russian, pioneering work was also published in Chinese. More recent reviews that comment on the physiological effects of BRs include those by Adam (1994), Sasse (1997), Yokota (1997), and Clouse and Sasse (1998).

The examples cited above illustrate that two important concepts were explicit from the beginning of research into the physiological effects of these potent plant growth regulators: firstly, that BRs can accelerate plant growth and maturation (which may or may not lead to absolute growth increases over time) and, secondly, that BR-induced effects cannot be considered in isolation, as BRs interact with other endogenous plant growth regulators and with environmental signals, particularly light quality. Other studies, both direct and indirect, suggested that BRs affect gene expression and can do so selectively, findings borne out by the results of molecular genetic approaches (see the chapter by S.D. Clouse and K.A. Feldmann, this volume).

2. Sites of Synthesis and Transport of BRs

Brassinosteroids occur at very low concentrations in plants, with pollen and immature seeds being the "richest" sources (see the chapter by S. Fujioka, this volume), but it is not yet known whether all tissues produce them. Endogenous BRs in roots are yet to be identified rigorously, but their presence can be inferred from the properties of BR-insensitive mutants (see the chapter by S.D. Clouse and K.A. Feldmann, this volume), and a putative BR receptor is expressed in roots of *Arabidopsis thaliana* (Li and Chory 1997). There is also some evidence from bioassay data from appropriate fractions of extracts of roots of pea and soybean (J. Sasse, unpublished).

Endogenous synthesis is clearly evident in *Catharanthus* cell cultures (see the chapter by A. Sakurai, this volume), while the results of inhibitor/replacement experiments in *Zinnia elegans* mesophyll cell cultures also imply endogenous synthesis in single cells (Iwasaki and Shibaoka 1991). In pollen, changes in levels of particular BRs and their conjugates can be correlated with the developmental stage (Asakawa et al 1996), implying that at least the later steps of biosynthesis occur there.

In pollen and seeds, short-distance effects of BRs can probably be assumed, but it is not yet clear whether the contribution of BRs to the control of stem elongation (see Section 3.2, this chapter) or vascular development (see Section 5.3, this chapter) depends on intracellular, or short-distance, effects or on long-distance transport of BRs from other organs. Endogenous long-distance transport of BRs (or their conjugates) is a possibility, as Takatsuto et al (1983) found that exogenous BRs supplied to roots of intact radish and tomato plants affected the elongation of hypocotyls and petioles, while epicotyls responded to BRs supplied at the bases of the hypocotyls of mung bean explants (Gregory and Mandava 1982).

Later studies with ^{14}C-labeled 24-epiBL confirmed movement from root to shoot, probably via the xylem; and transport from leaves, while slight, is also acropetal (e.g., Nishikawa et al 1994). Within cultured explants of tomato roots, it was clear applied BRs could move acropetally (Roddick et al 1993). Whether free BRs or their conjugates are transported in these cases is not yet clear. A specific mechanism may be required for BRs to cross cell membranes; with their hydrophilic side chains, it is difficult to envisage easy insertion into membranes in the manner and orientation attributed to most plant sterols.

3. Effects on Vegetative Growth

The effects on vegetative growth have been reviewed (cited in Section 1 and Sasse 1991a,b); and, overall, BRs should not be classified as auxins, GAs, or cytokinins. This is illustrated by the effects of BRs in 17 selected bioassays (Mandava et al 1981; Yopp et al 1981a). Several authors proposed that BR-induced effects might be mediated via auxin, that BR treatment might alter the levels of endogenous hormones, or that BR might enhance tissue sensitivity to auxin (reviewed in Mandava

1988). There were a few studies suggesting that different target tissues were affected, and that BR could promote elongation per se (Sasse 1990). This conclusion is now supported by results from molecular genetic studies (see the chapter by S.D. Clouse and K.A. Feldmann, this volume). However, complex interactions between BR and other plant hormones, sucrose, osmotica, and environmental signals certainly occur.

3.1 Whole Plants

Brassinosteroids have been detected in a few lower plants (see the chapter by S. Fujioka, this volume), and a recent study has shown that the growth rate of the alga *Chlorella vulgaris* can be enhanced dramatically by femtomolar levels of BL and its 24-epimer. This unicellular alga grows by expansion, followed by division, and the overall rate of these processes was markedly accelerated; cessation of growth occurred after 36 h, in contrast to the control culture, where slackening of the growth rate does not occur until 96 h (Bajguz and Czerpak 1996).

With higher plants, while some bioassays for BR-type activity utilize intact plants, and there are many reports of increased yields in crop plants after administration of BRs (see the chapter by Y. Kamuro and S. Takatsuto, this volume), there have been few detailed analyses of the effects of exogenous BRs on growth. One study of the growth of an intact monocot and dicot in their primary developmental stages (Braun and Wild 1984) showed a promotive effect of a BL isomer on photosynthetic capacity and biomass production. In mustard, fresh weights of the shoots increased, and both elongation and radial growth were stimulated. There was an increase of 40-50% in CO_2 fixation in vivo in wheat, and the levels of soluble proteins and reducing sugars also increased, but chlorophyll levels were only slightly affected. The rate of elongation of the more immature wheat leaf was markedly enhanced after 16 days when compared with control, but with time the differences between treated and control plants decreased. There was no significant increase in leaf area, but enhanced dry weight per leaf area suggested the thickness had increased. However, the leaf area was increased in bean seedlings after BR treatment (Meudt et al 1983) and rescue by BRs of a mutant of Arabidopsis blocked in an early stage of the biosynthesis of these hormones resulted in marked expansion of leaf blades (Kauschmann et al 1996).

Elongation was promoted in young stem tissues of many plants after treatment with moderate doses of BR (listed in Sasse 1991b), and this response is probably general. Interactions between BR treatment and the spectral quality of the light in which the plants were grown were observed in bean, with some inhibitory effects (Krizek and Mandava 1983a), and radish (Kamuro and Inada 1987). In this area too, mutants of Arabidopsis have provided interesting data (see the chapter by S.D. Clouse and K.A. Feldmann, this volume, and Section 6.2, this chapter). In darkness, alfalfa seedlings treated with micromolar levels of BL have shorter stems and roots, with swelling and twisting reminiscent of ethylene effects; this also occurred in plants grown in light (Hata et al 1986).

Inhibition of elongation, which may be mediated by ethylene, was also seen in intact cucumber seedlings after treatment with high doses of BL (Katsumi 1991), and a high dose of 24-epiBL inhibited shoot growth in wheat and mung bean as well (Roddick and Ikekawa 1992). The latter authors also found that lateral roots were particularly inhibited in wheat, mung bean, and maize. There is also some evidence from a study using the inhibitor uniconazole that endogenous BRs might inhibit the initiation of lateral roots in *Lotus japonicus* (Kawaguchi et al 1996). Inhibition of root growth also occurred in Arabidopsis (Clouse et al 1993), and it is notable that consistent inhibition of root growth in Arabidopsis allowed the identification of a BR-insensitive mutant (see the chapter by S.D. Clouse and K.A. Feldmann, this volume).

Brassinosteroids can also promote asymmetric growth in whole plants, such as the promotion of lamina inclination in rice (Takeno and Pharis 1982), and nanomolar levels fed to the roots caused curvature of the cotyledon petiole and of the hypocotyl in radish seedlings (Takatsuto et al 1983). BRs also induced epinasty in tomato leaves and petioles and this is mediated via endogenous ethylene biosynthesis. The biosynthetic precursor of ethylene, 1-aminocyclopropane-1-carboxylic acid, accumulated in leaves and petioles but not in sap. BRs stimulate ethylene biosynthesis between *S*-adenosylmethionine and 1-aminocyclopropane-1-carboxylic acid, and there are strict structural requirements for the steroids, while low levels of light inhibit the effect (references cited in Mandava 1988 and Arteca 1995). While it is tempting to ascribe all BR-induced growth inhibitions to the effects of induced endogenous ethylene biosynthesis, there may be species differences (Eun et al 1989), and such an explanation may not be warranted in seeds (see Section 5.1, this chapter).

The results described above imply complex interactions between endogenous and environmental signals that are difficult to analyze in whole plants, so it is not surprising that most physiological studies of the effects of plant growth regulators, including BRs, have been carried out in model systems. These might be explants containing many cell types, such as derooted seedlings, stem segments, and root sections, or "simpler" organs such as coleoptiles or pollen tubes. The use of cell or protoplast cultures has also proved useful. Techniques used included aging to reduce endogenous hormone levels, the measurement of dose/response curves and time courses, detailed kinetic studies, sequential or mixed treatments of BR and other hormones, the use of selected metabolic inhibitors, and comparisons of enzyme activities in tissue extracts.

3.2 Expansion Growth in Cell Cultures and Explants

While exogenous BRs often inhibit cell growth in culture (see Section 4, this chapter), single cells can elongate in response to BR in suitable media (Sala and Sala 1985). Coleoptiles are also sensitive; Miller (1972) had found oat coleoptiles elongated when treated with brassins, and BL fits between GA and auxin in the sequential response of wheat coleoptiles of different ages to exogenous hormones (Sasse 1985). Maize mesocotyls also respond, but auxin is more effective (Yopp et al 1981a).

Epicotyl sections from soybean are also sensitive to BL, and they illustrate a typical response to concentration and the time course of such a response (Clouse et al 1992). Lag times are usually longer than those for auxin, for example.

Promotive effects on the expansion of cotyledon explants have been shown for cucumber, where the maximum response was far less than that to cytokinin. In radish, the effects of the two regulators were additive, and there was accompanying stimulation of proton secretion with 24-epiBL treatment, in contrast to the effect of cytokinin (De Michelis and Lado 1986). That endogenous BRs are required for normal expansion of cotyledons (and hypocotyl) is indicated by anatomical evidence in some *cbb* mutants of Arabidopsis that can be rescued by BR administration (Kauschmann et al 1996).

Leaf sections have also been studied; in wheat, sections comprising younger tissue were more sensitive to exogenous BL than those of more mature tissue, in contrast to the effect of applied auxin (Sasse 1985). The unrolling of etiolated wheat leaf segments is also stimulated by BRs, and this observation has been developed as a convenient bioassay for them (Wada et al 1985).

Asymmetric expansion also occurs. In the most frequently used bioassay for BRs, which employs rice explants consisting of the second lamina and a lamina joint and sheath, marked inclination occurs after BR treatment, at nanomolar concentrations and below (Wada et al 1984). After a detailed study of the system, Cao and Chen (1995) proposed that the enhanced expansion of the adaxial cells of the pulvinus of the joint was due to BR-induced increases in water potential there and enhanced proton extrusion. Endogenous ethylene synthesis was also increased, and the effects were nullified in the presence of an inhibitor of such biosynthesis. So this nastic response may be mediated by endogenous ethylene; it is reminiscent of the retardation of hook opening by BL, where the asymmetric growth is also abolished by an inhibitor of ethylene biosynthesis (Yopp et al 1981a,b).

Most physiological studies of the effects of BRs have been done with dicot stem sections, usually comprising tissue from below the dividing zone. Early work has been reviewed (Mandava 1988) and detailed studies in cucumber (Katsumi 1991) and de-etiolating dwarf pea segments (Sasse 1990, 1991b) have been summarized.

In segments from the maturing stem of pea and squash, the presence of the epidermis is required for auxin-induced elongation, but both peeled and unpeeled sections of squash hypocotyl respond to BL (Tominaga et al 1994). Experiments with partially split stem sections from such tissue also suggested that both inner and outer tissues responded to BL (Sasse 1990; Tominaga et al 1994) (Fig. 1), and epidermal cells of the epicotyl of the dwarf pea mutant *lkb* clearly elongated after administration of BL to the lowest leaf (Yokota 1997). Sections of soybean taken from below the plumule also respond to BL (Clouse et al 1992), and in situ hybridization experiments have shown that expression of *BRU1*, a gene induced by BL, occurs across the stem, with particular labelling in the phloem and parenchyma cells surrounding the xylem elements (Oh et al 1998).

The zone most sensitive to exogenous BL in the de-etiolating pea stem consists of immature expanding tissue (from below the hook to the laterally enlarging zone of the stem). Sections from this zone do not respond to exogenous plant growth regulators in the usual manner when partially split (Fig. 1); there is only slight relief of tension, and the lack of a characteristic response to auxin or fusicoccin (FC) suggested that the epidermis was not yet limiting elongation. The morphology of intact sections from this zone can be dramatically altered by exogenous plant growth regulators (Fig. 2), and a particular effect of BL treatment was maintenance of the narrow diameter at the base, as well as the clear promotion of elongation (Sasse 1987). Notably, FC did not maintain a narrow basal diameter, nor did GA_1. Later experiments with azuki bean epicotyl segments confirmed that exogenous BL alone promoted elongation without lateral expansion, and that there was a relatively long lag period (Mayumi and Shibaoka 1995).

Elongation of this very plastic zone of the pea stem can be markedly affected by treatment with combinations of growth regulators; GA_1 plus BL or BL plus FC were particularly promotive. Similar responses occur in Azuki bean segments (Mayumi and Shibaoka 1995). These authors also showed that the three hormones BL, auxin, and GA promoted elongation synergistically when they were applied together.

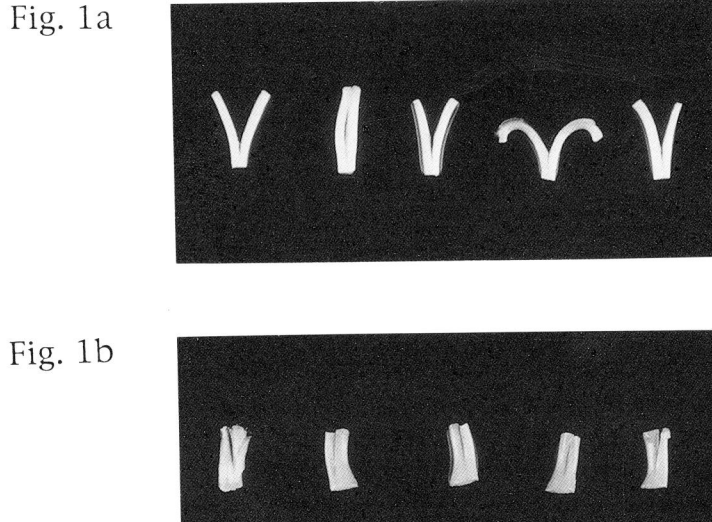

Fig. 1a

Fig. 1b

Fig. 1. Split sections from dwarf pea stems. **a.** Section from more mature stem treated for 4 h with, from left to right; water, 1 µM 4-chloroindole-3-acetic acid, 1 µM brassinolide (BL), 1 µM fusicoccin (FC), 1 µM gibberellin (GA_1). **b.** Less mature section treated for 3 h with, from left to right; water, 1 µM 4-chloroindole-3-acetic acid, 1 µM BL, 1 µM FC, 1 µM GA_1. Quantitative data in Sasse (1990)

Fig. 2a

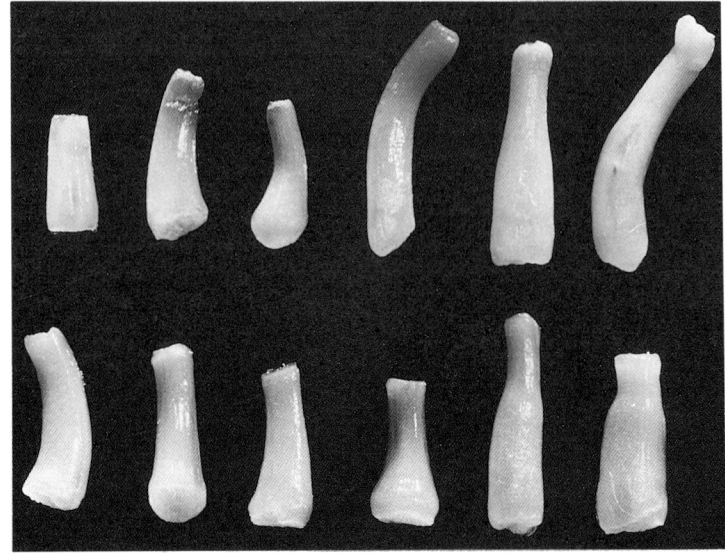

Fig. 2b

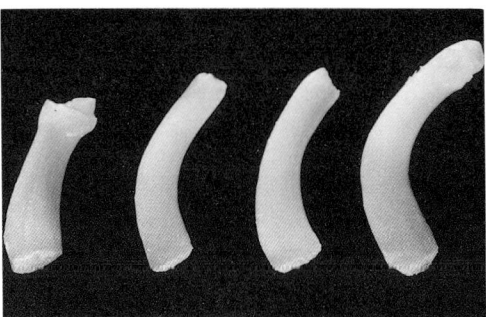

Fig. 2. Expansion growth of segments from the most brassinosteroid-responsive zone of the de-etiolating pea stem. Segments were grown in 2% sucrose for 24 h, with single or mixed growth regulators. **a.** From left to right, upper row; initial (after storage in 20% v/v glycerol at 6°C), control, 100 μM ethephon, 1 μM BL, 1 μM FC, 1 μM BL plus 1 μM FC: bottom row; control, 1 μM indole-3-acetic acid (IAA), 20 μM zeatin, 1 μM IAA plus 20 μM zeatin, 1 μM BL plus 1 μM IAA, 1 μM BL plus 20 μM zeatin. **b.** From left to right, control, 0.3 μM BL, 0.5 μM GA_1, 0.3 μM BL plus 0.5 μM GA_1. Quantitative data in Sasse (1987)

Brassinolide also interacted synergistically with auxin or cytokinin in their promotion of lateral expansion of the pea stem section, not only by a reduction in the expected elongation but also by enhancement of the expected lateral expansion (Fig. 2). Concomitant abscisic acid treatment also decreased elongation of the BL-treated segment, and kinetic analysis showed that the response capacity of the segment was affected (Sasse 1989). Ethephon treatment decreased BL-induced elongation by affecting the response capacity, and the elongation of this segment was reduced in the presence of other selected inhibitors, particularly of microtubule organization or RNA or protein biosynthesis. As well, maintenance of adequate cellulose biosynthesis was important for BL-induced elongation, in contrast to auxin-induced elongation, which is less sensitive to inhibition of cellulose biosynthesis (Sasse 1990).

If comparable interactions between hormones occur in vivo, BRs may be involved in the control of the overall form of the plant and its modification by environmental signals (see Section 6, this chapter). There are reports of BR treatment altering endogenous levels of auxin and abscisic acid levels (e.g., Eun et al 1989), and such changes might modify development and form, particularly of the stem and its tissues. Continued molecular approaches, the use of mutants, and the engineering of plants that overproduce BRs will be most valuable for increased understanding of overall stem development and morphology.

As with whole plants, effects of exogenous BRs on explants of root tissue are mainly inhibitory, with some promotion of elongation reported. These and effects on adventitious and lateral roots have been reviewed (Roddick and Guan 1991; Sasse and Sasse 1994). One case of promotion of adventitious rooting in cuttings seemed to depend on a particular light regimen (Sathimoorthy and Nakamura 1990), and another promotive treatment included 5 weeks of storage after application of the BR (Ronsch et al 1993). Other promotions have been reported when a synthetic BR analogue was used (see the chapter by Y. Kamuro and S. Takatsuto, this volume). The apparently conflicting results may well be influenced by concentration, pretreatments, and/or differing levels of stress in the parent plant.

3.3 Mechanism of Action in Expansion Growth

While the specificity of inhibitors is never ideal, and results must be interpreted with caution (e.g., uniconazole inhibits the biosynthesis of BRs as well as GAs), the use of selected inhibitors in dose/response and kinetic studies suggested that BR-induced expansion depended on mRNA and protein synthesis (reviewed in Mandava 1988 and Sasse 1991b). BR-induced expansion of root sections is accompanied by proton extrusion, membrane hyperpolarization, and the uptake of potassium ions. Interestingly, such responses also occur after treatment of root sections with sterols that do not promote expansion (reviewed in Mandava 1988), but energy-dependent proton extrusion may well be a necessary component of the BR-induced expansion, as an inhibitor of membrane-bound ATPase also inhibited BL-induced elongation in cucumber hypocotyl sections (Katsumi 1991).

Mayumi and Shibaoka (1995, 1996) confirmed the need for cellulose biosynthesis and the importance of microtubule organization in BL-induced elongation; their detailed studies of microtubule orientation suggested at least two processes were involved in BRs' effects, one of which was dependent on the microtubules and may require phosphorylation of proteins that possibly link microtubules to the plasmalemma. Also, for cells to expand, not only must the supply of new cell wall materials be maintained, but the existing wall must be modified to allow their incorporation, and several authors have shown that BR treatment alters the biophysical properties of the wall (e.g., Tominaga et al 1994). Concomitant with these changes are changes in the expression of genes for wall-modifying proteins; these are reviewed by Clouse and Sasse (1998) and in the chapter by S.D. Clouse and K.A. Feldmann, this volume.

Claims of the enhancing effect of BRs on the activities of some enzymes in crude homogenates from BR-treated plants should be assessed with caution, as most lack statistical analysis; some exceptions are a study of acid invertase in beet and oxidative enzymes in moth bean (see Sections 5.4 and 6.3.1, this chapter). So far there is no evidence for direct effects of BRs on enzyme-substrate complexes or rates of reaction, and the relationship of BRs to effects on membrane ATPases merits more detailed study. Nor is it clear whether BR-induced expansion and the accompanying uptake of water always depends on changes in osmotic potential within the cell's compartments; there are conflicting reports, and more detailed studies are needed. Exploration of effects on aquaporin synthesis or function may be worthwhile and further study of the effects of BRs on the permeability of cell membranes (cf. Katsumi et al 1987). The few reports of selective uptake or exclusion of metal ions in plants treated with BRs (see Section 5.4, this chapter) imply effects at the membrane level, either directly or in the production or function of transporters.

The reception and transduction of the BR signal is also a tantalizing question - is the receptor intracellular like those for many animal steroids, or is it membrane-bound, requiring a second messenger system? Do both or more types occur? The recent cloning of the *BRI*1 gene (Li and Chory 1997) suggests a putative membrane-bound receptor, but it is not yet known whether BL binds to it, either directly or through adaptor-type protein(s).

4. Effects on Cell Division

While detailed histological study of *cbb* mutants of Arabidopsis lacking functional BRs shows that their dwarfism is primarily due to reduced cell size rather than number (Kauschmann et al 1996), effects on cell division are known. Enhanced division due to brassins treatment was reported in bean plants, and interactions between high and low light intensities and brassins concentration were described. Only one combination of light intensity and brassins concentration produced cell division as well as elongation and swelling, but the earlier paper reporting brassins-promoted cell division in pith cells was consistent with this finding (Worley and Mitchell 1971, Krizek and Worley 1973). Later, increased activities of RNA and DNA polymerases in pinto and mung bean tissue were shown to correlate with the swelling or distinctive splitting due to treatment with high doses of BL (Kalinich et al 1985).

As mentioned in Section 3.1 of this chapter, the growth of *Chlorella* entails division, and the accelerative effect of BLs occurs at extremely low concentration, but most studies of cultures of higher plant cells showed inhibition of expansion and division after BR treatment. In the case of transformed tobacco cells, a marked decrease in auxin content was shown as well (Bach et al 1991). However, more recent work with protoplasts from Chinese cabbage has shown 24-epiBL-induced promotion of division and cluster and colony formation in the presence of 2,4-D and kinetin (Nakajima et al 1996).

5. Effects on Development

It is clear from work describing the dwarfed phenotypes and altered development of BR-deficient and BR-insensitive mutants of Arabidopsis and pea that BRs are essential for normal plant development (see the chapter by S.D. Clouse and K.A. Feldmann, this volume). Thus physiological studies on the effects of BRs are now of wider interest, and many early reports point to areas where the application of more modern approaches and techniques could be valuable for our understanding of the control of plant growth.

5.1 Seed Germination

Jones-Held et al (1996) have summarized work on the effects of exogenous BRs on seed germination and subsequent early seedling growth, and explored effects of a brief treatment of cress seeds with BL in some detail. Germination was unaffected, but root growth was inhibited within 48 h, and so was DNA synthesis; and it was unlikely that the inhibition of root growth involved ethylene. However, recent work exploring the complex conditioning and germination of witchweed and clover broomrape seeds showed that BRs can promote the rate of seed germination and overcome the inhibitory effect of light (Takeuchi et al 1995 and references cited therein). It is notable that only slight acceleration of germination occurs with 24-epiBL treatment of *Eucalyptus camaldulensis* seeds, but that there is a much more marked promotion when the seeds are germinated in saline conditions (Sasse et al 1995). As for endogenous BRs, during the germination of radish seeds levels of castasterone decreased and those of BL increased (Schmidt et al 1991), suggesting it may have a role in normal germination, or it must be available for successful extension, particularly of the shoot.

5.2 Meristems

The effects of BRs on extension growth of the stem (see Section 3, this chapter) and cell division in model systems (see Section 4, this chapter) have been discussed, but the effects on apical meristems are largely unknown. That the first gene found to be up-regulated by BL (*BRU1*) has sequence similarity to meri 5 (see the chapter by S.D. Clouse and K.A. Feldmann, this volume) hints that BRs may have a role in meristems, so in vivo studies and in situ hybridization could be rewarding. It may also be significant that a high dose of BL switched the apex of an inflorescence back to vegetative growth (see Section 5.5, this chapter).

Differentiation of cells derived from cambium, a lateral meristem, is also important for seedling development. Production of conducting tissue is crucial for the plant, and Fukuda (1996,1997) has recently reviewed xylogenesis. It is becoming increasingly obvious that, as well as auxin and cytokinin, BRs have an important role in this process, and it is known that BRs occur in cambial scrapings (see the chapter by S. Fujioka, this volume).

5.3 Vascular Differentiation

There were some observations of modified vascular tissue in early work studying the effects of brassins, but the first clear indication that exogenous BR accelerated tracheid differentiation came from a model system employing explants from *Helianthus tuberosus* (Clouse and Zurek 1991). In cultures of mesophyll cells from *Zinnia elegans*, the inhibition of endogenous GA and BR biosynthesis by treatment with uniconazole retarded tracheid differentiation, which could be restored by BL or 28-homoBL administration (but not by GA) (Iwasaki and Shibaoka 1991). BL is essential for entry into the final stage of tracheid differentiation, where secondary wall formation and cell death occur. The relationship between BL and the genes expressed at this stage is being investigated (Yamamoto et al 1997). In Arabidopsis, comparison of the *cpd* mutant (whose biosynthesis of BRs is blocked) with wild type has shown that the balance of xylem and phloem production is altered in the mutant, with a predominance of phloem at the expense of xylem (Szekeres et al 1996). Intense expression of *BRU1* in paratracheary parenchyma cells of soybean stem (Oh et al 1998) is consistent with a role for BRs in the differentiation of xylem, but effects on the production of cell types other than tracheids in woody tissues, e.g., fibres, remain to be explored.

5.4 Source/Sink Relationships

In their study of the effect of light quality on BL-induced growth of bean, Krizek and Mandava (1983b) found evidence for redirection of assimilate and proposed a mobilizing role for BR. Studies with ^{14}C-labeled assimilates in rice and *Vicia* also suggested enhanced partitioning to sink regions from source leaves and enhanced uptake of sucrose in BR-treated leaves. Stimulation of the proton ATPase was invoked to explain increased uptake of solutes in several systems, which included stomata (where closure was promoted), leaf tissue, and vascular bundles (reviewed in Adam 1994).

Work on the import of toxic ions by roots has revealed interesting selectivities for metal ions after BR treatment in conjunction with fertilizer; uptake of heavy metals such as Cs, Cd, and Pb was reduced in crops grown in contaminated soil (V. Khripach, personal communication). On the other hand, uptake of Mg from media containing very low levels of this nutrient was enhanced in spruce (Ronsch et al 1996), and reports of enhanced growth of crop plants in suboptimal conditions after BR treatment (see the chapter by Y. Kamuro and S. Takatsuto, this volume) also implies improved uptake of nutrients. Kuno (1997) found enhanced translocation into shoots of N and P after BL treatment of leaves of mulberry, but not of Mg, K, or Ca. Taken together, these data suggest that the source/sink relationship is yet another area of plant physiology where more investigation into the role(s) of BRs could be rewarding.

5.5 Reproduction

There are few reports so far on the effects of BRs on the reproductive stage of a plant's life cycle, even though pollen is a rich source of endogenous BRs. Suge examined the effect of relatively high doses of BL on sex expression in *Luffa cylindrica* and found that bisexual and pistillate flowers were produced on a staminate inflorescence, but the sepals were deformed at higher doses and the tops of the inflorescences developed into vegetative shoots that were similar to the main shoot. He found little effect on flower induction in the short- and long-day plants *Perilla* and *Raphanus*, except for a slight acceleration of maturation in unvernalized radish (Suge 1986). However, flower number could be increased in strawberry by treatment with a BR analogue, and increases in fruit weight followed (see the chapter by Y. Kamuro and S. Takatsuto, this volume).

Hewitt et al (1985) compared the effect of BL with that of other plant hormones on the growth of pollen tubes in vitro and found that it promoted elongation at nanomolar levels, an order of magnitude below those of auxin or GA. Also, parthenogenetic production of haploid seeds occurred in Arabidopsis and *Brassica juncea* after BL treatment (Kitani 1994). These data, together with the report that the *cpd* mutant of Arabidopsis is male sterile because its pollen failed to elongate during germination, but could set seed if BRs were provided (Szekeres et al 1996), suggest BRs do have at least an enabling role in embryo and/or seed production. As for the availability of appropriate BRs, Taylor et al (1993) suggested BRs could be stored in starch granules of maturing pollen and be released upon imbibition, while Asakawa et al (1996) showed that the level of a conjugated precursor decreased as pollen developed, with concomitant increases in free BRs.

It is not known whether BRs are important for continued fruit and seed development, but exogenous BR treatment can improve fruit set (Sugiyama and Kuraishi

1989) and seed filling (Hamada 1986) and it can retard abscission of fruitlet explants (Iwahori et al 1990). It may be significant that immature seeds are also relatively rich sources of BRs, and that they may be involved in fruit ripening through the production of an XET (see the chapter by S.D. Clouse and K.A. Feldmann, this volume).

5.6 Senescence

There are also only a few reports of effects of BRs on senescence. Exogenous BL accelerated senescence in *Xanthium* and *Rumex* leaf explants (Mandava et al 1981), and some similar findings were reported from Zhao's group, who used detached cucumber cotyledons and noted accelerated destruction of chloroplasts (Zhao et al 1990). Later, the group showed accelerated senescence in mung bean seedlings, accompanied by enhanced peroxidase activity and malondialdehyde accumulation, and decreased superoxide dismutase and catalase activity (He et al 1996). In contrast, Khripach's group, working on lipid peroxidation in pea (see Section 6.3.4, this chapter), reported decreased malondialdehyde levels after 24-epiBL treatment. However, Arabidopsis *det2* mutants lacking bioactive BRs show a delay in leaf and chloroplast senescence and flowering (Li et al 1996), which is consistent with a role for endogenous BRs in maturation.

6. Interactions with Environmental Signals

6.1 Tropisms

Effects in phototropism have not been studied, but Yopp et al (1981b) observed promotion of the geotropic response in horizontally placed sunflower and bean hypocotyls. If auxin-depleted sections were used, BL alone did not stimulate the response; a prior or simultaneous treatment with auxin was required to elicit the effect. Similarly, if bean first internode sections were gravitropically stimulated, a 2- to 3-second pretreatment with BL markedly promoted the upward curvature but added auxin was ineffective, suggesting that endogenous auxin was sufficient. The recovery rate when the curved segments were again placed vertically was also enhanced. In vertical segments treated with unilateral applications of auxin, BL was promotive only in the presence of auxin or if applied before the auxin. Such treatment did not promote uptake of auxin, increase its turnover, or affect auxin transport, but it did decrease the concentration of auxin within the tissue (Meudt 1987), in contrast to the results of Eun et al (see Section 3.2, this chapter).

High doses of BL produced transverse growth and twining of etiolated mesocotyls of maize, and these effects could not be mimicked by treatment with ethephon. Rice and wheat roots also twined in response to BR treatment (He et al 1991). The authors considered twining could be a specific BR response, but this study does not appear to have been pursued.

6.2 Light

Complex interactions between BRs and light have been observed in adventitious rooting (mentioned in Section 3.2, this chapter) and seed germination, including seeds such as clover broomrape whose germination is under phytochrome control (see Section 5.1, this chapter). Most interest, however, has been in photomorphogenesis. Early workers had shown that BL could overcome the inhibition of stem elongation of etiolated plants on exposure to light and confirmed the observations obtained with brassins in beans (see Sections 1 and 3.1, this chapter). Later work with monochromatic light (Kamuro and Inada 1987) confirmed that BL could overcome the marked inhibition of elongation in mung bean hypocotyls induced by red light and that elongation was unaffected under far-red; an interaction with phytochrome signalling was again postulated. There were also hints that BRs might help maintain the etiolated state (Sasse 1991a).

In recent years, results with mutants of Arabidopsis deficient in the biosynthesis of, or response to, BRs have supported these ideas and provided evidence for the essentiality of BRs for normal plant growth in both light and dark (see the chapter by S.D. Clouse and K.A. Feldmann, this volume). A characteristic of the BR synthesis and response mutants of Arabidopsis is their de-etiolation in the dark, but this was not obvious in the analogous *lkb* and *lka* mutants of pea (Nomura et al 1997). Also, while the *lip*1 mutant of pea does exhibit a de-etiolated phenotype in the dark (Frances and Thompson 1997), normalization by BR administration has not been achieved so far (J. Reid, personal communication) so an investigation of endogenous BR levels in that mutant could be useful. There may also be differences in tomato (Mustilli and Bowler 1997, see also the chapter by S.D. Clouse and K.A. Feldmann, this volume).

It is considered that the BR biosynthetic gene products act downstream from the light receptors in Arabidopsis (reviewed by Chory and Li 1997; Clouse and Sasse 1998), and Chory et al (1996) predict "that light negatively regulates BR synthesis or responsiveness in the hypocotyl, while simultaneously promoting BR synthesis or responsiveness in leaf cells." Also, light may modulate not only endogenous biosynthesis of BRs but also their interactions with other hormones, such as cytokinins, which may down-regulate BR biosynthesis via the *CPD* gene (Szekeres et al 1996).

The perception of light signals, their transduction and amplification, and their relationship with hormone-induced responses are most active areas of research, so many more exciting and challenging results can be expected in future. BRs are also predicted to play a role in photoperiodism and endogenous rhythms (Jackson and Thomas 1997), so interesting results can be expected in this area too.

6.3 BRs and Stress Effects

Enhanced yield in response to 22-,23-,24-triepiBL treatment of lettuce in suboptimally fertilized soil was reported by Meudt et al (1983), and Hamada (1986) considered an important action of BL was to enhance resistance to various other stresses, such as cold, fungal infection, herbicide injury, and salt. These responses to exogenous BRs have been confirmed (Takematsu et al 1986), and interest in this aspect of BR research is increasing. Several areas warrant more detailed investigation, not only for their practical application but also for increasing our understanding of the mechanisms of responses to BR.

6.3.1 Temperature Stress

Protective effects of BR treatment on growth, chlorophyll biosynthesis, and membrane stability in chilled rice, maize, and cucumber have been reported (He et al 1991; Katsumi 1991), while in a bromegrass cell culture system there was a slight increase in freezing tolerance after treatment with 24-epiBL that was less than that attainable with abscisic acid. However, the protective effect in heat treatment of the same cell culture was more comparable, but the two regulators differed in their modulation of hsp90 transcript levels (Wilen et al 1995). Kulaeva et al (1991) had also noted promotive effects of two BRs on the synthesis of heat shock proteins and granules in wheat leaves, but germination of moth bean seeds in 24-epiBL solutions, followed by exposure of the young seedlings to 48°C, only led to increased damage, as shown by solute leakage and lipid peroxidation (Upadhyaya et al 1991).

6.3.2 Water Stress

Mild drought stress in sugar beet is ameliorated by treatment with a synthetic BR (Schilling et al 1991), and effects in other crop plants have been explored. Two wheat varieties showed subtle differences in response to water stress; and in the one more sensitive to water stress, treatment with 28-homoBL increased grain yield, relative water content, and soluble protein concentration, while ion leakage was reduced. So it was again suggested that BR treatment enhances membrane stability. Similarly, with juvenile gram and in sorghum, a synergistic beneficial effect was observed with simultaneous 24-epiBL and abscisic acid treatment. Also, the stomatal transpiration rate was diminished (cf. Section 5.4, this chapter) (Singh et al 1993; Sairam 1994; Xu et al 1994a,b).

6.3.3 Salt Stress

As well as the early observations of ameliorative effects of BR treatment in rice growing under conditions approximating 50 mM salt (Hamada 1986, Takematsu et al 1986), Kulaeva et al (1991) reported a protective effect of pretreatment with BR on nuclear and chloroplast ultrastructure in barley leaf segments exposed to 500 mM salt. Germination of *Eucalyptus camaldulensis* seeds was accelerated in 150 mM salt in the presence of 24-epiBL (Sasse et al 1995), but with seedlings grown in salt under hydroponic conditions more damage was noted if 24-epiBL were supplied to the roots. However, when shoots were treated repeatedly, some ameliorative effects were observed, and a significant increase in axillary bud burst was seen (J. Sasse, unpublished).

6.3.4 Other Stresses

Transplantation stress in Pinus radiata seedlings after undercutting and lifting can be prolonged due to air gaps at the root/soil interface when replanted. New roots must be regenerated and good root/soil contact made before the water stress is relieved. Recovery, as shown by enhanced root growth 4 months later, was accelerated after a root soak in 1 µM 24-epiBL immediately after lifting (Sasse and Sands 1992).

Effects on herbicide safening, where BL treatment reduced the damage from treatment of rice with simazine, butachlor, or pretilachlor, while confirmed (Takematsu et al 1986), have not been pursued vigorously. Neither has the protection BL pre-treatment provided against fungal infections, nor where it synergistically enhanced and prolonged the effect of validamycin A (Hamada 1986). However, Szekeres et al (1996) reported that the BR biosynthetic mutant of Arabidopsis, *cpd*, showed very low expression of the pathogenesis-related genes *PR*1, *PR*2, and *PR*5, with significant induction of them if *CPD* were overexpressed in transgenic plants. Transcript levels of some other stress-related genes, such as chalcone synthase, alcohol dehydrogenase, and lipoxygenase, were elevated in the mutant, in contrast to other stress-related genes, which were unaffected. They suggested that the *CPD* function may regulate stress signalling negatively, possibly via modulation of lipoxygenase.

The effect of 24-epiBL on such oxidative lipid degradation has been studied at the biochemical level in pea, and this BR was shown to decrease the levels of breakdown products not only in normally aerated tissue but particularly in conditions of hypoxia or enhanced CO_2 levels. 24-EpiBL-treatment was compared with that of kinetin and found to be more effective (Ershova and Khripach 1996).

Overall, these data suggest that wider ranges of defined conditions should be employed for the exploration of the responses of BR-treated plants to stress, and that generalization from limited data sets may not be warranted. However, the impression that BR treatment has more marked effects on the growth of stressed plants, rather than those growing in optimal conditions, does seem supported; and we can expect further exploration of potential applications of BRs and BR analogs (see the chapter by Y. Kamuro and S. Takatsuto, this volume).

7. Effects on Insect and Fungal Development

Brassinosteroids even have effects in insects; two synthetic (22*S*,23*S*)-28-homoBRs inhibited the moult of the cockroach, and they bound to a preparation of ecdysteroid receptors from *Calliphora vicina*, inhibiting the response to 20-hydroxyecdysterone. Some mild agonist effects were also observed (Richter and Koolmann 1991). However, in the tick *Amblyomma hebraeum* two BRs could not displace ponasterone A from the ecdysteroid receptor but did accelerate salivary gland degeneration induced by ecdysteroids (Charrois et al 1996). In an epithelial cell line of *Chironomus tentans* BRs were cytotoxic (Spindler et al 1992), so interest continues in their potentialities as environmentally friendly insecticides (Luu and Werner 1996).

Acceleration of fungal development by treatment with BRs, particularly in the production of fruiting bodies, has been reported, and an endogenous role was postulated for them (reviewed in Adam et al 1991; Marquardt and Adam 1991). These studies do not seem to have been pursued, and we still do not know whether there are endogenous BRs in fungi. However, it is clear that representative fungi can metabolize BRs and, in at least one case, elaborate a more bioactive compound (Voigt et al 1993a,b).

8. Conclusion and Perspectives

The reports discussed in this and other chapters highlight the pleiotropic effects of BRs in plants and imply crosstalk with other signalling systems, not just of biotic and abiotic stresses and cyclic environmental changes, but also of other hormone-induced responses. They also illustrate that, together with the powerful techniques of molecular genetics, studies at the physiological level are very useful, but that many topics need much more research. Reliable analyses of endogenous BRs in very small tissue samples and subsamples are urgently needed to clarify many questions.

While it is clear BRs can induce de novo gene expression and have some distinct physiological effects, it may be that the results of BR administration in various experimental systems reflect, in many cases, acceleration of the existing developmental program of the particular system studied. Also, it is well recognized that hormones are important as messengers at different levels of organization in the plant. Several authors (e.g., Zucconi 1988) have emphasized the dangers of extrapolation from reductionist model systems where a single effector results in a response that correlates with concentration or sensitivity changes to the much more complex interacting networks of response in the intact plant. There, emergent phenomena such as (seeming) redundancy may develop. Thus, a plea for the use of techniques of systems analysis (Trewavas 1986) for the integration of our knowledge of factors controlling plant growth may well be worth reconsideration. Likewise, the concept of self-organized criticality (Bak 1996) may well provide tools for the plant physiologist interested in integrating experimental and observational data into an understanding of the functioning of plants within their environment. In particular, the suggestion of Barlow (1987) that BRs, as pollen hormones, could serve as a signal exerting "an enabling effect at the highest level of plant organization, the whole plant/community boundary" may also warrant further investigation.

Whatever models of plant growth may be developed in future, the BR story may also serve as a warning that it is unlikely that all the chemical factors influencing plant growth have been discovered, let alone their interactions with each other and endogenous and exogenous physical signals. It is clear that at the molecular, cellular, organ, and whole plant levels the effects of BRs cannot be ignored, and we can expect many more exciting results.

References

Adam G (1994) Brassinosteroide-eine neue PhytohormonGruppe. Naturwissenschaften 81:210-217

Adam G, Marquardt V (1986) Brassinosteroids. Phytochemistry 25:1787-1799

Adam G, Marquardt V, Vorbrodt HM, et al (1991) Aspects of synthesis and bioactivity of brassinosteroids. In: Cutler HG, Yokota T, Adam G (Eds) Brassinosteroids: Chemistry, Bioactivity, and Applications. ACS Symp Series 474. Amer Chem Soc, Washington, DC, pp 74-85

Arteca RN (1995) Brassinosteroids. In: Davies PJ (Ed) Plant Hormones Physiology, Biochemistry and Molecular Biology, 2nd Edition. Kluwer, Dordrecht, pp 206-213

Asakawa S, Abe H, Nishikawa N, et al (1996) Purification and identification of new acyl-conjugated teasterones in lily pollen. Biosci Biotech Biochem 60:1416-1420

Bach TJ, Roth PS, Thompson MJ (1991) Brassinosteroids specifically inhibit growth of tobacco tumor cells. In: Cutler HG, Yokota T, Adam G (Eds) Brassinosteroids: Chemistry, Bioactivity, and Applications. ACS Symp Series 474. Amer Chem Soc, Washington, DC, pp 176-188

Bajguz A, Czerpak R (1996) Effect of brassinosteroids on growth and proton extrusion in the alga *Chlorella vulgaris* Beijerinck (Chlorophyceae). J Plant Growth Regul 15:153-156

Bak P (1996) How nature works: the science of self-organised criticality. Springer-Verlag, New York

Barlow PW (1987) Requirements for hormone involvement in development at different levels of organization. In: Hoad GV, Lenton JR, Jackson MB, et al (Eds) Hormone Action in Plant Development. Butterworth, London, pp 39-51

Braun P, Wild A (1984) The influence of brassinosteroid on growth and parameters of photosynthesis of wheat and mustard plants. J Plant Physiol 116:189-196

Cao H, Chen S (1995) Brassinosteroid-induced rice lamina joint inclination and its relation to indole-3-acetic acid and ethylene. Plant Growth Regul 16:189-196

Charrois GJR, Mao H, Kaufman WR (1996) Impact on salivary gland degeneration by putative ecdysteroid antagonists and agonists in the ixodid tick *Amblyomma hebraeum*. Pesticide Biochem Physiol 55:140-149

Chory J, Li J (1997) Gibberellins, brassinosteroids and light-regulated development. Plant Cell Environ 20:801-806

Chory J, Catterjee M, Cook R, et al (1996) From seed germination to flowering, light controls plant development via the pigment phytochrome. Proc Natl Acad Sci USA 93:12066-12071

Clouse SD, Sasse JM (1998) Brassinosteroids: essential regulators of plant growth and development. Annu Rev Plant Physiol Plant Molec Biol 49: 427-451

Clouse SD, Zurek D (1991) Molecular analysis of brassinolide action in plant growth and development. In: Cutler HG, Yokota T, Adam G (Eds) Brassinosteroids: Chemistry, Bioactivity, and Applications. ACS Symp Series 474. Amer Chem Soc, Washington, DC, pp 122-140

Clouse SD, Zurek DM, McMorris TC, et al (1992) Effect of brassinolide on gene expression in elongating soybean epicotyls. Plant Physiol 100:1377-1383

Clouse SD, Hall AF, Langford M, et al (1993) Physiological and molecular effects of brassinosteroids on *Arabidopsis thaliana*. J Plant Growth Regul 12:61-66

Cutler HG (1991) Brassinosteroids through the looking glass: an appraisal. In: Cutler HG, Yokota T, Adam G (Eds) Brassinosteroids: Chemistry, Bioactivity, and Applications. ACS

Symp Series 474. Amer Chem Soc, Washington, DC, pp 334-345

De Michelis MI, Lado P (1986) Effects of a brassinosteroid on H+-extrusion in isolated radish cotyledons: comparison with the effects of benzyladenine. Physiol Plant 68:603-607

Ershova AN, Khripach VA (1996) Effect of epibrassinolide on lipid peroxidation in *Pisum sativum* at normal aeration and under oxygen deficiency. Russian J Plant Physiol 43:750-752

Eun J-S, Kuraishi S, Sakurai N (1989) Changes in levels of auxin and abscisic acid and the evolution of ethylene in squash hypocotyls after treatment with brassinolide. Plant Cell Physiol 30:807-810

Frances S, Thompson WF (1997) The dark-adaptation response of the de-etiolated pea mutant *lip*1 is modulated by external signals and endogenous programs. Plant Physiol 115:23-28

Fukuda H (1996) Xylogenesis: initiation, progression and cell death. Annu Rev Plant Physiol Plant Molec Biol 47:299-325

Fukuda H (1997) Tracheary element differentiation. Plant Cell 9:1147-1156

Gregory LE Mandava NB (1982) The activity and interaction of brassinolide and gibberellic acid in mung bean epicotyls. Physiol Plant 54:239-243

Grove MD, Spencer GF, Rohwedder WK, et al (1979) Brassinolide, a plant growth-promoting steroid isolated from *Brassica napus* pollen. Nature 281: 216-217

Hamada K (1986) Brassinolide in crop cultivation. In: McGregor P (Ed) Plant Growth Regulators in Agriculture. FFTC, Taiwan, pp 190-196

Hata S, Takagishi H, Egawa Y, et al (1986) Effects of compactin, a 3-hydroxy-3-methylglutaryl coenzyme A reductase inhibitor, on the growth of alfalfa (*Medicago sativa*) and seedlings and the rhizogenesis of pepper (*Capsicum annuum*). Plant Growth Regul 4:335-346

He R, Wang G, Wang X (1991) Effects of brassinolide on growth and chilling resistance of maize seedlings. In: Cutler HG, Yokota T, Adam G (Eds) Brassinosteroids: Chemistry, Bioactivity, and Applications. ACS Symp Series 474. Amer Chem Soc, Washington, DC, pp 220-230

He Y-J, Xu R-J, Zhao Y-J (1996) Enhancement of senescence by epibrassinolide in leaves of mung bean seedling. Acta Phytophysiol Sin 22:58-62

Hewitt FR, Hough T, O'Neill P, et al (1985) Effect of brassinolide and other growth regulators on the germination and growth of pollen tubes of *Prunus avium* using a multiple hanging drop assay. Aust J Plant Physiol 12:201-211

Iwahori S, Tominaga S, Higuchi S (1990) Retardation of abscission of citrus leaf and fruitlet explants by brassinolide. Plant Growth Regul 9:119-125

Iwasaki T, Shibaoka H (1991) Brassinosteroids act as regulators of tracheary-element differentiation in isolated *Zinnia* mesophyll cells. Plant Cell Physiol 32:1007-1014

Jackson S, Thomas B (1997) Photoreceptors and signals in the photoperiodic control of development. Plant Cell Environ 20:790-795

Jones-Held S, VanDoren M, Lockwood T (1996) Brassinolide application to *Lepidum sativum* seeds and the effects on seedling growth. J Plant Growth Regul 15:63-67

Kalinich JF, Mandava NB, Todhunter JA (1985) Relationship of nucleic acid metabolism to brassinolide-induced responses in beans. J Plant Physiol 120:207-214; 125:345-353

Kamuro Y, Inada K (1987) Effect of light conditions on brassinolide induced mung bean epicotyl elongation and radish growth. Proc Plant Growth Regul Soc Amer 14:221-223

Katsumi M (1991) Physiological modes of brassinolide action in cucumber hypocotyl growth. In: Cutler HG, Yokota T, Adam G (Eds) Brassinosteroids: Chemistry, Bioactivity, and

Applications. ACS Symp Series 474. Amer Chem Soc, Washington, DC, pp 246-254
Katsumi M, Tsuda A, Sakurai H (1987) Brassinolide-induced stimulation of membrane permeability and ATPase activity in light-grown cucumber hypocotyls. Proc Plant Growth Regul Soc Amer 14:215-220
Kauschmann A, Jessop A, Koncz C, et al (1996) Genetic evidence for an essential role of brassinosteroids in plant development. Plant J 9:701-713
Kawaguchi M, Imaizumi-Anraku H, Fukai S, et al (1996) Unusual branching in the seedlings of *Lotus japonicus*-gibberellins reveal the nitrogen-sensitive cell divisions within the pericycle on roots. Plant Cell Physiol 37:461-470
Khripach VA, Zhabinskii VN, Lakhvich FA (1993) Brassinosteroids. Nauka i Tecknika, Minsk
Kitani Y (1994) Induction of parthenogenetic haploid plants with brassinolide. Jpn J Genet 69:35-39
Krizek, DT, Mandava, NB (1983a) Influence of spectral quality on the growth response of intact bean plants to brassinosteroid, a growth-promoting lactone. I. Stem elongation and morphogenesis. Physiol Plant 57:317-323
Krizek, DT, Mandava, NB (1983b) Influence of spectral quality on the growth response of intact bean plants to brassinosteroid, a growth-promoting lactone. II. Partitioning of assimilate and chlorophyll content. Physiol Plant 57:324-329
Krizek DT, Worley JF (1973) The influence of light intensity on the internodal response of intact bean plants to brassins. Bot Gaz 134:147-150
Kulaeva ON, Burkhanova EA, Fedina AB, et al (1991) Effect of brassinosteroids on protein synthesis and plant-cell ultrastructure under stress conditions. In: Cutler HG, Yokota T, Adam G (Eds) Brassinosteroids: Chemistry, Bioactivity, and Applications. ACS Symp Series 474. Amer Chem Soc, Washington, DC, pp 141-157
Kuno K (1997) Effects of plant growth steroid brassinolide, on dry-weight growth and nutrient translocation in mulberry shoots. Nippon Sanshigaku Zasshi (J Sericultural Sci Jpn) 66:57-58
Li J, Chory J (1997) A putative leucine-rich repeat receptor kinase involved in brassinosteroid signal transduction. Cell 90:929-938
Li J, Nagpal P, Vitart V, et al (1996) A role for brassinosteroids in light-dependent development of *Arabidopsis*. Science 272:398-401
Luu B, Werner F (1996) Sterols that modify moulting in insects. Pestic Sci 46:49-53
Mandava NB (1988) Plant growth-promoting brassinosteroids. Annu Rev Plant Physiol Plant Molec Biol 39:23-52
Mandava NB, Thompson MJ (1984). Chemistry and functions of brassinolide. In: Nes WD, Fuller G, Tsai L-S (Eds) Isopentenoids in plants. Biochemistry and functions. Marcel Dekker, New York. pp 401-431
Mandava NB, Sasse JM, Yopp JH (1981) Brassinolide, a growth-promoting steroidal lactone II. Activity in selected gibberellin and cytokinin bioassays. Physiol Plant 53:453-461
Marquardt V, Adam G (1991) Recent advances in brassinosteroid research. In: Boerner H, Martin, D, Sjut V, et al (Eds) Chemistry of Plant Protection, Vol. 7: Herbicide Resistance - Brassinosteroids, Gibberellins, Plant Growth Regulators. Springer-Verlag, Berlin, pp 103-139
Mayumi K, Shibaoka H (1995) A possible double role for brassinolide in the re-orientation of cortical microtubules in the epidermal cells of Azuki bean epicotyls. Plant Cell Physiol 36:173-181
Mayumi K, Shibaoka H (1996) The cyclic reorientation of cortical microtubules on walls with a crossed polylamellate structure: effects of plant hormones and an inhibitor of protein

kinases on the progression of the cycle. Protoplasma 195:112-122

Meudt WJ (1987) Chemical and biological aspects of brassinolide. In: Fuller G, Nes WD (Eds) Ecology and Metabolism of Plant Lipids. ACS Symp Ser 325. Amer Chem Soc, Washington, DC, pp 53-75

Meudt WJ, Thompson MJ, Bennett HW (1983) Investigations on the mechanism of the brassinosteroid response. III. Techniques for potential enhancement of crop production. Proc Plant Growth Regul Soc Amer 10:312-318

Miller MM (1972) Studies on the characterisation of the effect of brassin utilizing the coleoptile and total development cycle of *Avena sativa* L. M.Sc. Thesis, University of Southern Illinois at Carbondale, Illinois, USA

Mitchell JW, Whitehead MR (1941) Responses of vegetative parts of plants following application of extract of pollen from *Zea mays*. Bot Gaz 102:770-790

Mustilli AC, Bowler C (1997) Tuning in to the signals controlling photoregulated gene expression in plants. EMBO J 16:5801-5806

Nakajima N, Shida A, Toyama S (1996) Effects of brassinosteroid on cell division and colony formation of Chinese cabbage mesophyll protoplasts. Jpn J Crop Sci 65:114-118

Nishikawa N, Toyama S, Shida A, et al (1994) The uptake and transport of ^{14}C-labelled epibrassinolide in intact seedlings of cucumber and wheat. J Plant Res 107:125-130

Nomura T, Nakayama M, Reid JB, et al (1997) Blockage of brassinosteroid synthesis and sensitivity causes dwarfism in *Pisum sativum*. Plant Physiol 113:31-37

Oh M-H, Romanov WG, Smith RC, et al (1998) Soybean *BRU*1 encodes a functional xyloglucan endotransglycosylase that is highly expressed in inner epicotyl tissues during brassinosteroid-promoted elongation. Plant Cell Physiol 39:124-130

Richter K, Koolman J (1991) Antiecdysteroid effects of brassinosteroids in insects. In: Cutler HG, Yokota T, Adam G (Eds) Brassinosteroids: Chemistry, Bioactivity, and Applications. ACS Symp Series 474. Amer Chem Soc, Washington, DC, pp 265-279

Roddick JG, Guan M (1991) Brassinosteroids and root development. In: Cutler HG, Yokota T, Adam G (Eds) Brassinosteroids: Chemistry, Bioactivity, and Applications. ACS Symp Series 474. Amer Chem Soc, Washington, DC, pp 231-245

Roddick JG, Ikekawa N (1992) Modification of root and shoot development in monocotyledon and dicotyledon seedlings by 24-epibrassinolide. J Plant Physiol 140:70-74

Roddick JG, Rijnenberg AL, Ikekawa N (1993) Developmental effects of 24-epibrassinolide in excised roots of tomato grown in vitro. Physiol Plant 87:453-458

Ronsch, H, Adam G, Matsche J, et al (1993) Influence of (22*S*,23*S*)-homobrassinolide on rooting capacity and survival of adult Norway spruce cuttings. Tree Physiol 12:71-80

Ronsch H, Adam G, Voigt B (1996) Retardation of needle chloroses by brassinosteroids in magnesium-deficient seedlings of Norway spruce. Proc Plant Growth Regul Soc Amer 23:62

Sairam RK (1994) Effects of homobrassinolide application on plant metabolism and grain yield under irrigated and moisture-stress conditions of two wheat varieties. Plant Growth Regul 14:173-181

Sakurai A, Fujioka S (1993) The current status of physiology and biochemistry of brassinosteroids. Plant Growth Regul 13:147-159

Sala C, Sala F (1985) Effect of brassinosteroid on cell division and enlargement in cultured carrot (*Daucus carota* L.) cells. Plant Cell Rep 4:144-147

Sasse JM (1985) The place of brassinolide in the sequential response to plant growth regulators in elongating tissue. Physiol Plant 63:303-308

Sasse JM (1987) Effects of brassinolide and other natural plant growth regulators on the morphology of pea stem tissue. Proc Plant Growth Regul Soc Amer 14:30-39

Sasse JM (1989) Using PEST to study the interactions of brassinolide and other natural plant growth regulators. Proc Plant Growth Reg Soc Amer 16:82-87

Sasse JM (1990) Brassinolide-induced elongation and auxin. Physiol Plant 80:401-408

Sasse JM (1991a) The case for brassinosteroids as endogenous plant hormones. In: Cutler HG, Yokota T, Adam G (Eds) Brassinosteroids: Chemistry, Bioactivity, and Applications. ACS Symp Series 474. Amer Chem Soc, Washington, DC, pp 158-166

Sasse JM (1991b) Brassinolide-induced elongation. In: Cutler HG, Yokota T, Adam G (Eds) Brassinosteroids: Chemistry, Bioactivity, and Applications. ACS Symp Series 474. Amer Chem Soc, Washington, DC, pp 255-264

Sasse JM (1997) Recent progress in brassinosteroid research. Physiol Plant 100:696-701

Sasse JM, Sands R (1992) Brassinosteroids and transplantation stress. Proc Plant Growth Regul Soc Amer 19:135-138

Sasse JM, Sasse JM (1994) Brassinosteroids and roots. Proc Plant Growth Regul Soc Amer 21:228-232

Sasse JM, Smith R, Hudson I (1995) Effect of 24-epibrassinolide on germination of seeds of *Eucalyptus camaldulensis* in saline conditions. Proc Plant Growth Regul Soc Amer 22:136-141

Sathimoorthy P, Nakamura S (1990) In vitro root induction by 24-epibrassinolide on hypocotyl segments of soybean (*Glycine max* (L) Merr.). Plant Growth Regul 9:73-76

Schilling G, Schiller C, Otto S (1991) Influence of brassinosteroids on organ relations and enzyme activities of sugar-beet plants. In: Cutler HG, Yokota T, Adam G (Eds) Brassinosteroids: Chemistry, Bioactivity, and Applications. ACS Symp Series 474. Amer Chem Soc, Washington, DC, pp 208-219

Schmidt J, Yokota T, Adam G, et al (1991) Castasterone and brassinolide in *Raphanus sativus* seeds. Phytochemistry 30:364-365

Singh J, Nakamura Sota Y (1993) Effect of epibrassinolide on gram (*Cicer arietinum*) plants grown under water stress in the juvenile stage. Indian J Agric Sci 63:395-397

Spindler K-D, Spindler-Barth M, Turberg A (1992) Action of brassinosteroids on epithelial cell line from *Chironomus tentans*. Z Naturforsch 47c:280-284

Steffens GL (1991) U.S. Department of Agriculture brassins project. In: Cutler HG, Yokota T, Adam G (Eds) Brassinosteroids: Chemistry, Bioactivity, and Applications. ACS Symp Series 474. Amer Chem Soc, Washington, DC, pp 2-17

Suge H (1986) Reproductive development of higher plants as influenced by brassinolide. Plant Cell Physiol 27:199-205

Sugiyama K, Kuraishi S (1989) Stimulation of fruit set of 'Morita' navel orange with brassinolide. Acta Hort 239:345-348

Szekeres M, Nemeth K, Koncz-Kalman Z, et al (1996) Brassinosteroids rescue the deficiency of CYP90, a cytochrome P450, controlling cell elongation and de-etiolation in *Arabidopsis*. Cell 85:171-182

Takatsuto S, Yazawa N, Ikekawa N, et al (1983) Structure-activity relationship of brassinosteroids. Phytochemistry 22:2437-2441

Takematsu T, Takeuchi Y, Choi CD (1986) Overcoming effects of brassinosteroids on growth inhibition of rice caused by unfavourable growth conditions. Shokucho (a journal from Japan Assoc Adv Phytoreg) 20:2-12

Takeno K, Pharis RP (1982) Brassinosteroid-induced bending of the leaf lamina of dwarf rice seedlings: an auxin mediated phenomenon. Plant Cell Physiol 23:1275-1281

Takeuchi Y, Ogasawara M, Konnai M, et al (1992) Application of brassinosteroids in agriculture in Japan. Proc Plant Growth Regul Soc Amer 19:343-352

Takeuchi Y, Omigawa Y, Ogasawara M, et al (1995) Effects of brassinosteroids on conditioning and germination of clover broomrape (*Orobanche minor*) seeds. Plant Growth Regul 16:153-160

Taylor PE, Spuck K, Smith PM, et al (1993) Detection of brassinosteroids in pollen of *Lolium perenne* L. by immunocytochemistry. Planta 189:91-100

Tominaga R, Sakurai N, Kuraishi S (1994) Brassinolide-induced elongation of inner tissues of segments of squash (*Cucurbita maxima* Duch.) hypocotyls. Plant Cell Physiol 35:1103-1106

Trewavas A (1986) Understanding the control of plant development and the role of growth substances. Aust J Plant Physiol 13:447-457

Upadhyaya A, Davis TD, Sankhla N (1991) Epibrassinolide does not enhance heat shock tolerance and antioxidant activity in moth bean. Hort Sci 26:1065-1067

Voigt B, Porzel A, Naumann H, et al (1993a) hydroxylation of the native brassinosteroids 24-epicastasterone and 24-epibrassinolide by the fungus *Cunninghamella echinulata*. Steroids 58:320-323

Voigt B, Porzel A, Undisz, K, et al (1993b) Microbial hydroxylation of 24-epicastasterone by the fungus *Cochliobolus lunatus*. Nat Prod Lett 3: 123-129

Wada K, Marumo S, Abe H, et al (1984) A rice lamina inclination test - a micro-quantitative bioassay for brassinosteroids. Agric Biol Chem 48:719-726

Wada K, Kondo H, Marumo S (1985) A simple bioassay for brassinosteroids: a wheat leaf-unrolling test. Agric Biol Chem 49:2249-2251

Wilen RW, Sacco M, Gusta LV, et al (1995) Effects of 24-epibrassinolide on freezing and thermotolerance of bromegrass (*Bromus inermis*) cell cultures. Physiol Plant 95:195-202

Worley JF, Mitchell JW (1971) Growth responses induced by brassins (fatty plant hormones) in bean plants. J Am Soc Hort Sci 96:270-273

Xu HL, Shida A, Futatsuya F, et al (1994a) Effects of epibrassinolide and abscisic acid on sorghum plants growing under soil water deficit. I. Effects on growth and survival. Jpn J Crop Sci 63:671-675

Xu HL, Shida A, Futatsuya F, et al (1994b) Effects of epibrassinolide and abscisic acid on sorghum plants growing under soil water deficit. II. Physiological basis for drought resistance induced by exogenous epibrassinolide and abscisic acid. Jpn J Crop Sci 63:676-681

Yamamoto R, Demura T, Fukuda H (1997) Brassinosteroids induce entry into the final stage of tracheary element differentiation in cultured *Zinnia* cells. Plant Cell Physiol 38:980-983

Yokota T (1997) The structure, biosynthesis and function of brassinosteroids. Trends Plant Sci 2:137-143

Yokota T, Takahashi N (1986) Chemistry, physiology and agricultural application of brassinolide and related steroids. In: M. Bopp (Ed) Plant Growth Substances 1985. Springer-Verlag, Berlin, pp 129-138

Yopp JH, Colclasure GC, Mandava NB (1979) Effects of brassin-complex on auxin and gibberellin mediated events in the morphogenesis of the etiolated bean hypocotyl. Physiol Plant 46:247-254

Yopp JH, Mandava NB, Sasse JM (1981a) Brassinolide, a growth-promoting steroidal lactone I. Activity in selected auxin bioassays. Physiol Plant 53:445-452

Yopp JH, Mandava NB, Thompson MJ, et al (1981b) Activity of brassinosteroid in selected bioassays in combination with chemicals known to synergize or retard responses to auxin and gibberellin. Proc Plant Growth Regul Soc Amer 8:138-145

Zhao Y-J, Xu R-J, Luo W-H (1990) Inhibitory effects of abscisic acid on epibrassinolide-induced senescence of detached cotyledons in cucumber seedlings. Chin Sci Bull 35:928-931

Zucconi F (1988) Epigenetic regulation in plants. Israel J Bot 37:131-144

8
Molecular Genetics of Brassinosteroid Action

STEVEN D. CLOUSE[1] AND KENNETH A. FELDMANN[2]

[1] Department of Horticultural Science, Box 7609
North Carolina State University, Raleigh, NC 27695, USA
[2] Department of Plant Sciences
University of Arizona, Tucson, AZ 85271, USA

1. Introduction

Physiological and chemical analyses have been critical in understanding the activity of plant hormones and will continue to play an essential role in these studies. During the past decade the application of molecular genetics has complemented physiological and biochemical approaches and has resulted in significant advances in dissecting the mechanisms of action of the major plant hormones. Numerous hormone-regulated genes have been cloned, and in many instances the regulatory sequences responsible for hormone modulation of gene expression have been identified (Deikman 1997, Hagen 1995). Transgenic plants with altered endogenous hormone levels have been generated to test predictions of physiological experiments using exogenously applied hormones (Klee and Estelle 1991), and hormone-deficient mutants with lesions in genes encoding biosynthetic enzymes have been combined with microchemical techniques, such as gas chromatography/mass spectroscopy (GC/MS), to elucidate several pathways of hormone biosynthesis (e.g., Sponsel 1995). The discovery and characterization of hormone-insensitive mutants has led to the identification of ethylene receptors in Arabidopsis and tomato, in addition to other essential components of the signal transduction pathway (Ecker 1995). The use of similar molecular genetic approaches in investigating brassinosteroid (BR) action has brought impressive results in a relatively short time.

Based on their widespread distribution in the plant kingdom and diverse physiological effects at nanomolar concentrations, most early researchers in the field argued that BRs should be considered essential endogenous regulators of plant growth

Key words: brassinosteroids, brassinolide, BR insensitive mutants, BR deficient mutants, signal transduction, biosynthesis, gene regulation, receptor kinase, *DWF*, *BRI*, *CPD*, *DET*2, *LK*, *DPY*, *Arabidopsis thaliana*

and development (Mandava 1988, Sasse 1991). The phenotype of recently identified BR-deficient and BR-insensitive mutants, all showing extreme dwarfism, altered leaf morphology, reduced male fertility, and de-etiolation of dark-grown seedlings, also suggested that BR activity was required for normal growth (Clouse et al 1996, Kauschmann et al 1996, Li et al 1996, Nomura et al 1997, Szekeres et al 1996). The observations that only BRs could rescue the deficient mutants to wildtype and that the insensitive mutant could respond to all other hormones except BR provided convincing genetic evidence that BRs are indeed essential for normal plant growth and development and must be considered along with auxin, cytokinins, gibberellins (GAs), abscisic acid, and ethylene in any comprehensive model of plant growth. The proliferation of recent reviews on BR action indicates a renewed general interest in these steroids as signaling molecules regulating plant development (Clouse and Sasse 1998, Clouse 1996, Clouse 1997, Ecker 1997, Fujioka and Sakurai 1997, Hooley 1996, Sasse 1997, Yokota 1997). This chapter examines the molecular genetic studies that have contributed to our understanding of BR biosynthesis and signal transduction, as well as the role of BR-regulated gene expression in developmental programs such as stem elongation and vascular differentiation.

2. Cloning of BR-Regulated Genes

While plant and animal hormones can directly affect physiological processes such as ion channel activity and phosphorylation states of structural proteins and metabolic enzymes (Blatt and Thiel 1993), the primary initial mode of action of many hormones is the modulation of specific gene expression via transcriptional and post-transcriptional mechanisms. BRs have proven to be no exception to this common rule. Early work with inhibitors of RNA and protein synthesis showed that continued RNA transcription and de novo protein synthesis were required for BR-promoted stem elongation (Mandava et al 1987), and Kulaeva et al (1991) found that BRs promoted de novo synthesis of numerous polypeptides under stress conditions. BR effects on specific mRNAs were demonstrated by two-dimensional gel electrophoresis of in vitro translated mRNA, with BR treatment of soybean stem segments, Arabidopsis seedlings, and Arabidopsis peduncle segments, resulting in altered transcript levels for over 50 different genes (Clouse and Zurek 1991, Clouse et al 1992). The cloning of cDNAs corresponding to mRNA transcripts whose levels are regulated by BR during elongation, stress response, and vascular differentiation have now been accomplished. *BRU*1 (for BR Upregulated no.1) was isolated by differential screening of elongating soybean epicotyls treated with or without 0.1 µM brassinolide, and enhanced expression of *BRU*1 was shown to be specific to BRs during the early phases of elongation (Zurek and Clouse 1994). Subsequent work (Oh et al 1998) showed that *BRU*1 encoded a xyloglucan endotransglycosylase (XET). The *TCH*4 gene of Arabidopsis (Xu et al 1995) along with the *LeXET* gene of tomato (Catala et al 1997) have been shown to be BR-regulated XETs, although both of these genes are also regulated by auxin. Dual regulation of mung bean ACC synthase mRNA levels by BR and auxin has also been reported (Arteca 1995). BRs have been shown to increase mRNA levels of heat shock protein 90 in bromegrass cell suspension cultures (Wilen et al 1995) and to affect the expression of genes controlling late stages of tracheary element differentiation of *Zinnia elegans* mesophyll cells, including those encoding phenylalanine ammonia lyase, cinnamic acid 4-hydroxylase, and a cysteine protease (Yamamoto et al 1997). The molecular mechanisms of BR-regulated gene expression are beginning to be addressed, using initially the *BRU1* and *TCH4* genes as models of posttranscriptional and transcriptional regulation, respectively.

2.1 *BRU*1 and Posttranscriptional Regulation

Brassinosteroids, when applied at 10 nM to 1 µM, promote the elongation of apical epicotyl segments of young, light-grown soybean seedlings in the absence of exogenous auxin (Clouse et al 1992). Differential colony hybridization of approximately 10,000 clones from a BR-treated soybean epicotyl cDNA library yielded 12 clones that were putatively BR-regulated. Further screening revealed that 11 of these were also auxin-regulated, while one, *BRU1*, appeared to be specifically regulated by BR. RNA transcript analysis of *BRU1* confirmed that while auxins and GAs could promote soybean epicotyl elongation only active BRs increased *BRU1* transcript levels (Zurek and Clouse 1994). This indicates that increased *BRU1* expression is not simply the result of elongation per se, but rather that the *BRU1* message increases in response to BR treatment. RNase protection assays were also used to monitor *BRU1* transcript levels in different tissues and developmental stages of soybean. The highest expression was observed in 14-day-old stem tissue, with epicotyl expression (both apical and basal) slightly exceeding that of hypocotyl. *BRU1* transcript levels in 14- and 28-day primary leaves were less than 7% of the level in 14-day epicotyls (Zurek and Clouse 1994). The *BRU1* cDNA hybridized to an mRNA of approximately 1,050 nucleotides that contained an open reading frame of 283 amino acids with a hydrophobic core of 21 contiguous amino acids near the *N*-terminus suggestive of a signal peptide. Sequence analysis revealed that *BRU1* had extensive homology to numerous XETs, and recombinant BRU1 protein was able to catalyze the incorporation of radiolabeled xyloglucan oligosaccharides into high molecular weight xyloglucan, implicating BRU1 as an active XET (Oh et al 1998).

The most common method used to distinguish between transcriptional and posttranscriptional levels of eukaryotic gene control is run-on transcription in isolated nuclei (Darnell 1982). When nuclei isolated from soybean epicotyls treated with or without 0.1 µM BR for various times were used for run-on transcription assays, it was evident that BR did not cause transcriptional induction of *BRU1* at any time examined (Zurek and Clouse 1994). Therefore, BR appears to be regulating *BRU1* expression at a posttranscriptional level. While the most common mechanism for gene regulation by steroid hormones in animals is transcriptional activation of specific genes that contain hormone-responsive elements in their promoters (Beato et al 1995), posttranscriptional mechanisms have been reported. For example, Brock and

Shapiro (1983) showed that estrogen stabilizes vitellogenin mRNA against cytoplasmic degradation, and Nielsen and Shapiro (1990) found that the untranslated region of the vitellogenin mRNA was responsible for estradiol-dependent mRNA stabilization in the presence of the estrogen receptor. Posttranscriptional modulation of c-*myc* expression by estrogen has also been reported (Santos et al 1988). In plants, abscisic acid, cytokinin, and ethylene all have been shown to stabilize specific mRNAs (Gallie 1993). It is also notable that the 3' untranslated region of the *BRU*1 message contains an AUUUUA motif that is similar to the AUUUA motif known to destabilize transcripts in both plants and animals (Green 1993). A working hypothesis is that BR might alter *BRU*1 mRNA stability via a *trans*-acting factor that binds to *cis*-acting sequences in the *BRU*1 message. One possibility that has precedence in other systems is that the *BRU*1 message is destabilized by a specific ribonuclease that recognizes a sequence in the 3' untranslated region. In the presence of elevated BR levels, the sequence is blocked by another protein, or the level or activity of the ribonuclease is altered. The investigation of this and other mechanisms of posttranscriptional regulation of *BRU*1 by BR is currently underway (J Jiang, S Clouse, unpublished results).

2.2 *TCH*4, Transcriptional Regulation and the BR Response Element

The Arabidopsis *TCH*4 gene is currently the primary model system for examining transcriptional regulation of gene expression by BRs. *TCH*4 was originally isolated in a screen for touch-inducible genes and, like *BRU*1, was subsequently shown by sequence analysis and assays of recombinant TCH4 protein to encode an XET. *TCH*4 is strongly expressed in expanding tissues, particularly in dark-grown hypocotyls and in organs that undergo cell wall modification such as vascular elements (Xu et al 1995). Besides mechanical stimulation, *TCH*4 also responds to changes in light and temperature and to hormones. In contrast to *BRU*1, both auxin and BR regulate *TCH*4 expression, but with quite different kinetics. Auxin treatment results in an increase in *TCH*4 expression within 10 minutes with a maximum at 30 minutes, while BR treatment shows detectable increases after 30 minutes with a maximum at 2 hours (Xu et al 1995). Besides *TCH*4, there are at least seven other XET or XET-related sequences in Arabidopsis, but only a subset of these are BR-regulated (Xu et al 1996).

In contrast to *BRU*1, *TCH*4 appears to be transcriptionally regulated by BR since a promoter:β-glucuronidase (GUS) fusion of the Arabidopsis *TCH*4 gene lacking any 5' untranslated sequence (*TCH*4:GUS) gave patterns of BR-regulated GUS expression similar to that observed for *TCH*4 expression in BR-treated nontransgenic plants (J Braam, personal communication). The BR response element is being pursued in a -958 to -1 promoter fragment of *TCH*4 by generating an ordered set of 5' deletions using exonuclease III/mung bean nuclease followed by ligation of the promoter deletions into the binary vector pBI101 containing the GUS reporter gene.

These constructs have been vacuum infiltrated into Col-0 Arabidopsis plants, and T3 homozygous lines are being analyzed for GUS expression by histochemical staining and Northern analysis in the presence and absence of BR (R Torisky, J Braam, S Clouse, unpublished results). Results to date indicate that the BR response lies between -128 and -288, and a linker scanning mutagenesis experiment is underway to more precisely define the sequence of the *cis*-acting element that is required for transcriptional activation of the *TCH*4 gene by BR. Once the BR response element is in hand, a series of highly informative experiments become possible, including studying transcription factors that bind to the response element (see below), and creating yeast strains with the response element driving a minimal promoter-reporter gene construct. This then becomes a tool into which various candidates for the BR receptor/signal transduction components can be introduced and their activity monitored, reminiscent of the elegant experiments performed in uncovering the mechanism of mammalian steroid hormone receptors (Evans 1988).

3. Molecular Mechanisms of BR-Regulated Cell Expansion

Cell expansion is critical for growth, differentiation, and morphogenesis in all plant organs. The control of cell expansion requires coordinated alterations in wall mechanical properties, cell hydraulics, biochemical processes, and gene expression (Cosgrove 1997). Current views of primary cell walls in dicotyledenous and non-Poaceae monoctoyledenous plants hold that cellulose microfibrils are tethered into a network by noncovalent attachment to hemicelluloses (primarily xyloglucans); and this network, in turn, is embedded in a pectic gel matrix (Carpita and Gibeaut 1993). For turgor-driven cell expansion to proceed, the cell wall must transiently yield by slippage or breakage of the hemicellulose tethers. Incorporation of new wall polymers must proceed concomitantly with cell expansion to prevent thinning and weakening of the walls. Thus, the possible hormonal regulation of the synthesis and activity of enzymes involved in wall modification and biosynthesis, such as XETs, glucanases, expansins, sucrose synthase, and cellulose synthase, becomes an obvious model to explain the molecular basis of hormone-modulated cell elongation. In support of this model, BR regulation of genes encoding XETs and expansins has now been demonstrated (Clouse 1997), and BRs have been shown by biophysical measurements to promote wall loosening in soybean epicotyls (Zurek et al 1994) and *Brassica chinensis* and *Cucurbita maxima* hypocotyls (Tominaga et al 1994, Wang et al 1993).

4. Molecular Mechanisms of BR-Modulated Cell Differentiation

The differentiation of vascular tissue, particularly the xylem, has been extensively studied for several decades, and an essential role for both auxins and cytokinins in this process has been clearly demonstrated (Aloni 1995). Experiments have shown that BRs are also likely to be critically involved in xylogenesis, both in planta and in the two major in vitro model systems used to study this developmental pathway. When tuber explants of *Helianthus tuberosus* are cultured with auxin and cytokinin, approximately 30% of the parenchyma cells differentiate into tracheary elements after 72 hours. The addition of nanomolar concentrations of brassinolide at the start of culture results in a 10-fold increase in the number of tracheary elements present at 24 hours, suggesting that brassinolide accelerates the rate of auxin- and cytokinin-promoted differentiation (Clouse and Zurek 1991). An acceleration of tracheary element differentiation in the presence of auxin and cytokinin is also seen when 0.2 nM brassinolide is added to isolated mesophyll cells of *Zinnia elegans* (Iwasaki and Shibaoka 1991). Moreover, uniconazole (an inhibitor of GA and BR biosynthesis) completely blocks tracheary element differentiation in *Zinnia elegans*, which is overcome by BR but not by GA treatment.

Zinnia elegans mesophyll cells have proven to be an excellent model system for identifying biochemical and molecular markers associated with the well-known morphological changes that accompany xylem differentiation, and the overall process has been divided into three major stages (reviewed in Fukuda 1997). Stage I is a dedifferentiation process that is initiated by wounding, auxin, and cytokinin. During Stage I, actin filaments are rearranged and tubulin gene expression and tubulin synthesis are elevated. Wound induction of genes encoding phenylpropanoid biosynthetic enzymes such as phenylalanine ammonia lyase (*ZePAL1*, *ZePAL2* and *ZePAL3*) and cinnamate-4-hydroxylase (*ZC4H*) is also observed. Stage II commences after 24 to 36 hours of culture and results in restriction of developmental potential similar to that seen when meristimatic cells differentiate into procambial initials. Stage II is associated with increases in certain subcellular organelles and microtubules, heightened calmodulin and calmodulin-binding protein activity, and the expression of Stage II-specific genes such as *TED2*, *TED3*, *TED4*, and *ZCAD1*. Phenylpropanoid pathway gene expression abates during Stage II. The transition between Stage II and Stage III (which occurs 48 to 54 hours after the initiation of culture) represents an irreversible checkpoint in tracheary element differentiation. During Stage III secondary wall synthesis predominates, and the phenylpropanoid pathway genes *ZePAL1*-*ZePAL3* and *ZC4H* are again induced, this time for lignin biosynthesis. A variety of genes involved in programmed cell death are also induced (e.g., the cysteine protease *ZCP4*), and Stage III terminates with autolysis.

Recent evidence strongly supports the hypothesis that endogenous BRs are essential for the transition from Stage II to Stage III in *Z. elegans* tracheary element differentiation (Yamamoto et al 1997). Uniconazole, as mentioned above, inhibits BR biosynthesis and prevents tracheary element differentiation. However, uniconazole had no effect on genes expressed specifically in Stage I or Stage II but completely abolished expression of those genes induced during Stage III. Moreover, in the presence of uniconazole the phenylpropanoid pathway genes were expressed normally during Stage I but failed to turn on during Stage III. Finally, BR rescued Stage III gene expression in the presence of uniconazole and allowed lignification and autolysis to proceed. This is the first report of BRs controlling a specific step in differentiation by the regulation of gene expression.

Evidence is also available suggesting that BRs play a role during xylem differentiation in planta. Studies of the spatial expression of *BRU*1 in developing soybean epicotyls revealed a marked expression of this BR-regulated XET gene in paratracheary parenchyma cells surrounding vessel elements (Oh et al 1998). Moreover, *TCH*4, the BR-regulated XET in Arabidopsis, showed significant expression in the vascular tissue of roots and shoots (Xu et al 1995). XETs have previously been proposed to play a role in vascular differentiation (Fry et al 1992), and the *BRU*1 and *TCH*4 studies support this hypothesis. The abnormalities in vascular differentiation observed in the BR-deficient *cpd* mutant (Szekeres et al 1996) and the identification of endogenous BRs in cambial scrapings of *Pinus silvestris* (Kim et al 1990) and a *Eucalyptus* species (T Yokota, unpublished results) lend further support to the involvement of BRs in xylem differentiation in vivo.

5. BR Biosynthesis Mutants

The identification of mutants at specific steps in biochemical and developmental pathways remains one of the most powerful means that molecular geneticists have at their disposal for dissecting these complex pathways. This strategy has turned out to be an especially expeditious route for identifying components of the brassinosteroid biosynthetic pathway. All of the mutants placed in this pathway have to date possessed a dwarf growth habit. BR dwarfs are distinguished from GA dwarfs in that the latter dwarfs tend to possess an inflorescence that is shorter than BR dwarfs; and, most importantly, many of the GA dwarfs fail to germinate in the absence of exogenously supplied GA. It is thus relatively easy to identify a collection of putative BR dwarfs and, by testing them with intermediates in the BR pathway, pinpoint a step at which the mutant is defective. In this way, a relatively large number of dwarfs have been mapped to the BR pathway.

Several research groups have been involved in the identification of BR dwarfs in Arabidopsis. The largest collection of mutant loci has been isolated by Feldmann and colleagues, which they refer to as *dwf*1-*dwf*8 (*dwarf*). Takahashi et al (1995) isolated an allele of *dwf*1 that was named *dim*1 (*diminuto*), while Kauschmann et al (1996) characterized three dwarf loci that were referred to as *cbb*1-*cbb*3 (*cabbage*), with *cbb*1 again being allelic to *dwf*1. Szekeres et al (1996) identified a BR-rescued dwarf that they referred to as *cpd* (*constitutive photomorphogenesis and dwarfism*), and Li et al (1996) isolated a BR dwarf in a screen for de-etiolation mutants (*det*2; *de-etiolated*). Clouse et al (1996) characterized a BR-insensitive mutant, *bri*1

Table 1. Designations for BR dwarf mutants of *Arabidopsis thaliana*

Locus	Mutant alleles	Other designations	Reference
DWF1 [a]	*dwf1*-1		Feldmann et al 1989
	dwf1-2	*dim*1 [b]	Takahashi et al 1995
	dwf1-3/5		Feldmann KA [h]
	dwf1-6	*cbb*1 [c]	Kauschmann et al 1996
BRI1 [d]	*bri1*-1		Clouse et al 1996
	bri1-2	*cbb*2	Kauschmann et al 1996
	bri1-3/7	*dwf2*-1/5	Feldmann and Azpiroz 1994
	bri1-101/118	*bin*1-1/18	Li et al 1996
CPD [e]	*cpd*-1		Szekeres et al 1996
	cpd-2	*cbb*2	Kauschmann et al 1996
	cpd-3/5	*dwf3*-1/3	Feldmann and Azpiroz 1994
DWF4	*dwf4*-1/4		Azpiroz et al 1998
DWF5	*dwf5*-1/4		Feldmann KA, Dilkes BP, Che S [h]
	dwf5-5	*le* [f]	M. Koornneef, personal comm.
DET2 [g]	*det2*-1		Li et al 1996
	det2-101/110	*dwf6*-1/10	Feldmann KA, Dilkes BP [h]
DWF7	*dwf7*-1		Feldmann KA, Choe S, Tanaka A [h]
DWF8	*dwf8*-1		Feldmann KA, Choe S [h]

[a] *dwarf*
[b] *diminuto*
[c] *cabbage*
[d] *brassinosteroid-insensitive*
[e] *constitutive photomorphogenesis and dwarfism*
[f] *lepida*
[g] *de-etiolated*
[h] unpublished results

(*brassinosteroid insensitive*), and Li and Chory (1997) identified a further 18 alleles of this locus. The eight known Arabidopsis BR loci and allelic designations are shown in Table 1.

All of the BR dwarfs have several common characteristics that are best described at the outset of this discussion (see also color plate, page 190). These dwarfs exhibit a short robust stature, short dark-green leaves, and reduced fertility or sterility. In addition, most of the BR dwarfs display prolonged development. In the light, the dwarfs can be classified as small (*bri1* and *cpd*) or standard (*dwf1*, *det2*, *dwf4*-*dwf8*) dwarfs. Further, the *dwf4*, *det2*-1, and *dwf8* alleles are the smallest of the standard dwarfs (Table 2). All of the mutants exhibit varying degrees of de-etiolation when grown in the dark. The alleles of *bri1*, *cpd*, and *dwf4* possess the shortest hypocotyls in the dark.

5.1 Early Steps in BR Biosynthesis

The known active BR intermediates are ultimately derived from mevalonate via the isoprenoid pathway. The first committed step to BR biosynthesis remains to be identified. However, several mutants have been identified that map to the early part of this pathway. The *ste1* mutation interferes with the conversion of episterol to 24-methylenecholesterol. The *ste1* mutant was identified in a brute force screen as a plant with modified sterol composition; the mutant contains higher levels of Δ^7-sterols than the wildtype (Gachotte et al 1995). The altered sterol profile in *ste1* was partially complemented by the yeast *ERG1* gene, which encodes a Δ^7-sterol-C5-desaturase (Gachotte et al 1995). This enzyme introduces the double bond between C5 and C6. Interestingly, *ste1* displayed no detectable phenotype. However, as the mutant does not completely lack Δ^5-sterols and as the *ste1* allele has not been sequenced, it is possible that this is a leaky allele and that enough Δ^5-sterol enters the BR pathway so that it is not limiting and thus no dwarf phenotype is exhibited.

Three additional mutants, *dwf1*, *dwf5*, and *dwf7*, have been identified that are also defective in early steps of the BR pathway (S Choe, S Fujioka, KA Feldmann, unpublished results). All three mutants exhibit a standard dwarf phenotype in the light (20-25% of the wild-type height at 5 weeks of age) (Table 2). In addition, when grown in the dark, the hypocotyls are intermediate in length between the most severe mutants, such as *bri* and *cpd*, and wildtype.

Table 2. Height of light-grown wild-type *Arabidopsis thalinana* and BR dwarfs at 5 week of age

Genotype	Height (cm)
Wildtype (WS)	25.8
dwf1-1	5.47
bri1-4	1.4
cpd-4	1.92
dwf4-1	2.77
dwf5-1	7.0
det2-101	5.1
dwf7-1	4.5
dwf8-1	4.8 (30.3 [a])

WS, Wassilewskija ecotype
[a] Wildtype height, Enkheim ecotype

5.1.1 dwf5

The *dwf5* mutant is unusual among the dwarfs in that all four alleles (two in Wassilewskija ecotype and two in Enkheim ecotype) are fertile (S Choe, KA Feldmann, unpublished results). The fertility appears to be the result of the anthers being brought in proximity to the stigmatic surface. Whether fertility is restored because of longer filaments or shorter gynoecia, as compared to other dwarfs, is presently being investigated. *DWF5* was mapped to the lower arm of chromosome 1 near another dwarf mutant, *le (lepida)*, previously isolated by M. Koornneef; an allelism test has confirmed that *le* and *dwf5* map to the same locus. Biochemical analysis of endogenous sterol levels in *dwf5* showed that mutants failed to accumulate campestanol, campesterol, or 24-methylenecholesterol as is typical in the wildtype (KA Feldmann, S Fujioka, unpublished results). This result suggests that the block occurred before 24-methylenecholesterol. Feeding studies were less useful in pinpointing the defective step in the pathway as many of the BR intermediates that occur in the early part of the pathway are not effective at inducing elongation, even in the wildtype. However, a synthetic BR, 22α-hydroxycampesterol, and the BR intermediates that occur later in the pathway rescued the dwarfism in *dwf5* suggesting that the steps from campesterol to brassinolide are intact (KA Feldmann, S Fujioka, unpublished results).

5.1.2 dwf7

The *dwf7* gene maps to the top of chromosome 3 (A Tanaka, unpublished result) and is presently defined by only one mutant allele. Biochemical analysis of endogenous sterol levels in *dwf7* showed that mutants failed to accumulate campestanol, campesterol, or 24-methylenecholesterol, similar to *dwf5* (S Choe, KA Feldmann, S Fujioka, unpublished results). This result suggests that the block occurred before 24-methylenecholesterol. Again, similar to *dwf5*, feeding studies were less useful in pinpointing the defective step in the pathway. However, 22α-hydroxycampesterol, and the BR intermediates that occur later in the pathway rescued the dwarfism in *dwf7* suggesting that in this mutant the steps from campesterol to brassinolide are intact (S Choe, KA Feldmann, S Fujioka, unpublished results).

Characterization of the biosynthetic defects in *dwf5* and *dwf7* will be advanced when internal standards for biochemical analysis of these upstream intermediates are available and when the genes are cloned and characterized. Only one step is predicted to occur between the Δ^7-sterol-C5-desaturase (STE1) and 24-methylenecholesterol, the Δ^7-reductase. While *DWF5* or *DWF7* may encode this step, the fact that another dwarf locus resides in the early part of the BR pathway suggests that the *ste1* mutant is a leaky allele. Additional mutants will need to be identified to verify the first committed step in BR biosynthesis.

5.1.3 dwf1 and lkb

The *dwf1* mutants have been isolated by several investigators. This mutant was first described by Feldmann et al (1989) as the first T-DNA-tagged mutant in plants. The

DWF1 gene was cloned and characterized (Genebank U12400), but its sequence gave no clues about its function in plants at that time. Takahashi et al (1995) isolated a second T-DNA-tagged allele (*dim1* = *dwf1*-2). Their characterization of the *DWF1* (*DIM1*) gene indicated that it contained a nuclear localization signal, and they speculated that the gene product was not involved in light signaling or in the biosynthesis or signaling of phytohormones. Further characterization of *DWF1* shows that it encodes an oxido reductase (Mushegian and Koonin 1995; BP Dilkes, KA Feldmann, unpublished results). Recently, Kauschmann et al (1996) also isolated an allele of *dwf1* (*cbb1* = *dwf1*-6) and showed that the dwarf phenotype was rescued by 28-homobrassinolide, 24-epicastasterone, and 24-epibrassinolide. Other BR intermediates were not available for testing at that time.

We have followed up these observations and tested *dwf1*-1 on various BR intermediates including 22α-hydroxycampesterol. The rescue by 22α-hydroxycampesterol and the BR intermediates late in the pathway show that *dwf1* is blocked before campesterol (BP Dilkes, KA Feldmann, unpublished results). Further, Klahre et al (1997) showed that *dwf1*-2 accumulated 24-methylenecholesterol. They also used deuterium labeled 24-methylenecholesterol to show that it was not converted to campesterol in this mutant. Since this is predicted to be a two-step reaction, an isomerization step to 24-methyldesmosterol and a reduction to campesterol, Klahre et al (1997) added labeled 24-methyldesmosterol to the cultures to show that the reduction to campesterol was affected. These combined data strongly suggest that *DWF1* acts at the reduction step between 24-methylenecholesterol and campesterol.

Brassinosteroid dwarfs (*lk*, *lka* and *lkb*) have also been isolated from the common garden pea (*Pisum sativum*). The *lka* mutant is insensitive to exogenously supplied brassinolide and likely functions in BR signal transduction (Nomura et al 1997). The dwarfism in the *lk* and *lkb* mutants could be rescued by brassinolide. The *lkb* mutant has drastically reduced levels of brassinolide, castasterone, and 6-deoxocastasterone as well as campestanol and campesterol and increased levels of 24-methylenecholesterol (Yokota et al 1997). This suggests that *LKB* is the pea homolog of the *DWF1* gene in Arabidopsis. The *lk* mutant is discussed further below.

5.2 Campesterol to Campestanol

The conversion of campesterol to campestanol is mediated by three reactions including dehydrogenation, isomerization, and reduction (Li et al 1997, Fujioka et al 1997). A mutant in the reductase step was identified by Li et al (1996). The *det2* mutant, which maps the bottom of chromosome 2, was initially identified in a screen for de-etiolated mutants. Evidence that *DET2* encodes a biosynthetic enzyme includes the fact that BRs, but not other growth regulators, rescued the *det2* phenotype to wild type, as did the transformation of *det2* with human 5α-reductases driven by the CaMV 35S promoter (Li et al 1996, 1997). Moreover, recombinant DET2 protein was able to reduce several 3-oxo-$\Delta^{4(5)}$ mammalian steroids when expressed in human embryonic kidney cells but failed to reduce 3β-hydroxy-$\Delta^{5(6)}$ steroids, providing convincing evidence that the DET2 enzyme performs the same function as human steroid

5α–reductases. Further, Fujioka et al (1997) measured the endogenous levels of BR intermediates in *det2* and found that (24*R*)-24-methylcholest-4-ene-3-one accumulated and campestanol was reduced to about 10% of that found in the wild type (see the chapter by A. Sakurai, this volume).

Azpiroz et al (1998) identified ten additional alleles of *det2* (*dwf6*) from T-DNA populations or dwarfs obtained from the Arabidopsis Biological Resource Center at Ohio State University (ABRC). Interestingly, all of these dwarf mutants were larger in stature than *det2-1*. Part of this difference in phenotype may be due to the omission of weak alleles in the Li et al (1996) report and the under-representation of strong alleles, due to reduced fertility, from the ABRC. In addition, the ten alleles characterized by Azpiroz et al (1998) were in the Wassilewskija and Enkheim ecotypes, whereas the Li et al (1996) alleles were in the Columbia ecotype.

The *lk* dwarf of pea also had greatly reduced levels of castasterone, 6-deoxocastasterone, and campestanol. The authors speculate that *LK* has the same function as the *DET2* gene of Arabidopsis (Yokota et al 1997).

In the conversion of campesterol to campestanol there appears to be at least two other steps for which a mutant has yet to be identified. It is interesting to note that there are more than a dozen alleles of *det2* and none for the other two steps between campesterol and campestanol. Also of interest, low stringency screening conditions for sequences related to *DET2* indicate that there is a second gene with homology to *DET2* in the Arabidopsis genome, which is not unexpected given that three alleles of *det2* showed measurable levels of campestanol, suggesting that a second reductase is present (Fujioka et al 1997). The characterization of this gene may be necessary to understand how the BR pathway is regulated in different developmental or environmental states.

5.3 Campestanol to Brassinolide

5.3.1 *dwf4*: Rate-Limiting Step in the BR Pathway

The *dwf4* mutant is a standard BR dwarf (Table 2). When grown in the dark, *dwf4* plants have a de-etiolated growth habit: short hypocotyl and open and somewhat expanded cotyledons. While *dwf4* mutants possess viable pollen, the plants remain sterile or nearly sterile due to failure of the anthers to reach the stigmatic surface. The *DWF4* locus was mapped to the lower arm of chromosome 4 (Azpiroz et al 1998) and is defined by four mutant alleles. Interestingly, while all four alleles appear to contain loss of function mutations, the two alleles isolated from the Wassilewskija ecotype are smaller than the two alleles isolated from the Enkheim ecotype (Azpiroz et al 1998).

Sequence analysis of *DWF4* showed that it encodes a cytochrome P450 with 43% identity to *CPD* (described below) and as such is the second member of the CYP90 family. This sequence similarity suggests that *DWF4* functions as a hydroxylase. Also, similar to *CPD*, *DWF4* did not cluster with the Group A plant cytochrome P450s (Choe et al 1998).

Feeding studies with BR intermediates showed that only 22α-hydroxylated BRs rescued the *dwf*4 phenotype. The *dwf*4 seedlings failed to respond to even high concentrations of campestanol or 6-oxocampestanol but were rescued by 6-deoxocathasterone and cathasterone, respectively. Moreover, *dwf*4 seedlings were rescued to wild-type phenotype by 22α-hydroxycampesterol but not by campesterol. These results suggest that *DWF4* is able to mediate the hydroxylation of at least three substrates: campestanol, 6-oxocampestanol, and campesterol (Choe et al 1998).

5.3.2 *cpd*

The next step in the pathway was confirmed by a small dwarf that was rescued by teasterone but failed to respond to cathasterone (Szekeres et al 1996). The *cpd* mutant possessed a light-grown phenotype in the dark in that it had a short hypocotyl, no apical hook, open cotyledons, and extended leaf primordia; and after prolonged growth in the dark, it contained numerous rosette leaves. In contrast to the wildtype, the cotyledons of *cpd* differentiated trichomes and stomata (Szekeres et al 1996). Another indicator of a light-grown phenotype was the expression of certain light-regulated genes in *cpd*. In the light, *cpd* plant height was 3-5% that of the wildtype.

The *CPD* locus was mapped to the top of chromosome 5, and the locus was cloned from a T-DNA insertion mutant. CPD was found to encode a new class in the cytochrome P450 gene superfamily (CYP90) with homology to specific domains of steroid hydroxylases (Szekeres et al 1996). CPD clusters with CYP85 from tomato (Bishop et al 1996) and CYP88 from maize (Winkler and Helentjaris 1995).

Szekeres et al (1996) used BR feeding studies to show that *CPD* was likely to encode a 23α–hydroxylase in that the *cpd* mutant was rescued to wild-type phenotype by BR intermediates hydroxylated at the C23 position, including teasterone, 3-dehydroteasterone, typhasterol, and castasterone. We (S Choe, KA Feldmann, unpublished results) have since shown that *cpd*-4 is also rescued by 23-hydroxylated BR intermediates in the late C6 oxidation pathway.

The naturally occurring *dpy* mutant of tomato is an intermediate dwarf with severely altered leaf morphology, including the downward curling and dark green color typical of the Arabidopsis BR mutants. We (R Cerny, S Clouse, unpublished results) found that spraying with brassinolide completely rescued the mutant phenotype to wild type [color plate, page 190]. More detailed studies showed that 6-deoxoteasterone, along with all subsequent intermediates of the late C6 oxidation pathway, rescued *dpy*, while cathasterone, 6-deoxocathasterone, and all of its precursors failed to do so (R Cerny, S Fujioka, S Clouse, unpublished results). This suggests that *DPY* is the tomato homolog of *CPD* in Arabidopsis, and cloning by transposon tagging is underway to determine the extent of sequence homology between the tomato and Arabidopsis genes (R Cerny, G Bishop, S Clouse, unpublished results). Color plate (page 190) shows that *dpy* also exhibits the de-etiolation response characteristic of Arabidopsis BR mutants.

5.3.3 *dwf8*

The *dwf8* mutant is a standard size dwarf and exhibits extremely reduced fertility. At 5 weeks of age light-grown *dwf8* plants possess a height that is 15% (4.8 cm) that of the respective wildtype (Table 2). When grown in the dark, *dwf8* seedlings display an intermediate de-etiolated response. Initial rescue experiments have been somewhat difficult to interpret. Light-grown *dwf8* seedlings fail to respond to either 6-deoxoteasterone or teasterone but do respond to 3-dehydro-6-deoxoteasterone and 3-dehydroteasterone. However, the latter BR intermediates do not completely rescue the dwarf phenotype (S Choe, KA Feldmann, unpublished results). Moreover, GC-SIM (selected ion monitoring) analysis of *dwf8* tissues indicates that typhasterol accumulates suggesting that *DWF8* mediates the hydroxylation of typhasterol to castasterone (S Fujioka, personal communication). The precise role of *DWF8* will likely become clear with further biochemical analysis using labeled intermediates.

5.4 Conclusions from Mutant Analyses

Several questions pertaining to BR biosynthesis remain unanswered that can best be addressed by mutant analyses. What additional steps are necessary for brassinolide biosynthesis? Do BR intermediates before brassinolide have intrinsic bioactivity in Arabidopsis? Is the BR pathway similar to the networked pathway for GA biosynthesis?

The first question is being addressed by saturating for BR-complemented dwarfs. The second question, that of bioactivity of BR intermediates, will hopefully be addressed by the identification of mutants in the final steps leading to brassinolide biosynthesis. We have noted that the severity of the dwarf phenotypes is correlated with their position in the pathway; the mutants near the end of the pathway are much smaller than those early in the pathway. The only exception to this is *dwf8*, but this may be a weak allele. This increasing severity in phenotype may mean that null mutants in the final steps of the pathway may be lethal. The third question, that of networking, is being addressed by characterizing double mutants. If the BR pathway consists exclusively of the branched pathways without interconnections, then a double mutant between a null allele of *dwf4* and that of *cpd* should be of a *dwf4* stature, given that *dwf4* would be predicted to be epistatic to *cpd*. In fact, the double mutant has an additive phenotype (R Azpiroz, KA Feldmann, unpublished results). In the same way a double mutant between *dwf1* and *dwf4* would be expected to look like *dwf1*, but again the double mutant is additive. In every case where we have verified the double mutants, they have been additive (KA Feldmann, BP Dilkes, S Choe, F Tax, unpublished results). This must mean that the BR pathway is more interconnected than current models show or that there are duplicate genes for many of these steps.

Even with this large collection of dwarfs, the absence of BR pathway intermediates would have prevented this research from advancing in a rapid manner. These molecules have been prepared and made available by S. Fujioka, A. Sakurai, and

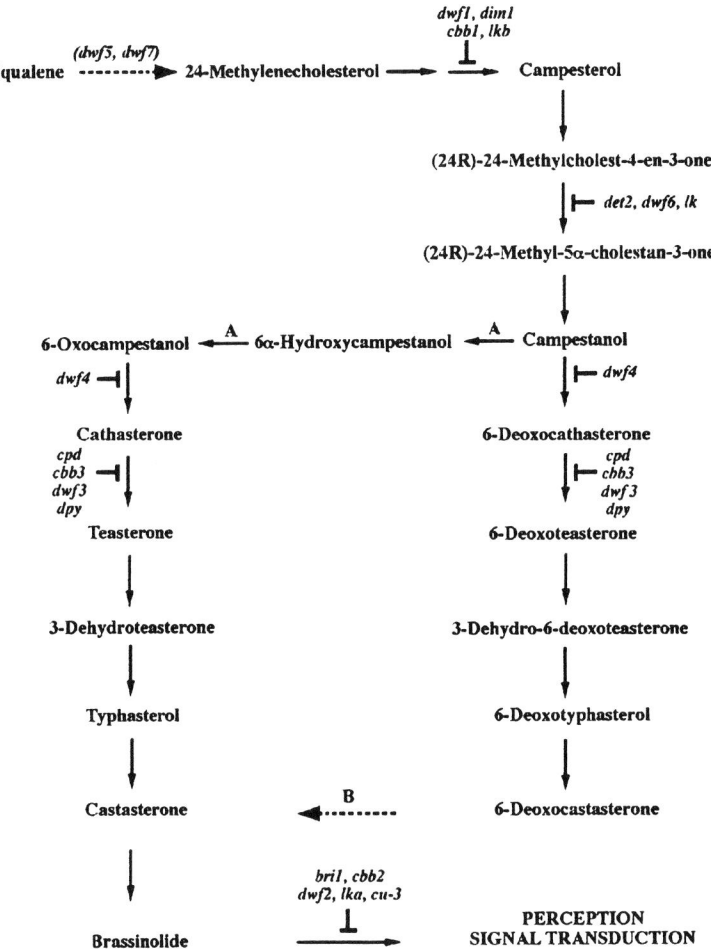

Fig. 1. Site of action of BR mutants. The proposed site of action for Arabidopsis (*dwf1-dwf7*, *dim1*, *cbb1-cbb3*, *det2*, *cpd*, *bri1*), pea (*lkb*, *lk*, *lka*), and tomato (*dpy*, *cu-3*) mutants is indicated based on the most recent evidence. **A** represents the early C6 oxidation pathway, while **B** designates the late C6 oxidation pathway. Dashed lines indicate multiple steps

their collaborators. One BR intermediate that has been especially useful is 22α-hydroxycampesterol (Choe et al 1998). As described earlier, the hydroxyl group is added to campesterol by *DWF4*, and 22α-hydroxycampesterol can rescue the *dwf4* phenotype. Similarly, the *cpd* phenotype can be rescued by 22α,23α-dihydroxycampesterol; 22α-hydroxycampesterol serves as a substrate for *CPD*. Presumably the same three enzymes that mediate the conversion of campesterol to campestanol also mediate the conversion of 22α,23α-dihydroxycampesterol to teasterone and 6-deoxoteasterone. As many of the BR intermediates that occur in the early part of the pathway are not useful in demonstrating a rescued phenotype, likely due to the rate-limiting step at *DWF4*, 22α-hydroxycampesterol presents a way of circumventing this limiting step. As such, the failure to rescue a dwarf phenotype by 22α-hydroxycampesterol indicates that the lesion is between campesterol and campestanol or in one of the last five steps in the BR pathway leading to brassinolide. 22α-Hydroxycampesterol rescued dwarfs are easy to place in the pathway, as *dwf4* and *cpd* alleles have discernible phenotypes in contrast to mutants such as *dwf*1, *dwf*5, and *dwf*7. Utilizing these observations, we are now isolating additional BR dwarfs and efficiently placing them in the pathway. In this way, it seems likely that a mutant will soon be identified for every step in the BR pathway. The current status of BR mutant analysis is summarized in Fig. 1.

6. BR Signal Transduction

6.1 BR-Insensitive Mutants and the Search for a BR Receptor

The value of hormone-insensitive mutants in dissecting signal transduction pathways in plants has been clearly demonstrated for ethylene (Ecker 1995) and abscisic acid (Finkelstein and Zeevaart 1994), and this approach has been fruitful for BRs as well. BR-insensitive mutants in Arabidopsis were first identified by the ability of mutant plants to elongate roots in the presence of inhibitory concentrations of BR with respect to wild type (Clouse et al 1993). One mutant, named *bri*1, showed severe pleiotropic effects on development including dwarfism, de-etiolation, male sterility, and altered leaf morphology, which suggested that the BRI1 protein played an important role in BR signal perception or transduction (Clouse et al 1996). Numerous alleles of *bri*1 with identical phenotype have been isolated in independent screens (KA Feldmann, personal communication; Kauschmann et al 1996, Li and Chory 1997) and, as described in Table 1, these have been renamed *bri*1-2 (*cbb*2), *bri*1-3 through *bri*1-7 (*dwf*2-1 through *dwf*2-5), and *bri*1-101 through *bri*1-118 (*bin*1 through *bin*18). The phenotype of *bri*1-1 is shown in color plate (pg 190).

Several laboratories attempted the positional cloning of *BRI*1, with Li and Chory (1997) reporting the first sequence. Based on the structural similarity of animal and plant steroid signaling molecules, it was reasonable to assume that plants would have members of the intracellular superfamily of steroid receptors; and it was speculated that BRI1 might be a steroid-like receptor (Clouse 1996, Clouse et al 1996). However, BRI1 sequence analysis revealed strong homology not to steroid receptors but to leucine-rich receptor kinases, which function at the cell surface to transduce extracellular signals (Li and Chory 1997). All required components of a receptor kinase could be clearly identified in the sequence including a signal peptide, extracellular domain, membrane-spanning domain, and intracellular kinase domain. Sequence analysis of five mutant alleles of *BRI*1 showed that four of the alleles had mutations in critical areas of the kinase domain, and one had a mutation in a unique 70 amino acid island in the putative extracellular ligand binding domain (Li and Chory 1997).

The identification of an important BR signaling mutant as a receptor-like kinase is significant given the importance of this class of proteins in controlling both animal and plant development. In plants, all receptor-like kinases identified to date are of the serine/threonine type in their kinase domain but have divergent extracellular ligand-binding domains, as would be expected if these receptors are responding to diverse ligands. Besides *BRI*1 numerous leucine-rich repeat receptor-like kinases have been identified in plants. The *ERECTA* gene (Torii et al 1996) has an important role in plant morphogenesis, *CLAVATA*1 (Clark et al 1997) controls shoot and floral meristem size in Arabidopsis, *SERK* is expressed specifically during somatic embryogenesis in carrot cells (Schmidt et al 1997), *Xa*21 is a disease-resistance gene in rice (Song et al 1995), and *RLK*5 (Walker 1993) and *TMK*1 (Chang et al 1992) are Arabidopsis leucine-rich repeat receptor-like kinases of unknown function. Other receptor-like kinases in plants that do not have leucine-rich repeat type extracellular domains include *Ath.lecRK*1 (Herve et al 1996), which has a lectin-like extracellular domain; *SRK* (Goring and Rothstein 1992) whose extracellular domain is homologous to S-locus glycoproteins involved in self-incompatibility in Brassica; *PR*5K (Wang et al 1996) with an extracellular domain similar to pathogenesis-related proteins; and *CrRLK*1 (Schulze-Muth et al 1996) a *Catharanthus roseus* receptor-like kinase whose extracellular domain is not homologous to any known sequence.

In animals, receptor kinases play critical roles in controlling cell proliferation and differentiation. In contrast to plants, animal receptor kinases have tyrosine kinase activity rather than serine/threonine activity. Leucine-rich repeats are found in the extracellular domain of tyrosine receptor kinases (Schneider and Schweiger 1991), but more frequently cysteine-rich repeats and immunoglobulin repeats are found in such important receptors as the insulin receptor, epidermal growth factor receptor (EFG), and the platelet-derived growth factor (PDGF) receptor (reviewed in Heldin 1995). Animal tyrosine receptor kinases dimerize upon ligand binding and autophosphorylate tyrosine residues. This activates the kinase domain, which phosphorylates intracellular targets such as G proteins, phospholipase C, and transcription factors such as proto-oncogenes, thus amplifying and proliferating the initial signal (Heldin 1995). Plant receptor kinases may act in a similar manner since most of the receptor-kinases described above have been shown to autophosphorylate on threonine and serine when recombinant kinase domain is incubated with labeled ATP. Recombinant BRI1 kinase domain also apparently autophosphorylates (M Oh, S Clouse, unpublished data). A potential substrate for plant receptor-like kinases, called KAPP, has been identified that interacts with RLK5 (Stone et al 1994) and CLAVATA (Williams et al 1997) in vitro. The protein consists of a phosphatase type 2C domain and a kinase interaction domain. A potential substrate for BRI1 is described below, which if verified would be the first transcription factor shown to interact with a plant receptor-like kinase.

Although obviously an important component of the BR signal transduction pathway, the role of BRI1 as the BR receptor has not yet been confirmed by direct binding studies. In plants no ligands have been identified for plant receptor kinases, and in animals all known ligands for such receptors are polypeptides or glycoproteins, not small molecules such as steroids (Walker et al 1996). The BR receptor might be a distinct polypeptide that binds to BRI1 in the presence of BR, or an unidentified ligand might exist that is also required for BR activity. The demonstration of direct binding of BR to BRI1 will not necessarily exclude the possibility that there are also intracellular BR receptors. In animals both intracellular and extracellular steroid receptors co-occur, with the intracellular receptor mediating gene expression and the extracellular receptor modulating non-genomic responses such as calcium ion flux and phosphorylation status of various proteins (Mendoza et al 1995).

The BR-insensitive mutants in species other than Arabidopsis have now been putatively identified. The naturally occurring tomato mutant *cu-3* appears to be BR-insensitive (R Cerny, R Gardner, S Clouse, unpublished results) and exhibits many of the phenotypic features of Arabidopsis *bri1* including extreme dwarfism, dark green cabbage-like leaves, male sterility, delayed senescence, and de-etiolation (color plate, page 190). It also shows the same insensitivity to BR in root elongation assays as *bri1* and, similar to *bri1*, shows normal sensitivity to abscisic acid, GA, cytokinins, and indoleacetic acid; and hypersensitivity to 2,4-dichlorophenoxy acetic acid (R Cerny, R Gardner, S Clouse, unpublished results). Based on phenotype and physiological responses, it is possible that *cu-3* is a lesion in a receptor-like kinase homolog of *BRI1*. Having the sequence of a second putative BR receptor from another species would be valuable when looking for conserved motifs, particularly in the extracellular ligand-binding domain. The *lka* mutant of pea also has a dwarf phenotype, and because it shows reduced response to exogenous BR (compared to the BR-deficient dwarf *lkb*) it is also predicted to be BR-insensitive (Normura et al 1997).

6.2 Downstream Components of BR Signal Transduction

During the attempted positional cloning of *BRI1*, a candidate gene was identified that mapped to the proper location on the bottom of chromosome 4 and appeared to partially rescue the *bri1* phenotype when introduced into the mutant as a CaMV35S promoter:cDNA construct (T Altmann, S Clouse, unpublished results). The cDNA had extensive sequence homology to a barley gene that was itself proposed to be an estrogen receptor homolog (Speulman and Salamini 1995). Moreover, recombinant protein encoded by this cDNA bound specifically to a region of the *TCH*4 promoter thought to contain the putative BR response element (M Oh, R Torisky, J Braam, S Clouse, unpublished results). Such a gene has properties expected for an intracellular steroid receptor, but sequence analysis of five *bri1* alleles showed no mutation in this gene and *BRI1* was subsequently shown by Li and Chory to be a leucine-rich repeat receptor-like kinase, as detailed above (Li and Chory 1997). This gene, while not *BRI1*, still remained of interest since it was the first putative transcription factor shown to bind specifically to the promoter of a BR-regulated gene and was thus named *TCH4-BF*1. It has recently been found that the kinase domain of BRI1 apparently phosphorylates recombinant TCH4-BF1 in vitro suggesting the intriguing possibility that TCH4-BF1 plays a role in BR signaling pathways (M Oh, S Clouse, unpublished results).

The genomic sequence of *TCH4-BF*1 reveals an open reading frame of 228 amino acids encoding a protein of 26,122 daltons; and four introns of 82, 225, 583, and 257 bp are interspersed throughout the reading frame. There is no obvious signal peptide, but a sequence rich in lysine and arginine at the *N*-terminus is reminiscent of nuclear localization signals; and numerous consensus sequences for serine or threonine phosphorylation are present in *TCH4-BF*1. Message levels of *TCH4-BF*1 do not appear to be regulated by BR. BLASTP analysis shows that the most closely related sequence in the database is the barley ES43 protein, with 61% identity and 77% similarity over a continuous stretch of 163 amino acids. The *ES*43 clone was identified in a screen for plant genes homologous to the DNA binding domain of animal steroid hormone nuclear receptors (Speulman and Salamini 1995). While there is homology between ES43 and the first zinc finger of the estrogen receptor, including conserved spacing of cysteine residues indicative of a zinc finger, the overall structure is not as convincing. There is only one zinc finger, as opposed to two in steroid receptors, and the zinc finger in ES43 occurs at the carboxy terminus rather than in the middle of the protein as in steroid receptors. Furthermore, there is no obvious ligand binding domain. When a BLASTP search is run on ES43 or TCH4-BF1, no steroid receptor homologs are retrieved, but numerous members of an important group of zinc-finger-like transcription factors called PHD finger proteins do show homology.

The PHD finger transcription factors play critical roles in the development of multicellular organisms by regulating the expression of homeotic genes (Aasland et al 1995). In *Drosophila* there are two classes of PHD finger transcriptional regulators, the polycomb and trithorax groups, which are required for the determined state of imaginal discs. Lesions in *PCL* (polycomb group) or *ASH*1 (trithorax group) genes result in typical homeotic mutations where legs are replaced by antenna, anterior wings by posterior wings, and so on (Lonie et al 1994, Tripoulas et al 1996). Human homologs of the trithorax group such as *ALR* and *ALL*-1 have been cloned, and these PHD finger proteins appear to be involved in the transcriptional regulation of genes controlling cell proliferation and differentiation since their mutation leads to oncogenesis (Prasad et al 1997). PHD fingers were first identified in plants (in fact PHD is an acronym for Plant Homeo Domain) but no extensive analysis of plant PHD finger proteins has been undertaken (Beckmann and Cashmore 1993). All PHD fingers contain a conserved Cys4-His-Cys3 motif in a 50-80 amino acid expanse that is thought to bind two zinc atoms (Aasland et al 1995). TCH4-BF1 contains all seven of the canonical Cys and His residues in the PHD finger. The structural homology and DNA binding properties of TCH4-BF1 strongly suggest that TCH4-BF1 functions as a PHD finger transcription factor. If BRI1 is indeed the BR receptor, then intermediaries to carry the signal from the cell surface to the promoter of BR-regulated genes such as *TCH*4 will be required.

Phosphorylation/dephosphorylation is one of the most common forms of posttranslational protein modification and is found extensively as a critical component of signal transduction pathways. For example, binding of vertebrate epidermal growth factor to its cognate tyrosine receptor kinase results in receptor dimerization and autophosphorylation (Heldin 1995). The activated receptor interacts with and phosphorylates an intracellular transcription factor, Stat3, which is then translocated to the nucleus where it activates the transcription of specific epidermal growth factor responsive genes (Park et al 1996). This scenario is played out in numerous signal transduction pathways involving membrane-bound receptors and extracellular ligands. Receptor-mediated phosphorylation of a transcription factor may result in its activation by increased nuclear localization, direct enhancement of DNA binding, or activation of the transcriptional domain, usually as a result of binding to other nuclear proteins in the transcription complex (Hill and Treisman 1995). The identification of the *bri*1 mutant (Clouse et al 1996) and subsequent cloning of the *BRI*1 gene (Li and Chory 1997) were significant advances in our understanding of BR action and signal transduction mechanisms and suggest that a receptor kinase is a critical component of BR signal transduction.

The identification of a putative PHD finger transcription factor that binds specifically to the promoter of a BR-regulated gene and that is apparently phosphorylated by BRI1 in vitro will initiate studies of the nature of BRI1 substrates and their role in downstream signal transduction. The fact that all 23 BR-insensitive mutants so far identified are alleles of the same locus suggests that either the BR signal transduction pathway involves only a few members, and of these only BRI1 is unique, or that other components of the pathway are common to a signal transduction pathway that is essential for viability such that mutation in these common intermediates would result in death. One hypothesis is that the portion of the BR signal transduction pathway involved in elongation is quite short and involves BR-mediated dimerization of BRI1 followed first by autophosphorylation and subsequently by direct phosphorylation of TCH4-BF1. The activated transcription factor then migrates to the nucleus to active genes involved in elongation such as *TCH*4 and possibly expansins, and so on (see Fig. 2). To test this hypothesis, a thorough in vitro and in vivo analysis of the interaction of BRI1 and TCH4-BF1 and examination of the spatial expression of *BRI*1 and *TCH4-BF*1 in developing tissues where *TCH*4 is expressed, is currently underway (M Oh, J Braam, T Altmann, S Clouse, unpublished results).

7. Future Prospects

After a long period of neglect by the general plant community, BRs are now receiving a great deal of international attention. The dramatic phenotype of BR-deficient and BR-insensitive mutants provides clear evidence for the critical role of BR activity in the growth and development of plants. The identification of BR biosynthetic and insensitive mutants in tomato and pea, extend the importance of these compounds from the experimental plant *Arabidopsis thaliana* to crop plants. Given that BRs are found throughout the plant kingdom, it is likely that greater understanding

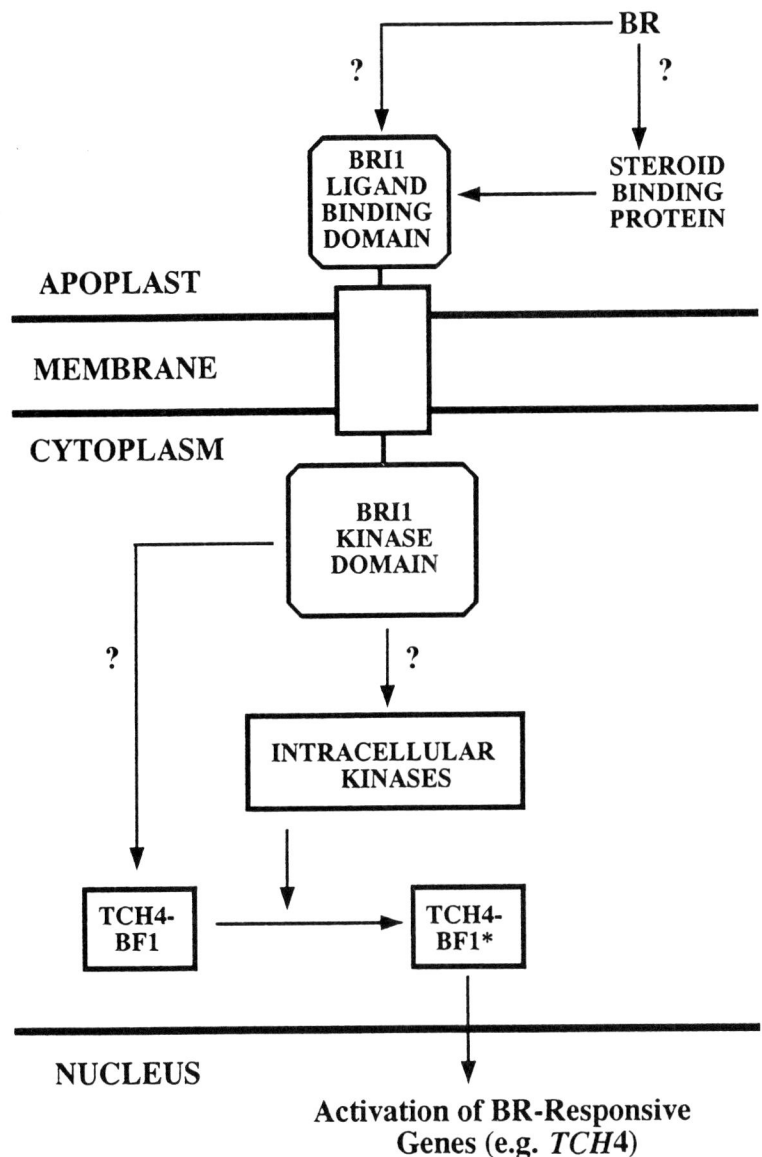

Fig. 2. A hypothetical scheme for BR signal transduction in *Arabidopsis*. BRI1, a putative leucine-rich receptor kinase, may bind BR directly (or BR bound to a steroid-binding protein that functions at the cell surface), which causes autophosphorylation of the BRI1 kinase domain. The BRI1 kinase domain then phosphorylates TCH4-BF1, a putative PHD finger transcription factor, either directly or via a separate intracellular kinase. The activated TCH4-BF1 (designated TCH4-BF1*) is localized to the nucleus where it binds to the promoter of *TCH4* and other BR-responsive genes, leading to enhanced gene expression

of the molecular mechanisms of BR action could have practical impact on generating transgenic crop plants of many species with altered growth properties.

From the perspective of basic developmental biology, the application of molecular genetics to the study of plant steroid hormone action has already yielded a putative BR receptor, a possible intermediate transcription factor in BR signal transduction and cloned genes regulated by BR at the transcriptional and posttranscriptional levels. Moreover, the fact that *BRI*1 is a member of a large family of receptor-like kinases in plants suggest that there may be significant cross-talk in signal transduction pathways, perhaps by sharing common substrates for phosphorylation. Studies on the possible interaction of BRI1 with other developmentally associated receptor-like kinases, such as CLAVATA1 and ERECTA, would be informative. Of critical importance is determining whether the BRI1 extracellular domain binds BR directly and the application of interactive cloning techniques may uncover additional substrates for the BRI kinase domain. The impending identification of a BR response element in the *TCH*4 promoter will clarify the terminal end of BR signal transduction and will provide a valuable molecular tool for a wide range of further studies.

The combined microchemical and molecular genetic analysis of the BR biosynthesis pathway has been highly successful, and future studies will address the subcellular localization of BR biosynthesis and how environmental and developmental signals regulate the endogenous levels of BR throughout the plant life cycle by affecting the genes and gene products involved in biosynthesis and metabolism. Epistasis studies using BR mutants crossed with other hormone and developmental mutants, should help to place BRs in the overall pattern of development as will studies of the ectopic expression of BR biosynthetic genes in transgenic plants.

References

Aasland R, Gibson T, Stewart A (1995) The PHD finger: implications for chromatin-mediated transcriptional regulation. TIBS 20: 56-59

Aloni R (1995) The induction of vascular tissues by auxin and cytokinin. In: P Davies (Ed) Plant Hormones: Physiology, Biochemistry and Molecular Biology. 2nd ed. Kluwer, Dordrecht, pp 531-546

Arteca RN (1995) Brassinosteroids. In: P Davies (Ed) Plant Hormones: Physiology, Biochemistry and Molecular Biology. 2nd, ed., Kluwer, Dordrecht, pp 206-213

Azpiroz R, Wu Y, LoCascio JC, et al (1998) An Arabidopsis brassinosteroid-dependent mutant is blocked in cell elongation. Plant Cell 10: 219-230

Beato M, Herrlich P, Schutz G (1995) Steroid hormone receptors: many actors in search of a plot. Cell 83: 851-857

Beckmann H, Cashmore A (1993) HAT3.1, a novel Arabidopsis homeodomain protein containing a conserved cysteine-rich region. Plant J 4: 137-151

Bishop GJ, Harrison K, Jones JDG (1996) The tomato *Dwarf* gene isolated by heterologous transposon tagging encodes the first member of a new cytochrome P450 family. Plant Cell 8: 959-969

Blatt M, Thiel G (1993) Hormonal control of ion channel gating. Annu Rev Plant Physiol

Plant Mol Biol 44: 543-568

Brock M, Shapiro D (1983) Estrogen stabilizes vitellogenin mRNA against cytoplasmic degradation. Cell 34: 207-214

Carpita N, Gibeaut D (1993) Structural models of the primary cell walls in flowering plants: consistency of molecular structure with the physical properties of the walls during growth. Plant J 3: 1-30

Catala C, Rose J, Bennett A (1997) Auxin-regulation and spatial localization of an endo-1,4-β-D-glucanase and a xyloglucan endotransglycosylase in expanding tomato hypocotyls. Plant J 12: 417-426

Chang C, Schaller G, Patterson S, et al (1992) The *TMK1* gene from *Arabidopsis* codes for a protein with structural and biochemical characteristics of a receptor protein kinase. Plant Cell 4: 1263-1271

Choe S, Dilkes BP, Fujioka S, et al (1998) The *DWF4* gene of Arabidopsis encodes a cytochrome P450 that mediates multiple 22α hydroxylation steps in brassinosteroid biosynthesis. Plant Cell 10: 231-244

Clark S, Williams R, Meyerowitz E (1997) The *CLAVATA1* gene encodes a putative receptor kinase that controls shoot and floral meristem size in *Arabidopsis*. Cell 89: 575-585

Clouse SD (1996) Molecular genetic studies confirm the role of brassinosteroids in plant growth and development. Plant J 10: 1-8

Clouse SD (1997) Molecular genetic analysis of brassinosteroid action. Physiol Plant 100: 702-709

Clouse SD, Zurek D (1991) Molecular analysis of brassinolide action in plant growth and development. In: HG Cutler, T Yokota, G Adam (Eds) Brassinosteroids Chemistry, Bioactivity, and Applications. ACS Sym Ser 474, Amer Chem Soc, Washington, DC, pp 122-40

Clouse S, Sasse J (1998) Brassinosteroids: Essential Regulators of Plant Growth and Development. Annu Rev Plant Physiol Plant Mol Biol 49: 427-451

Clouse SD, Zurek DM, McMorris TC, et al (1992) Effect of brassinolide on gene expression in elongating soybean epicotyls. Plant Physiol 100: 1377-1383

Clouse SD, Hall AF, Langford M, et al (1993) Physiological and molecular effects of brassinosteroids on *Arabidopsis thaliana*. J Plant Growth Regul 12: 61-66

Clouse SD, Langford M, McMorris TC (1996) A brassinosteroid-insensitive mutant in Arabidopsis thaliana exhibits multiple defects in growth and development. Plant Physiol 111: 671-678

Cosgrove D (1997) Relaxation in a high-stress environment: the molecular basis of extensible cell walls and enlargement. Plant Cell 9: 1031-1041

Darnell J (1982) Variety in the level of gene control in eukaryotic cells. Nature 297: 365-371

Deikman J (1997) Molecular mechanisms of ethylene regulation of gene transcription. Physiol Plant 100: 561-566

Ecker J (1995) The ethylene signal transduction pathway in plants. Science 268: 667-675

Ecker J (1997) BRI-ghtening the pathway to steroid hormone signaling events in plants. Cell 90: 825-827

Evans R (1988) The steroid and thyroid hormone receptor superfamily. Science 240: 667-675

Feldmann KA, Azpiroz R (1994) Primary dwarfs. In: J Bowman (Ed) Arabidopsis: An Atlas of Morphology and Development, Springer-Verlag, New York, pp 82-85

Feldmann KA, Marks MD, Christianson ML et al (1989) A dwarf mutant of Arabidopsis generated by T-DNA insertion mutagenesis. Science 243: 1351-1354

Finkelstein R, Zeevaart J (1994) Gibberellin and abscisic acid biosynthesis and response. In:

E Meyerowitz, C Somerville (Eds) Arabidopsis. Cold Spring Harbor Laboratory Press, Cold Spring Harbor, NY, pp 523-554

Fry S, Smith R, Renwick K, et al (1992) Xyloglucan endotransglycosylase, a new wall-loosening enzyme activity from plants. Biochem J 282: 821-828

Fujioka S, Sakurai A (1997) Brassinosteroids. Natural Product Rep 14: 1-10

Fujioka S, Li J, Choi YH, et al (1997) The Arabidopsis *deetiolated2* mutant is blocked in early brassinosteroid biosynthesis. Plant Cell 11: 1951-1962

Fukuda H (1997) Tracheary element differentiation. Plant Cell 9: 1147-1156

Gachotte D, Meens R, Benveniste P (1995) An Arabidopsis mutant deficient in sterol biosynthesis: heterologous complementation by *ERG*3 encoding a Δ^7-sterol-C-5 desaturase. Plant J 8:407-416

Gallie D (1993) Posttranscriptional regulation of gene expression in plants. Annu Rev Plant Physiol Plant Mol Biol 44: 77-105

Goring D, Rothstein S (1992) The S-locus receptor kinase gene in a self-incompatible *Brassica napus* line encodes a functional serine/threonine kinase. Plant Cell 4: 1273-1281

Green P (1993) Control of mRNA stability in higher plants. Plant Physiol 102: 1065-1070

Hagen G (1995) The control of gene expression by auxin. In: P Davies (Ed) Plant Hormones: Physiology, Biochemistry and Molecular Biology. 2nd ed. Kluwer, Dordrecht, pp 228-245

Heldin C (1995) Dimerization of cell surface receptors in signal transduction. Cell 80: 213-224

Herve C, Dabos P, Galaud J, et al (1996) Characterization of an *Arabidopsis thaliana* gene that defines a new class of putative plant receptor kinases with an extracellular lectin-like domain. J Mol Biol 778-788

Hill C, Treisman R (1995) Transcriptional regulation by extracellular signals: mechanisms and specificity. Cell 80: 199-211

Hooley R (1996) Plant steroid hormones emerge from the dark. Trends Genet 12: 281-283

Iwasaki T, Shibaoka H (1991) Brassinosteroids act as regulators of tracheary-element differentiation in isolated Zinnia mesophyll cells. Plant Cell Physiol 32: 1007-1014

Kauschmann A, Jessop A, Koncz C, et al (1996) Genetic evidence for an essential role of brassinosteroids in plant development. Plant J 9: 701-713

Kim S-K, Abe H, Little CHA, et al (1990) Identification of two brassinosteroids from the cambial region of Scots pine (*Pinus silvestris*) by gas chromatography-mass spectrometry, after detection using a dwarf rice lamina inclination bioassay. Plant Physiol 94: 1709-1713

Klahre U, Fujioka S, Yokota T, et al (1997) Characterization of the *diminuto* mutant and genes regulated by brassinosteroids. Proc Plant Growth Reg Soc Amer 24: 99

Klee H, Estelle M (1991) Molecular genetic approaches to plant hormone biology. Annu Rev Plant Physiol Plant Mol Biol 42: 529-551

Kulaeva ON, Burkhanova EA, Fedina AB, et al (1991) Effect of brassinosteroids on protein synthesis and plant-cell ultrastructure under stress conditions. In: HG Cutler, T Yokota, G Adam (Eds) Brassinosteroids Chemistry, Bioactivity, and Applications. ACS Sym Ser 474, Amer Chem Soc, Washington, DC, pp 141-155

Li J, Chory J (1997) A putative leucine-rich repeat receptor kinase involved in brassinosteroid signal transduction. Cell 90: 929-938

Li J, Nagpal P, Vitart V, et al (1996) A role for brassinosteroids in light-dependent development of *Arabidopsis*. Science 272: 398-401

Li J, Biswas MG, Chao A, et al (1997) Conservation of function between mammalian and

plant steroid 5α-reductases. Proc Natl Acad Sci USA 94: 3534-39
Lonie A, D'Andrea R, Paro R, et al (1994) Molecular characterisation of the polycomblike gene of *Drosophila melanogaster*, a trans-acting negative regulator of homeotic gene expression. Development 120: 2629-2636
Mandava NB (1988) Plant growth-promoting brassinosteroids. Annu Rev Plant Physiol Plant Mol Biol 39: 23-52
Mandava N, Thompson M, Yopp J (1987) Effects of selected inhibitors of RNA and protein synthesis on brassinosteroid-induced responses in mung bean epicotyls. J Plant Physiol 128: 53-65
Mendoza C, Soler A, Tesarik J (1995) Nongenomic steroid action: independent targeting of a plasma membrane calcium channel and a tyrosine kinase. Biochem Biophys Res Comm 210: 518-523
Mushegian A, Koonin E (1995) A putative FAD-binding domain in a distinct group of oxidases including a protein involved in plant development. Prot Science 4: 1243-44
Nielson D, Shapiro D (1990) Estradiol and estrogen receptor-dependent stabilization of a minivitellogenin mRNA lacking 5,100 nucleotides of coding sequence. Mol Cell Biol 10: 371-376
Nomura T, Nakayama M, Reid JB, et al (1997) Blockage of brassinosteroid synthesis and sensitivity causes dwarfism in *Pisum sativum*. Plant Physiol 113: 31-37
Oh M-H, Romanov W, Smith R, et al (1998) *BRU1* encodes a xyloglucan endo-transglycosylase that is expressed in inner and outer tissues of elongating soybean epicotyls. Plant Cell Physiol 39: 124-130
Park O, Schaefer T, Nathans D (1996) In vitro activation of Stat3 by epidermal growth factor receptor kinase. Proc Natl Acad Sci USA 93: 13704-13708
Prasad R, Zhadanov A, Sedkov Y, et al (1997) Structure and expression pattern of human *ALR*, a novel gene with strong homology to *ALL*-1 involved in acute leukemia and to *Drosophila trithorax*. Oncogene 15: 549-560
Santos G, Scott G, Lee W, et al (1988) Estrogen-induced post-transcriptional modulation of c-*myc* expression in human breast cancer cells. J Biol Chem 263: 9565-9568
Sasse JM (1991) The case for brassinosteroids as endogenous plant hormones. In: HG Cutler, T Yokota, G Adam (Eds) Brassinosteroids Chemistry, Bioactivity, and Applications. ACS Sym Ser 474, Amer Chem Soc, Washington, DC, pp 158-166
Sasse JM (1997) Recent progress in brassinosteroid research. Physiol Plant 100: 696-701
Schmidt E, Guzzo F, Toonen M, et al (1997) A leucine-rich repeat containing receptor-like kinase marks somatic plant cells competent to form embryos. Development 124: 2049-2062
Schneider D, Schweiger M (1991) A novel modular mosaic of cell adhesion motifs in the extracellular domain of the neurogenic *trk* and *trk*B tyrosine kinase receptors. Oncogene 6: 1807-1811
Schulze-Muth P, Irmler S, Schroder G, et al (1996) Novel type of receptor-like protein kinase from a higher plant (*Catharanthus roseus*). J Biol Chem 271: 26684-26689
Song W, Wang G, Chen L, et al (1995) A receptor kinase-like protein encoded by the rice disease resistance gene, Xa21. Science 270: 1804-1806
Speulman E, Salamini F (1995) A barley cDNA clone with homology to the DNA-binding domain of the steroid hormone receptors. Plant Science 106: 91-98
Sponsel V (1995) The biosynthesis and metabolism of gibberellins in higher plants. In: P Davies (Ed) Plant Hormones: Physiology, Biochemistry and Molecular Biology. 2nd ed. Kluwer, Dordrecht, pp 66-97

Stone J, Collinge M, Smith R, et al (1994) Interaction of a protein phosphatase with an *Arabidopsis* serine-threonine receptor kinase. Science 266: 793-795

Szekeres M, Nemeth K, Koncz-kalman Z, et al (1996) Brassinosteroids rescue the deficiency of CYP90, a cytochrome P450, controlling cell elongation and de-etiolation in Arabidopsis. Cell 85: 171-182

Takahashi T, Gasch A, Nishizawa N, Chua H-H (1995) The *DIMINUTO* gene of Arabidopsis is involved in regulating cell elongation. Genes Dev 9:97-107

Tominaga R, Sakurai N, Kuraishi S (1994) Brassinolide-induced elongation of inner tissues of segments of squash (*Cucurbita maxima* Duch.) hypocotyls. Plant Cell Physiol 35: 1103-06

Torii K, Mitsukawa N, Oosumi T, et al (1996) The *Arabidopsis ERECTA* gene encodes a putative receptor protein kinase with extracellular leucine-rich repeats. Plant Cell 8: 735-746

Tripoulas N, LaJeunesse D, Gildea J, et al (1996) The *Drosophila ash*1 gene product, which is localized at specific sites on polytene chromosomes, contains a SET domain and a PHD finger. Genetics 143: 913-928

Walker J (1993) Receptor-like protein kinase genes of *Arabidopsis thaliana*. Plant J 3: 451-456

Walker J, Stone J, Collinge M, et al (1996) Receptor-like protein kinases. In: P Shewry, N Halford, R Hooley (Eds) Protein Phosphorylation in Plants. Oxford University Press, Oxford, pp 227-238

Wang TW, Cosgrove DJ, Arteca RN (1993) Brassinosteroid stimulation of hypocotyl elongation and wall relaxation in pakchoi (*Brassica chinensis* cv Lei-choi). Plant Physiol 101: 965-68

Wang X, Zafian P, Choudhary M, et al (1996) The PR5K receptor protein kinase from *Arabidopsis thaliana* is structurally related to a family of plant defense proteins. Proc Natl Acad Sci USA 93: 2598-2602

Wilen RW, Sacco M, Gusta LV, et al (1995) Effects of 24-epibrassinolide on freezing and thermotolerance of bromegrass (*Bromus inermis*) cell cultures. Physiol Plant 95: 195-202

Williams R, Wilson J, Meyerowitz E (1997) A possible role for kinase-associated protein phosphatase in the *Arabidopsis* CLAVATA1 signaling pathway. Proc Natl Acad Sci USA 94: 10467-10472

Winkler RG, Helentjaris T (1995) The maize *Dwarf*3 gene encodes a cytochrome P-450-mediated early step in gibberellin biosynthesis. Plant Cell 7:1307-1317

Xu W, Purugganan MM, Polisenksy DH, et al (1995) Arabidopsis *TCH*4, regulated by hormones and the environment, encodes a xyloglucan endotransglycosylase. Plant Cell 7: 1555-1567

Xu W, Campbell P, Vargheese A, et al (1996) The *Arabidopsis* XET-related gene family: environmental and hormonal regulation of expression. Plant J 9: 879-889

Yamamoto R, Demura T, Fukuda H (1997) Brassinosteroids induce entry into the final stage of tracheary element differentiation in cultured *Zinnia* cells. Plant Cell Physiol 38: 980-983

Yokota T (1997) The structure, biosynthesis and function of brassinosteroids. Trends Plant Science 2: 137-143

Yokota T, Nomura T, Kitasaka Y et al (1997) Biosynthetic lesions in brassinosteroid-deficient pea mutants. Proc Plant Growth Reg Soc Amer 24: 99

Zurek DM, Clouse SD (1994) Molecular cloning and characterization of a brassinosteroid-regulated gene from elongating soybean (*Glycine max* L.) epicotyls. Plant Physiol 104: 161-170

Zurek DM, Rayle DL, McMorris TC, et al (1994) Investigation of gene expression, growth kinetics, and wall extensibility during brassinosteroid-regulated stem elongation. Plant Physiol 104: 505-513

Color Plate. Phenotype of brassinosteroid dwarf mutants. **a** Comparison of 5-week-old wild-type *Arabidopsis thaliana* and mutant plants representing the eight brassinosteroid dwarf loci. Bars represent 1 cm. In the dark, dwarf seedlings also display short hypocotyls when compared to wild-type plants. From left, Ws-2 wildtype, *dwf*5-1, *dwf*7-1, *dwf*1-1, *dwf*6-1, *dwf*8-1, *dwf*4-1, *dwf*3-1, and *dwf*2-1. **b** Close-up of an 8-week-old *bri*1-1 (*brassinosteroid-insensitive*) mutant of Arabidopsis showing dark green, curled leaves. **c** The *dpy* mutant of tomato displays a curled leaf phenotype with dark, rugose leaves and suppression of axillary shoots. When treated with 0.1 µM brassinolide twice per week, the leaves lose their curl and are rescued to the wild-type phenotype. In the dark the *dpy* and *cu-3* mutants exhibit a de-etiolated phenotype with short hypoctyls, open cotyledons, and elimination of the apical hook

9
Structure-Activity Relationship

Carme Brosa

Institut Químic de Sarrià CETS, Universitat Ramon Llull, Via Augusta 390
E 08017-Barcelona, Spain

1. Introduction

The brassinosteroids (BRs) represent a class of endogenous plant growth regulators with ubiquitous occurrence in the plant kingdom. They possess high growth-promoting activity, and have been evaluated for use in improving crop yield, quality, stress tolerance, and disease resistance. The results reported with several plants have successfully demonstrated the suitability of these steroids for the enhanced production of field crops and vegetables (Cutler et al 1991, Brosa 1997). Although these findings are encouraging, further detailed studies are required before the full potential use of BRs can be realized.

Much effort is being done in the field of BRs from the physiological, biochemical, and molecular points of view as well as in the synthesis of new BR analogs. In this sense, the clarification of the structure-activity relationship is an important problem that remains to be solved and will help give a better understanding of the bioactivity and mode of action of such interesting compounds. The search for new BRs with a good bioactivity-cost relationship is an active area for improving the benefits from these potent plant growth regulators in agriculture.

In this chapter the different structure-activity relationships (SAR) established in the literature are discussed. It includes our prospects for finding a new way to define the SAR and a novel approach to the mode of action of BRs at the structural level.

Key words: structure-activity relationship, molecular modeling, conformational analysis, rice lamina inclination test, activity, BR inhibitor KM-01, brassinosteroids-receptor interaction, brassinolide, 24-epibrassinolide, (22S, 23S)24-epibrassinolide, 28-homobrassinolide, (22S, 23S)28-homobrassinolide, 24-epicastasterone, (22S, 23S) 24-epicastasterone, 28-homocastasterone

2. Early Structure-Activity Relationships

Different qualitative SAR have been established taking into account the activity data obtained in specific bioassay systems developed for other hormones and related growth substances such as auxins, gibberellins, and cytokinins. These SAR results are more or less stringent depending upon the bioassay used but also on the structural features of the set of BRs tested.

The first approach to defining an SAR was made by Wada and Marumo (1981) just 2 years after the isolation and identification of brassinolide (BL, **1**) (Fig. 1) (Grove et al 1979), the most potent natural BR found to date. The authors, after synthesizing some BRs with modified A-ring but having a 22S,23S glycol and 24S ethyl substituent on the side chain, and evaluating the activity by the rice lamina inclination test (RLIT), suggested that the 2α,3α-dihydroxy-7-oxa-6-ketone moiety was essential to elicit plant growth-promoting activity. Subsequently, Thompson et al. (1982) examined a number of BRs using the bean first and second internode bioassays. The study was focused on the influence in the activity of modified side chains having either a 22R,23R and a 22S,23S diol as well as a methyl (R or S) or ethyl (S) substituent at C-24, or lacking it. Based on the activity data from the two bioassays used, more extended structural requirements were established: i) a *trans* A/B ring system, ii) a 6-ketone or a 7-oxa-6-ketone in B-ring, iii) a 2α,3α-diol, iv) *cis* hydroxy groups at C-22 and C-23, and v) a methyl or ethyl substituent at C-24, the last two points being independent of the stereochemistry. They found a dissimilar pattern of biological activity in both bioassays used.

Similar structural requirements were found by Takatsuto et al (1983a) using the RLIT and by Cerana et al (1984) on elongation of maize root segments, but the first authors later pointed out the significance of an S configuration of the alkyl group at C-24 and of a RR configuration for the vicinal diol at C-22,C-23 (Takatsuto and Ikekawa 1984). In this study, they also suggested that the configuration of the alkyl group at C-24 was more important than that of the C-22,C-23-vicinal diol. However, the relative activities that they found in the RLIT were different than those found in the bean bioassays.

A stricter SAR was postulated on the basis of radish and tomato tests. Accordingly, a 22R,23R glycol was required, the 6-ketone at the B-ring was discarded as an essential group to elicit activity, and only a methyl or no alkyl group at C-24 was accepted (Takatsuto et al 1983b). Again, reversed results in the relative activity were observed in both bioassays and were also different when compared to the activity data obtained in the RLIT.

The presence of a suitable side chain having two hydroxy groups at C-22,C-23 was proved to be indispensable for high activity. Evaluation of some BR analogs having cholestane, cholane, and androstane side chains showed a strong decrease in activity when using the RLIT (Kondo and Mori 1983). On the contrary, the presence of the methyl groups at C-26 and C-27 may not be essential for biological activity (Takatsuto et al 1984).

Yokota and Mori (1992) reported in an extensive review that the structural re-

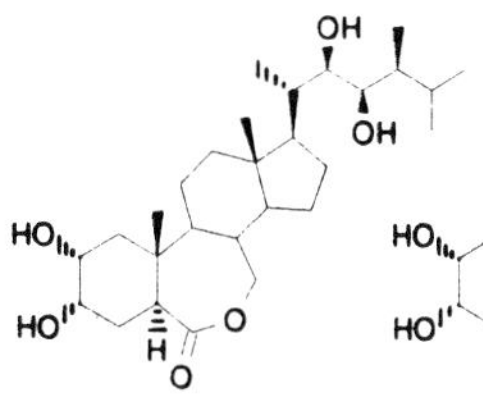

Fig. 1. Brassinosteroid analogs

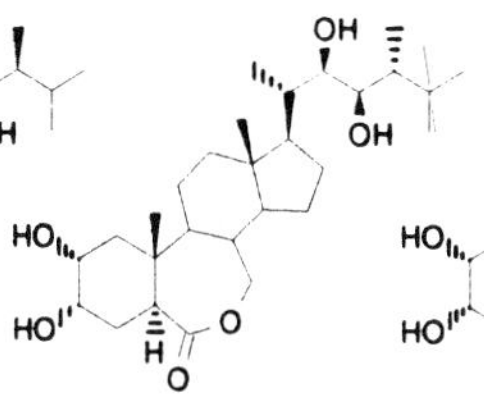

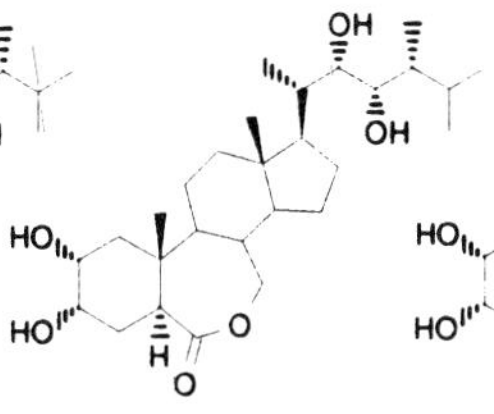

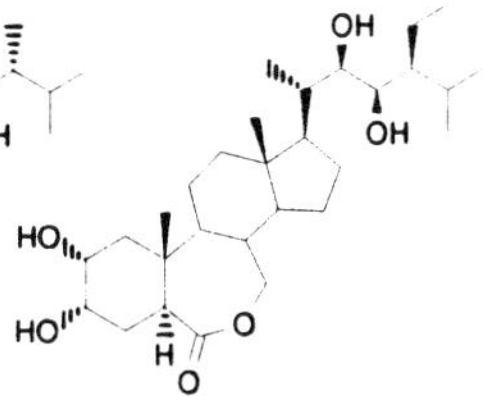

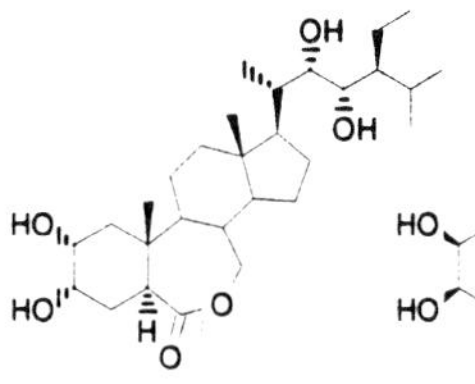

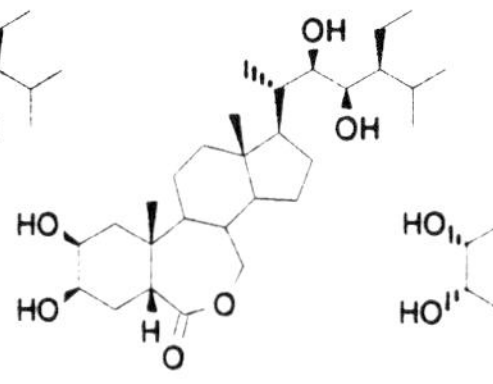

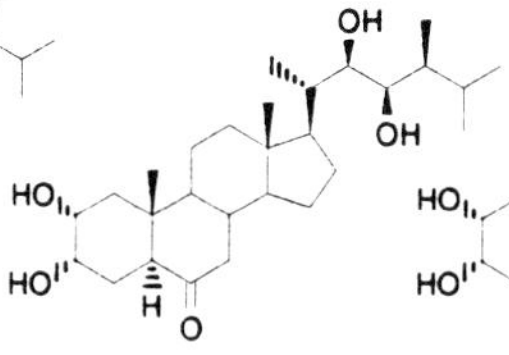

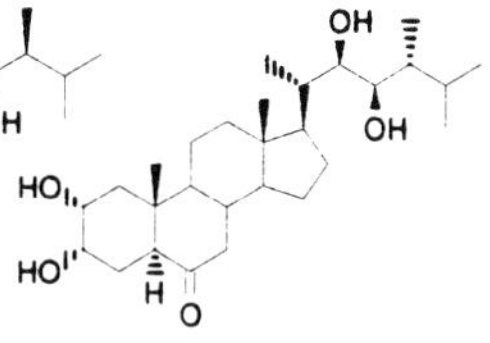

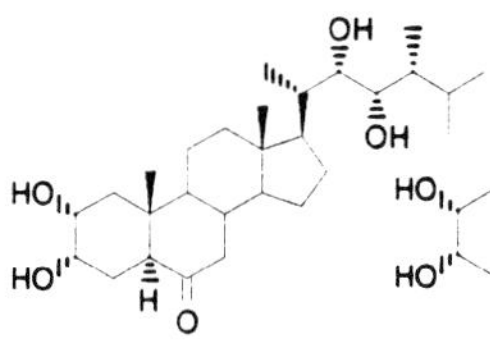

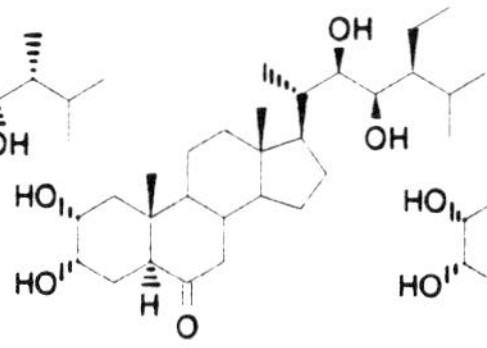

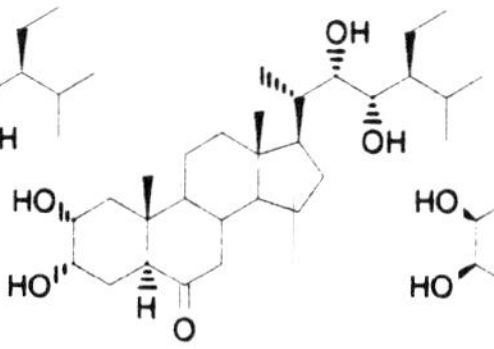

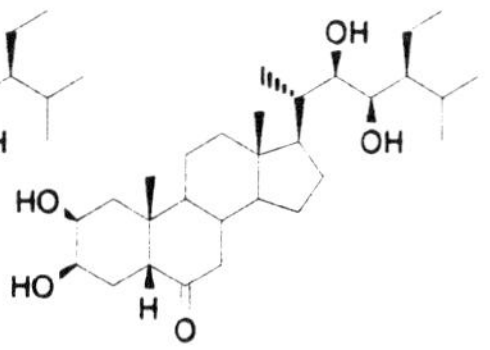

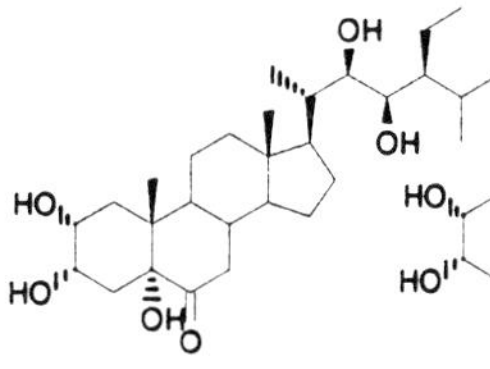

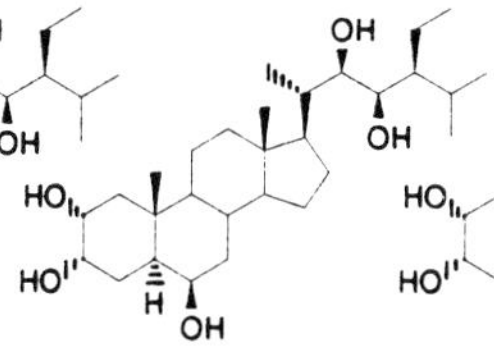

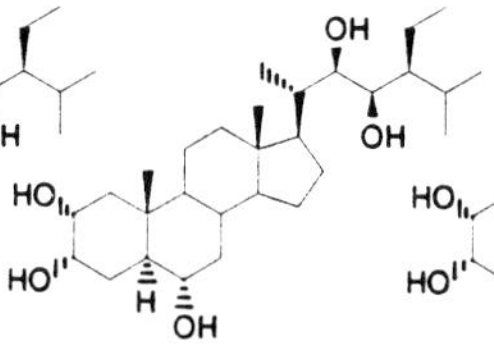

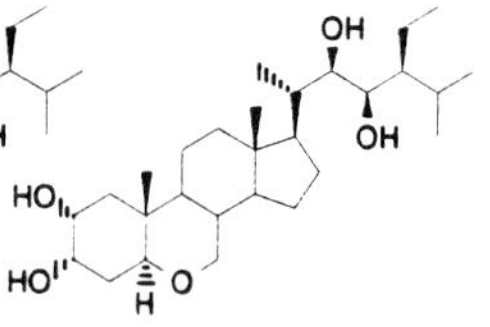

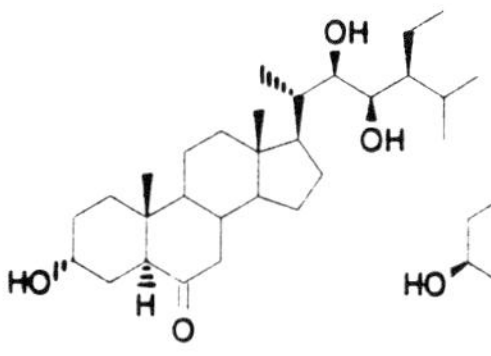

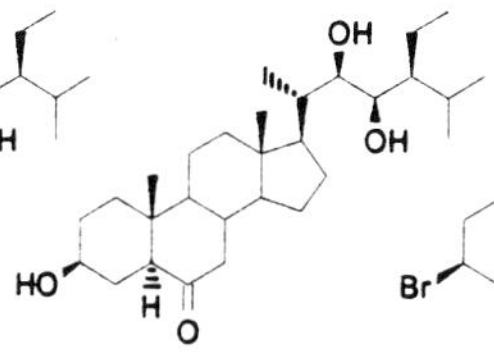

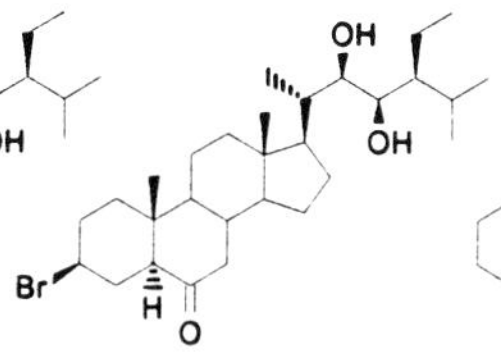

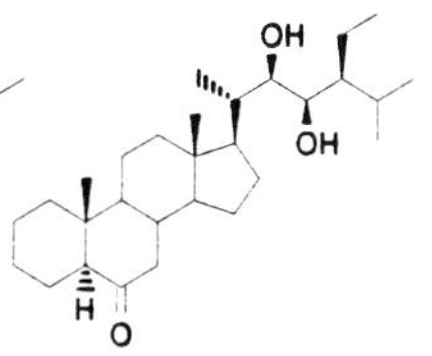

quirements at C-24 are much less rigorous in the RLIT. However, the authors included substituents like methyl, ethyl, methylene, or ethylidene at C-24 as well as an additional methyl group at C-25 as active compounds. The relative activity values reported ranged from 5 to 300.

With respect to the A-ring, the 28-homo-BR analog having a $3\alpha,4\alpha$-diol instead of $2\alpha,3\alpha$ was slightly less active than 28-homo-BL (HBL, **4**) in the RLIT. Hence these functionalities were also included in the structural model for high activity by Takatsuto et al (1987).

More contradictions to SAR were found by Kohout et al (1991) after synthesizing and evaluating by the bean first and second internode bioassays some compounds having pregnane and androstane side chains and some even with no side chain (Kohout et al 1987, Kohout and Strnad 1989, 1992, Kohout et al 1996). With the aim of preparing compounds with oxygenated functionalities at the side chain other than diols, they synthesized some esters and amides of the androstane and cholane type. While some of them elicited high activity in the bean second internode bioassay, their activity in the bean first internode bioassay was marginal. The activity pattern of other BR analogs was proved to be reversed in the two tests, assessing the difficulty of defining the specificity of their hormonal action.

In conclusion, the different SAR reported in the literature and the contradictions found in them reveals the weakness of the current proposals with regard to the structural requirements for high BR activity. The difference in the response of a BR observed among bioassays suggests that for each one different BR actions are measured. As was pointed out by Geuns (1983), BRs are active in various bioassays designed specifically for different plant growth regulators, so the specificity and usefulness of such bioassays must be treated with caution.

In spite of the particularity of each SAR, the structural requirements postulated up to now for high BR activity can be summarized as follows: i) $2\alpha,3\alpha$-diol in A-ring, ii) 7-oxalactone better than 6-ketone in B-ring, iii) A/B *trans* fused ring junction, iv) a *cis* C-22,C-23 diol preferentially with $22R,23R$ configuration, and v) a C-24 methyl or ethyl substituent.

In addition to the heterogeneous results depending on the bioassay, a closer look into the structural requirements established through a small number of BR analogs with few structural modifications shows that they are far from being general from two points of view.

First, the BR functionalities should not be considered independently as shown in the requirements postulated in the literature since we have found that they are closely related. Figure 2 shows the activity data of nine BRs measured using our modified RLIT (C. Brosa, L. Soca, and E. Terricabras, 1998, manuscript in preparation) based on the procedure developed by Takeno and Pharis (1982). From these data, it can be observed that compounds with a lactone in the B-ring (**2, 3, 4,** and **5**) elicit higher activity than their corresponding 6-ketone analogs (**8, 9, 10,** and **11**) according to the above requirements. Also, in agreement with them, compounds with $22S,23S$ configuration at the diol side chain (**3, 5, 9,** and **11**) are less active than the corresponding $22R,23R$ ones (**2, 4, 8,** and **10**), but the values strongly depend not only on the

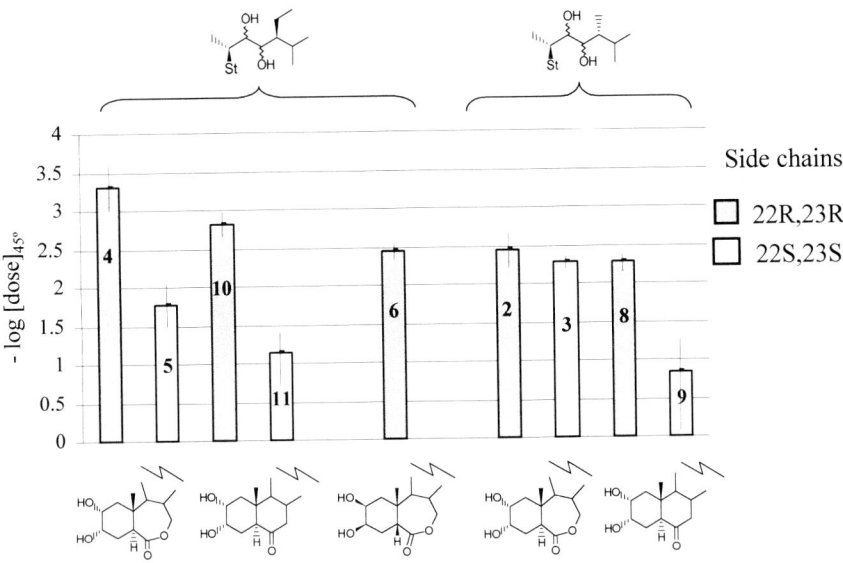

Fig. 2. Brassinosteroid activity expressed as -log [dose]$_{45}$ in the rise lamina inclination test. *Numbers on bars* correspond to compounds in Fig. 1

configuration of the diol at C-22,C-23 but also on the type of alkyl substituent at C-24. Thus, for a similar functionality in the skeleton, compounds with stigmastane side chain (**4**, **5**, **10**, and **11**) are more active than those with an ergostane side chain (**2**, **3**, **8**, and **9**) except for **3** which is more active than **5**. Moreover, regarding the configuration of the diol, the lactone **2** is less active than the ketone analog **10** when it is 22R,23R, thereby not meeting the requirements. When the configuration changes from 22R,23R to 22S,23S, however, the activity also changes and now becomes higher for **3** than for **11**. These results clearly indicate the existence of a relationship between functionalities and thus a strong interdependence among requirements ii), iv), and v). Furthermore, similar relationships are observed for the rest of the requirements.

The second point is that the postulated requirements limit the scope of applicability to the functionalities involved in the BRs tested set. For instance, they were developed without taking into account analogs having 2β,3β diol and/or A/B *cis* junction. No BRs with these functionalities were examined. Therefore, the question is why the stereochemistry of the glycol at C-2,C-3 and that of the A/B ring junction was fixed in the requirements as 2α,3α and A/B *trans*, respectively, if there was no proof of the activity elicited by compounds having the opposite ones. After synthesizing such compounds and evaluating their activity we obtained a very high activity for **6** with 2β,3β diol and A/B *cis* junction (Fig. 2) (Brosa et al 1994, 1996b).

The present discussion indicates the weakness of the postulated requirements in the literature for high BR activity. Thus, a more accurate way to define the SAR needs to be found, in which the overall change in the three-dimensional structure of the molecule should be taken into account.

3. New Approach for Structure-Activity Relationships Definition

Brassinosteroids may act at the molecular level through a receptor/ligand complex to regulate the expression of specific genes (Li et al 1996, Li and Chory 1997, see the chapter by S.D. Clouse and K.A. Feldmann, this volume). Therefore, the active BRs should have a single defined three-dimensional (3D) *"active conformation"* (AC) able to bind to the receptor. On this AC, the atoms directly involved in binding with the BR receptor ought to have the same spatial situation in all active molecules. Thus, the more complementary is the AC of a defined BR to the 3D structure of the receptor, the more active it should be. Therefore, a SAR that considered the AC would be a more rigorous way to establish the structural requirements, and might shed some light on the conformation of the BR bound to its receptor.

The establishment of a quantitative structure-activity relationship (QSAR) and the knowledge of which parts of the BR molecule are the most important ones in expressing activity would be useful for providing more information about the BR-receptor binding at the structural level. This also should lead, together with the extensive work being done at the molecular level (Clouse and Zurek 1991, Clouse 1996; see the chapter by S.D. Clouse and K.A. Feldmann, this volume), to a major advance in the understanding of BR action. Moreover, the QSAR will enable us to predict the activity of new analogs and will eventually be of help in the design of the most suitable BRs for agricultural application with the best synthetic cost-activity ratio. Since BL (**1**) is the most active natural BR found, we can assume that its AC will be the one that best fits the receptor and that it can thus be taken as reference.

One should take into account that the so-called AC is not necessarily the preferred conformation adopted in the steroid-receptor complex. Nevertheless, and due to the fact that the structure of the receptor is unknown at the moment, the information gained by comparing the AC of all the active BRs will be useful in defining a good SAR. Moreover, one can assume that the energy needed for the AC of various BRs to adopt the preferred conformation will not differ too much from one active BR to another.

Based on the above considerations, our strategy to go deeper into the knowledge of the BR mode of action at the structural level involves molecular modeling techniques, which also allows us to establish a QSAR. A detailed explanation of the methodology is shown in Fig. 3.

Thus, to establish a QSAR, a broad set of BRs having sufficient structural modifications, together with their corresponding strictly homogeneous activity data with statistical parameters, has been used (some of them are shown in Fig. 1). Considering only the active BRs of this set, a modified active analog approach (Klebe et al

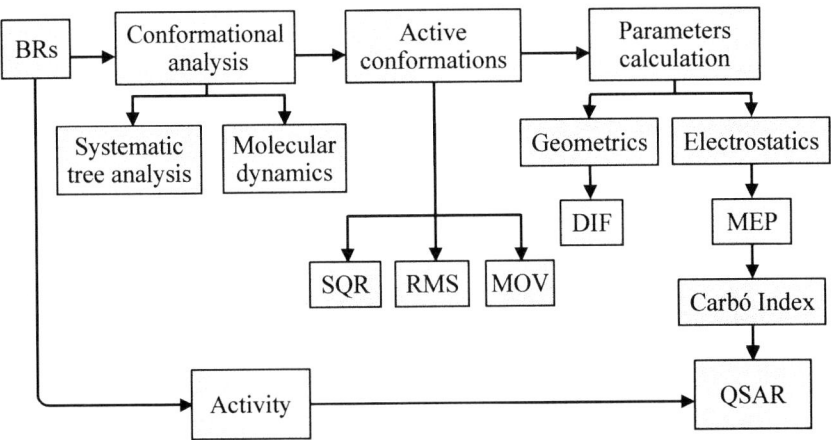

Fig. 3. Quantitative structure-activity relationship (QSAR) scheme. BRs, brassinosteroids; SQR, square residue; RMS, root mean square; MOV, molecular overlay volume similarity indices; DIF, sum of the absolute difference values for each distance between the oxygen atoms of brassinolide (BL, **1**) and that of the rest of BRs; MEP, molecular electrostatic potential

1994) has been used to obtain the AC for each BR. This study involves a conformational analysis and a 3D structure comparison. The conformational analysis has been carried out by a systematic tree analysis, corroborated by molecular dynamics. The AC of each compound has been selected among the conformers of each compound found using three different similarity indices of the 3D structures: the square residue (SQR), the root mean square (RMS), and the molecular overlay volume (MOV). For each AC, a set of geometric and electrostatic parameters has been calculated, and the correlation of these parameters with the activity has been investigated to establish the QSAR. A more detailed explanation of the methodology used in each of these steps as well as the results obtained is discussed in the following sections.

3.1 Activity Evaluation

The activity of BRs was evaluated first in the RLIT using intact rice seedlings as described by Takeno and Pharis (1982) but using *Bahia* as cultivar due to its higher sensitivity (Brosa 1997). Some modifications in the assay protocol were employed to improve the sensitivity and specificity. Figure 4a shows the increase observed in the sensitivity at a dose of 1 µg per plant for some BRs when the plants are maintained at 30° in the dark for 2 days instead of 2 days with a photoperiod of 16 hours light/8 hours dark after BR injection. This increase seems to be associated with the etiolated conditions used after the BR treatment.

This bioassay presents a drawback when doses higher than 2 to 5 µg per plant are applied. A lack of assimilation by the plant is observed as a white spot in the plant node, due probably to the high insolubility of some BRs in the biological medium.

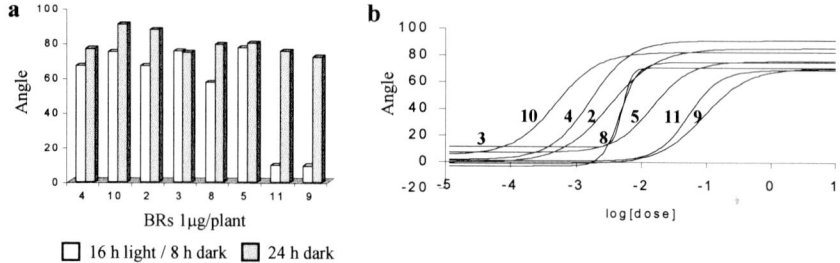

Fig. 4a-b. a Activity of a set of BRs in the rise lamina inclination test at a dose of 1 μg/plant and photoperiods of 16 h light/8 h dark (*in white*) and 24 h dark (*in gray*) after bassinosteroids injection. **b** Dose-dependent curves for **2** to **5** and **8** to **11** in the rice lamina inclination test with a photoperiod of 24 h dark

Table 1. Activity data and DIF and CI values similarity indices for BRs

Compounds	Activity	DIF	CI
1	5.93	0	1
4	3.29	0.9	0.82
10	2.80	2.6	0.85
2	2.45	0.8	0.97
6	2.45	4.2	0.76
8	2.27	2.9	0.81
3	2.08	2.6	0.74
17	1.91	2.2*	0.71
5	1.78	3.5	0.75
13	1.53	5.5	0.76
12	1.37	4.9	0.60
14	1.36	4.1	0.58
18	1.34	4.8*	0.63
11	1.14	12.7	0.74
15	1.13	7.4	0.48
20	1.00	1.1*	0.68
9	0.84	5.0	0.52
16	0.47	6.0	0.67
19	0.29	6.1*	0.68

Activity of BRs in the rice lamina inclination test expressed as -log $[dose]_{45°}$.
DIF, sum of the absolute difference values for each distance between the oxygen atoms of brassinolide (BL, **1**) and that of the rest of BRs; CI, Carbó electrostatic similarity index.
* DIF values lower with respect to the others because the lack of functionality at C-2 and/or C-3.

This fact has prevented determination of the maximum activity that can be obtained by low active BRs in their dose-dependent curves.

Following this new procedure, new dose-dependent curves have been obtained for BRs, some of them being shown in Fig. 4b. Not all the BRs reach the same maximum in the induced angle, being in a range from 65° to 100°. This suggests that not all the BRs act in the same manner at the doses where the activity is maximum. In fact, the macroscopic effect observed when rice seedlings are treated with BRs is the bending of its lamina. Various factors such as its binding affinity to the receptor, efficiency of transport through the plant, and so on are probably involved. If the affinity of binding to the receptor were related to the maximum bending elicited, the evaluation of this affinity could help in providing information about the specificity of the receptor with respect to the structural features of BRs.

The use of doses at 50% of the maximum activity is the parameter commonly accepted for SAR studies (Kubinyi 1993). In our case, 50% maximal activity corresponds, in most cases, to an angle of 45°. Since the complete dose-dependent curves for low activity compounds have not been obtained yet, the parameter used to compare the activity of BRs in our QSAR is expressed as minus logarithm of the dose at which the bending produced is 45° (-log $[dose]_{45°}$).

Table 1 shows the activity data for all the BRs shown in Fig. 1. Except BL (**1**) and castasterone (**7**), all of them have been synthesized by us (Ferrer et al 1990, Brosa et al 1992, 1994, 1996a, 1996b, 1996c, 1998a).

3.2 Conformational Analysis

Brassinosteroids present two flexible structural sites: the seven member B-ring in the case of lactones and the side chain. After having demonstrated the independence of the side chain conformation with respect to the geometry of the B-ring, the study was performed in two steps, taking advantage of the great distance between the two flexible points.

The first step was the conformational analysis of the B-ring without taking into account the side chain. This was carried out using different minimization over different starting geometry using the AM1[*1] and PM3[*2] semiempirical methods. The results were validated by performing a molecular dynamics simulation (C. Brosa, I. Zamora unpublished results, 1998) using the CVFF[*3] force-field included in the molecular modeling software package Discover[*4]. Two possible conformations for the seven member B-ring, with an energy difference of approximately 5 kcal/mol were found. Table 2 shows the torsion angles calculated for both conformers compared with the X-ray data for (22S,23S) 28-homobrassinolide (SHBL, **5**) (Kutschabsky et al 1990). It is worth pointing out that the torsion angles for the lower energy

---

[*1] Austin Model 1
[*2] Parametric Model 3
[*3] Consistent Valence Force Field
[*4] Discover 2.9.7. (1995) Biosym/MSI, 9685 Scranton Road, San Diego

Table 2. Torsion angles for the seven menber B ring of brassinolide (BL, **1**) obtained by molecular modeling and those of (22S,23S) 28-homobrassinolide (SHBL, **5**) from X-ray structure

Torsion Angle	Conformer A (°)	Conformer B (°)	X-ray (°)
C5-C6-O6-C7	86.06	14.52	4.5
C6-O6-C7-C8	-85.99	62.09	70.00
O6-C7-C8-C9	18.96	-92.19	-82.60
C7-C8-C9-C10	60.32	69.91	54.50
C8-C9-C10-C6	-89.32	-60.05	-52.30
C9-C10-C5-C6	60.28	76.91	77.90
C10-C5-C6-O6	-54.97	-78.63	-75.70

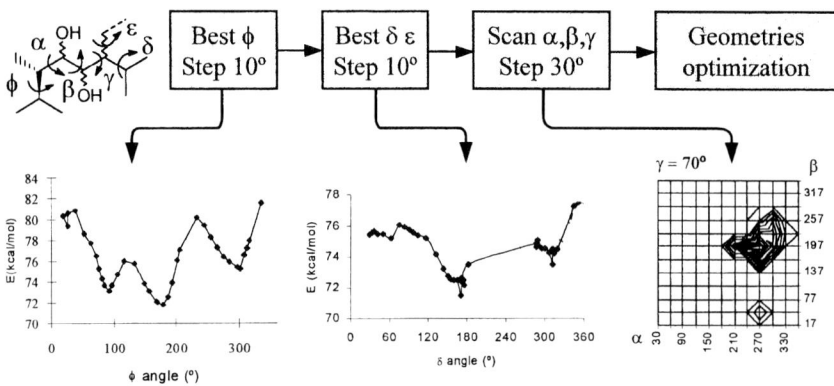

Fig. 5. Systematic tree analysis with some of the results obtained in the conformational analysis of BL (**1**). *Best*, angle rotation followed by energy optimization

conformer of B-ring, conformer B, are highly similar to those obtained by X-ray analysis.

Afterwards, the conformational analysis of the most common side chains present in BRs of campestane, ergostane, and stigmastane type having either a 22R,23R or a 22S,23S glycol was performed considering conformer B of the B-ring.

To perform this conformational study, we have considered all the bonds that can rotate through a defined angle (e.g., φ, α, β) to adopt a minimal energy conformation. Since the number of degrees of freedom is too large to perform an exhaustive analysis, a systematic tree analysis was planned in four steps for each type of side chain following the methodology shown in Fig. 5. In this figure, some of the results obtained for BL (**1**) as an example are presented. The order of how the angles have been taken in this methodology has been carefully selected to minimize steric interactions between substituents during the analysis.

Table 3. Number of conformers (NC) with minima energy for different BR side chains in a range of 5 kcal/mol and 10 kcal/mol

Side chains					
NC for 5 Kcal/mol	35	36	36	48	111
NC for 10 Kcal/mol	51	65	55	82	111

The force field MM2[*1] and the functions *best* and *scan* included in the MAD[*2] program have been used for the conformational analysis, and the geometry optimization has been performed with the HYPERCHEM[*3] program using the MM+[*4] force field.

Thus, starting with a *best* of the ϕ angle, which means rotation and optimization with a step of 10°, we obtained a set of angles with the corresponding energy. Considering only the ϕ angles with a ΔE lower than 5 kcal/mol, a *best* of δ and ϵ angles with a step of 10° was performed. With the dihedrals found within a range of 5 kcal/mol, a *scan* without minimization of α and β angles was performed for each γ angle with a step of 30° obtaining a set of contour maps. With the aim of avoiding some important conformer being discarded from the analysis, the study was repeated by extending the range to 10 kcal/mol. Finally, selecting the values of the angles with minimal energy level, followed by an optimization, we obtained the geometries or possible conformers in a range of 10 kcal/mol with respect to the absolute minimum value found for each compound. Although the number of possible conformers is different depending on the cut-off chosen, the selection of the AC gave the same result for both analyses.

Table 3 shows the number of possible conformers found for some of the side chain studies in a range of 5 and 10 kcal/mol, respectively. Whereas in the first case the number of possible conformations found are from 35 to 111 depending on the type of side chain, in the second case their numbers are from 51 to 111, and the side chain with 22S,23S-hydroxyl group and 24S-ethyl substituent presents the largest flexibility.

Similar to the conformational study of the B-ring, the analysis of the side chains has also been validated by molecular dynamic simulation, leading to the same results.

[*1] Molecular Mechanics 2
[*2] Molecular Advanced Design (1993) Oxford Molecular Ltd, Magdalen Center, Oxford Science Park, Sandford on Thames, Oxford, OX4 4G4, England
[*3] Hyperchem (1992) Autodesk Inc., Salsalito, CA USA
[*4] Molecular Mechanics Plus

3.3 Active Conformation Approach

In this section, the methodology followed to select the AC of a defined BR from all those found in the conformational analysis is described. This AC is expected to be the best fit with the receptor. The methodology consists of finding first the AC for BL (**1**), which is taken as reference, and then finding the AC for the rest of BRs by comparing their conformations with the AC found for BL (**1**).

The AC of BL (**1**) is obtained by comparing all its possible conformers with all the possible conformers of each of the other active BRs. Thus, the AC of BL (**1**) will be that which results in similarity to at least one conformer of all active BRs. In this study, the BR analogs **2**, **3**, **4**, **6**, **8**, and **10**, have been considered as the active ones (see Table 1). Once the AC of BL (**1**) is selected, the comparison between this conformer and all the possible conformers of a defined BR enables one to select its AC as the more similar conformation to that of BL (**1**).

On this AC approach three criteria were used to get a good discrimination over all of the possible conformers with minimum energy found in the conformational analysis: SQR, RMS, and MOV. All of them are based on different molecular similarity indices between BL (**1**) as reference and the rest of the active BRs. While the SQR and the RMS are based on geometrical similarity, the MOV takes into account the shape of the molecule. Figure 6a-c shows a graphical representation of the three criteria used.

In the first criterion, which considers the oxygen atoms' spatial orientation, the comparison with BL (**1**) was performed by means of SQR similarity index (SQR_{ij}). This involves the comparison of the distances (d) between the oxygen atoms in the molecule and between these atoms and defined planes obtained for each conformer (i) of BL (**1**) with those obtained for each conformer (j) of the other active compounds. The distances between oxygen atoms provide an idea of the spatial orientation of the groups that supposedly interact with the receptor. The distances between these atoms and defined planes describe the geometry of these with respect to the steroidal skeleton. The lower the SQR value for the conformers being compared, the greater will be its similarity. Figure 6a shows, by way of example, the conformer number 5 for BL (**1**) and the conformer number 27 for HBL (**4**). One of the planes as well as some of the distances measured are indicated in both cases. Equation 1 is used to calculate the SQR.

$$SQR_{ij} = \sum_{k=1}^{n°dist} \left(d_k^{1,i} - d_k^{BR,j}\right)^2 \qquad \text{Eq. 1}$$

Following the second criterion, the RMS similarity index (RMS_{ij}) was obtained after overlapping the C and D rings of each BL (**1**) conformer (i) with each one of the other compounds (j). Once they are superimposed, the distances between side-chain-related atoms and the A- and B-ring-related ones for each conformer are calculated. The RMS of these distances is the parameter used to define the similarity index: the lower the value the higher similarity. Figure 6b shows the overlap of the conformer 5 for BL (**1**) with two of the conformers (27 and 40) of HBL (**4**). Thus, whereas the

9 Structure-Activity Relationship

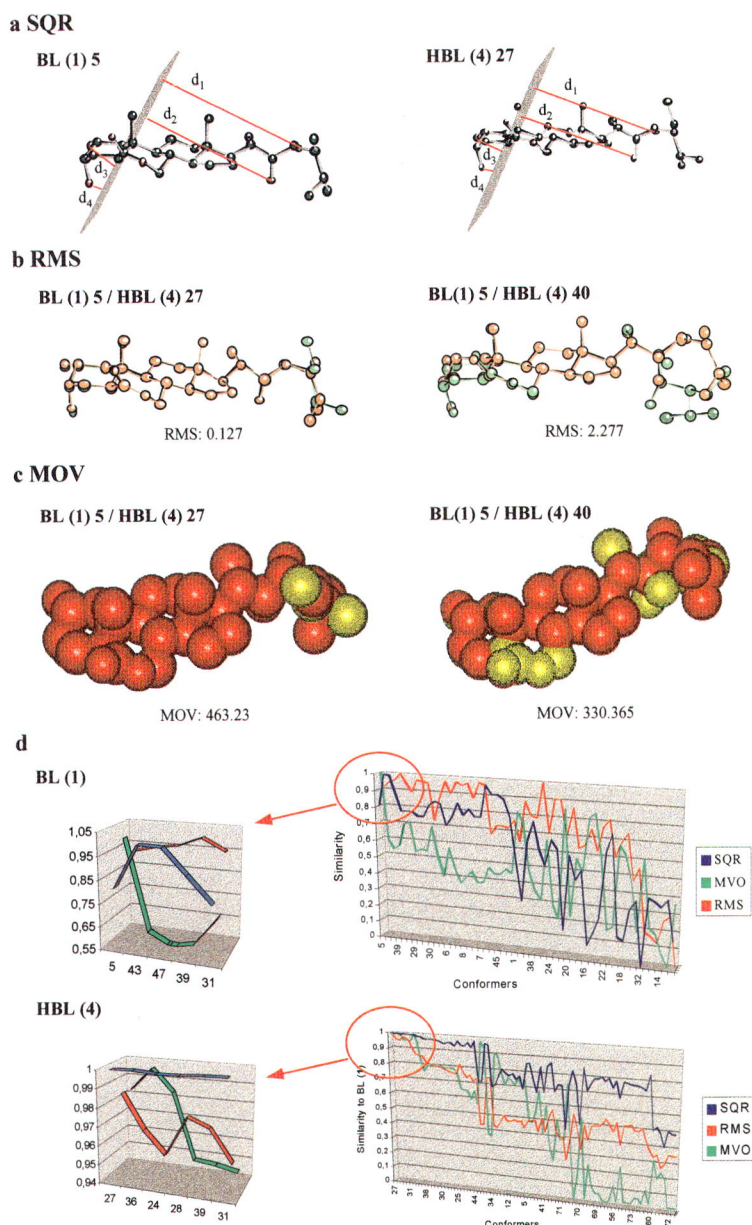

Fig. 6a-d. Conformation analysis for BL (**1**) (conformer 5) and 28-homobrassinolide (HBL, **4**) (conformers 27 to 40) following three criteria. **a** SQR. **b** RMS. **c** MOV. **d** Autoscaled results obtained by applying the three criteria over all the possible conformers for BL (**1**) and HBL (**4**)

conformers 5 for **1** and 27 for **4** show a good overlap (RMS = 0.127), the conformer number 40 for **4** with respect to the conformer 5 of **1** differs in the ring A and the side chain (RMS = 2.277). Equation 2 is used to calculate the RMS.

$$RMS_{ij} = \sqrt{\frac{\sum_{k=1}^{n^\circ dist} (d_k^{ij})^2}{n^\circ dist.}} \qquad \text{Eq. 2}$$

The last criterion used is the MOV. Once the C and D rings are superimposed, the overlap volume is calculated over all the atoms on the side chain only to obtain a better discrimination due to the high skeleton similarity for all the BRs studied. In this case, the higher the MOV value the greater the similarity. In Fig. 6c the conformers overlaid to calculate the MOV are the same as in the other criteria, giving the same conclusion as using the RMS criterion. Thus, a higher similarity is obtained between the conformers 5 for BL (**1**) and 27 for HBL (**4**) (MOV = 463.23) than between 5 for **1** and 40 for **4** (MOV = 330.365). Summarizing, the AC for each compound will be the one that minimizes the first and second criteria and maximizes the third one.

Figure 6d represents, by way of example, the autoscaled results obtained by applying the three criteria over all the possible conformers found in the conformational analysis for BL (**1**) as well as for HBL (**4**). Conformer number 5 has been chosen as the AC of BL (**1**) because it is the most similar by all the criteria to at least one conformer of all the active BRs used in this analysis, **2**, **3**, **4**, **6**, **8**, and **10**.

As has been indicated before, once the AC of BL (**1**) is chosen, the methodology is repeated for each active and nonactive BR using the AC of BL (**1**) as reference. Thus, the conformer number 27 for HBL (**4**) (Fig. 6d) is the most similar to the conformer number 5 for BL (**1**) in all the criteria used, and therefore it has been chosen as the AC for HBL (**4**).

In parallel with our work, an approach to molecular modeling has been reported by McMorris et al (1994). The distances between a reference point, C-16, on the rigid steroid skeleton and various atoms on the flexible side chain of BL (**1**), 24-epibrassinolide (EBL, **2**), HBL (**4**), and 22S,23S diastereoisomers of the last two (**3**, and **5**) were determined for the energy-minimized conformation calculated using the PC Model molecular modeling software for the Silicon Graphics 4D workstation. They found that all the distances for SHBL (**5**) differed substantially more from those corresponding to BL (**1**) than in the case of the other compounds, in accordance with the lower activity elicited by **5** in the soybean elongation assay. Therefore, they suggested that some relationship between these distances and the activity may be present. This finding is in agreement with what has been observed by us. Nevertheless, the simplifications made in the molecular modeling study should be treated with caution if one takes into account that there is more than one possible minimum energy conformation in a range of 5 kcal/mol for the side chain of all of these compounds. Moreover, there is a great difference between some of the conformers found, mainly in the spatial situation of the side chain atoms as is shown in

Fig. 6b, the accurate selection of the best conformer being one of the most critical steps in the molecular modeling analysis.

Also, a side chain conformation analysis of **1** and **2** was recently reported based on NMR spectroscopy (Stoldt et al 1997). In this case, the conformer found for BL (**1**) strongly differs to the one proposed by us (Fig. 6).

3.4 Calculation of Geometric and Electrostatic Parameters

Once the active conformation for all compounds was found, the QSAR required an alignment rule. To obtain a preliminary idea of which functionalities were important for activity, a Free-Wilson analysis (Kubinyi and Kehrhahn 1976) was performed. The *RR* configuration for the side chain hydroxy groups and α,α for A-ring diol were shown to be the more significant groups, accounting for 35% and 25%, respectively, of the total activity. These findings are in accordance with the two binding subsites in the receptor postulated by Takatsuto et al (1983b), but they considered the affinity of the A ring diol stronger than that of the side chain. Although they did not mention any interaction of the B-ring with the receptor, the functionalities present in this ring must influence it in some way. In fact, the activity diminishes when there is a 6-ketone, or practically disappears when there is a 6-oxalactone (Takatsuto et al 1987), 6-oxa (Brosa et al 1997), or no functionality (Yokota et al 1983) instead of the 7-oxalactone present in BL (**1**) (Table 1). Thus, the B-ring has been taken into account for a complete description of the activity, although its contribution was found to be less important in the Free-Wilson analysis (11%) (Brosa et al 1996b).

Using the AC of 20 BRs, a set of geometrical and electrostatic parameters were calculated; and in preliminary work (Brosa et al 1996b) a linear correlation between the activity and some parameters based on distances of atomic charges and Van der Waals radii was found. However, with the information gained by this correlation it was not possible to determine which atoms were really involved at these distances. A more accurate study indicated that the distances between the oxygen atoms at C-2,

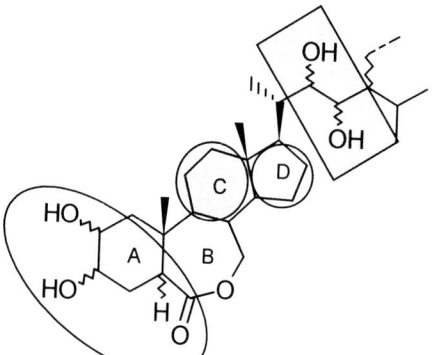

Fig. 7. Pairs of atoms that correlate with the activity. The oxygen atoms at C-2, C-3, or C-6 and the hydrogen or carbon atoms at the D ring (*white areas*), and the oxygen atoms at the C-22 or C-23 and the hydrogen or carbon atoms at the C ring (*shadow areas*)

BL (1)

Dist.	O-2	O-3	Ocarb	O22	O23
O-2	-	2.8	6.2	11.1	12.4
O-3	2.8	-	5.1	10.8	12.6
Ocarb	6.2	5.0	-	10.7	12.3
O-22	11.0	10.8	10.7	-	2.6
O-23	12.6	12.9	12.4	2.5	-

HBL (4)

BL (1)

Dist.	O-2	O-3	Ocarb	O22	O23
O-2	-	2.8	6.2	11.1	12.4
O-3	2.7	-	5.1	10.8	12.6
Ocarb	6.4	4.9	-	10.7	12.3
O-22	12.4	12.4	10.6	-	2.6
O-23	12.1	12.1	11.3	2.6	-

SHCS (11)

Fig. 8. Distances between oxygen atoms at C-2, C-3, C-6, C-22, and C-23 for 28-homobrassinolide (HBL, **4**) and (22S,23S)-28-homocastasterone (SHCS, **11**) (*white area*) and for brassinolide (BL, **1**) (*shadowed area*)

C-3, or C-6 and the hydrogen or carbon atoms in the D-ring (between nonshadow areas, Fig. 7), as well as the distances between the oxygen atoms at the C-22 or C-23 and the hydrogen or carbon atoms in the C-ring (between shadow areas, Fig. 7) matched with those found in the correlation. Because there are no structural modifications on the C- and D-rings in all the BRs studied, one could assume that the activity depends on the spatial situation of the oxygen atoms concerned.

Thus, the greater the difference of the distances between the oxygen atoms of a defined BR with respect to the related distances for BL (**1**), the lower will be its activity. Figure 8 summarizes these findings. Thus a BR will be more active the closer its values in the white area are to the ones for BL (**1**) in the shadowed area. In this figure there are two BR examples: HBL (**4**) eliciting high activity (3.29), and (22S,23S) 28-homocastasterone (SHCS, **11**) with much lower response (1.14). Whereas the values are quite similar to those of BL (**1**) in the first one, they differ significantly in the second.

In considering this, the geometry similarity index [DIF] was defined as the sum of the absolute difference values for each distance between the oxygen atoms of BL (**1**) and that of the rest of BRs. Therefore, the lower the DIF value, the more similar is the compound to BL (**1**). It can be observed in Table 1 that all the active compounds present a DIF value lower than 5, and in general they correlate well with the activity. Some examples of this correlation will be discussed later. Although there are some exceptions, it points to the importance of these oxygen atoms in the activity.

However, this DIF has its drawbacks. On one hand, it only considers the oxygen atoms' spatial position, not the type of functionality involved. For instance, the carbonyl oxygen of the lactone or ketone at the B-ring is considered similar to any of the hydroxyl oxygens both at the A-ring and the side chain, but they may well act in a different way in binding to the receptor. On the other hand, the information gained in the active conformation approach about the spatial situation of the oxygen atoms is lost when the absolute value of the distance is used for calculating the DIF value.

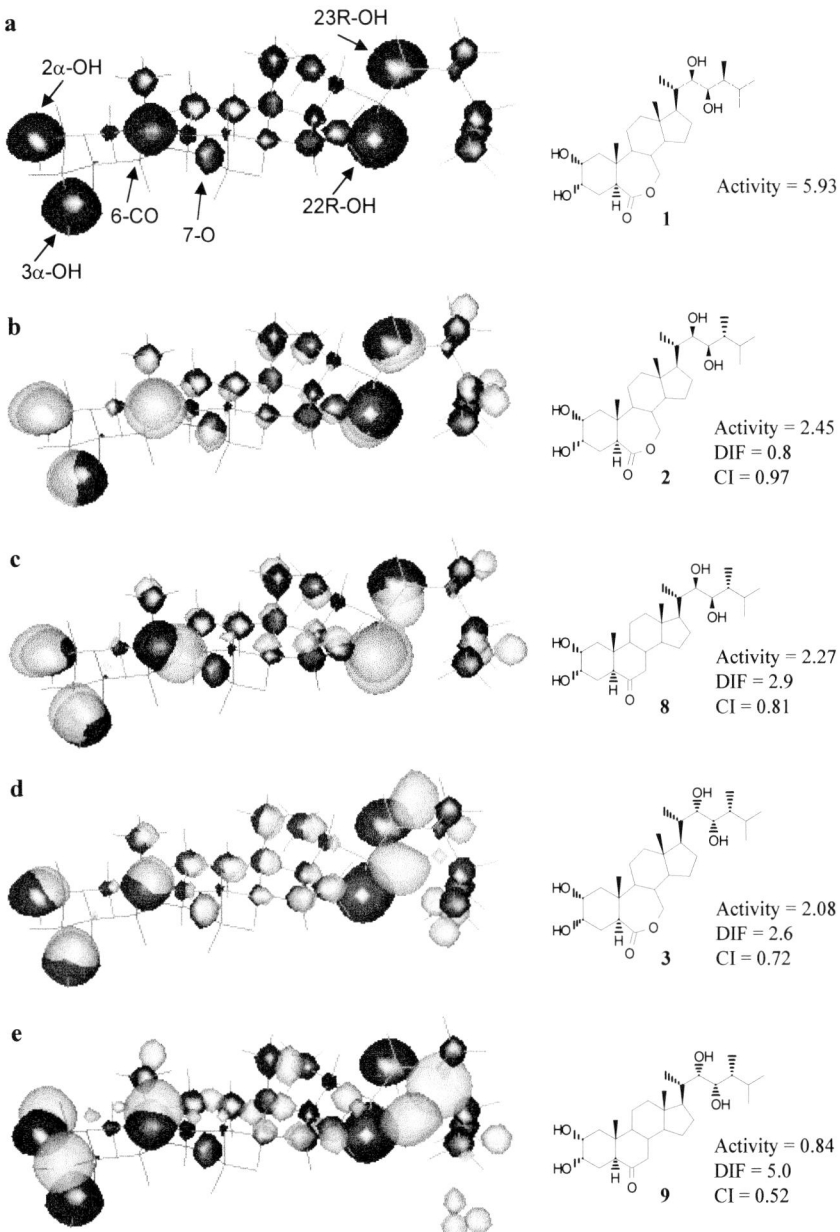

Fig. 9. a MEP map for BL (**1**). **b-e** MEP maps overlapping between BL (**1**) (*in black*) and **2**, **8**, **3**, and **9** (*in grey*). Activity [-log (dose)$_{45°}$] in rice lamina inclination test, DIF values, and CI values

For instance, the distance between C3-O and C22-O for EBL (**2**) (11.1 Å) is the same as that of (22S, 23S) 24-epibrassinolide (SEBL, **3**) (11.1 Å), although the C22-O has a different spatial situation in the two molecules (Fig. 9b,d). Moreover, the distances between defined atoms are not different enough (an average of 0.5 Å) between two compounds to allow good discrimination.

Having determined by means of our preliminary QSAR (Brosa et al 1996b) that the atomic charges play an important role in describing the activity, the alignment of the structures using electrostatic parameters seemed to be appropriate and the molecular electrostatic potential (MEP) was selected. The alignment of the structures was performed by superimposing the MEP map of BL (**1**) on that of the other compounds by means of the molecular similarity program ASP[*1], using the gaussian function (Good et al 1992) option with three terms to fit the distance inverse dependence. Different Carbó similarity indices (Carbó et al 1980) were obtained from MEP - electrostatic, shape, and lipophilicity – and tried to be correlated with the activity. From this analysis, only the electrostatic Carbó similarity index (CI) was found to be related to the activity, and it was used to quantify the similarity between two molecules.

Table 1 shows the CI value obtained for some of the BRs studied. With some exceptions, the activity correlates with the CI and is higher when the CI value is higher than 0.75. When the CI value is lower than 0.75, the correlation with the activity becomes worse and it is much difficult to discriminate each compound due to its structural similarity with respect to the total of the molecule because all of them have the same skeleton. One should take into account that the CI of two BRs differing in all the typical functional groups is, at least, 0.5.

Figure 9a-e graphically represents the MEP map at -30 isovalues for BL (**1**) (in black) and the overlap with those for EBL (**2**), 24-epicastasterone (ECS, **8**), SEBL (**3**), and (22S,23S)-24-epicastasterone (SECS, **9**) (in grey). Their CI, DIF value, and activity are also indicated.

Looking at Fig. 9b, all electrostatic potential sites corresponding to the four hydroxy groups and the lactone of the B-ring of EBL (**2**) (in grey) overlap perfectly with those of BL (**1**) (in black), hence with a very high CI and in agreement with its high activity. Only the alkylic part of the side chain does not overlap at all, as in the other examples shown. While the sites corresponding to the 2α-OH and the carbonyl of the lactone **2** lie in front of **1**, the black area being in back of the grey one, the 22R,23R glycol of **1** lies in front of **2**, the black area being over the grey one. In the case of ECS (**8**) (Fig. 9c) and comparing it with EBL (**2**), only the region of the B-ring does not superimpose well with the lactone of BL (**1**), thus decreasing its CI value as well as its activity. With **3** (Fig. 9d) a great difference is observed in both 22S and 23S hydroxy groups and the activity and the CI value fall again. Finally, the very low activity of SECS (**9**) (Fig. 9e), a compound where neither the region of the A-ring and side chain nor of the B-ring overlays well with BL (**1**), is in agreement

---

[*1] ASP V3.11 (1997) Oxford Molecular Ltd. Magdalen Center, Oxford Science Park, Sandforf on Thames, Oxford, OX4 4GA, England

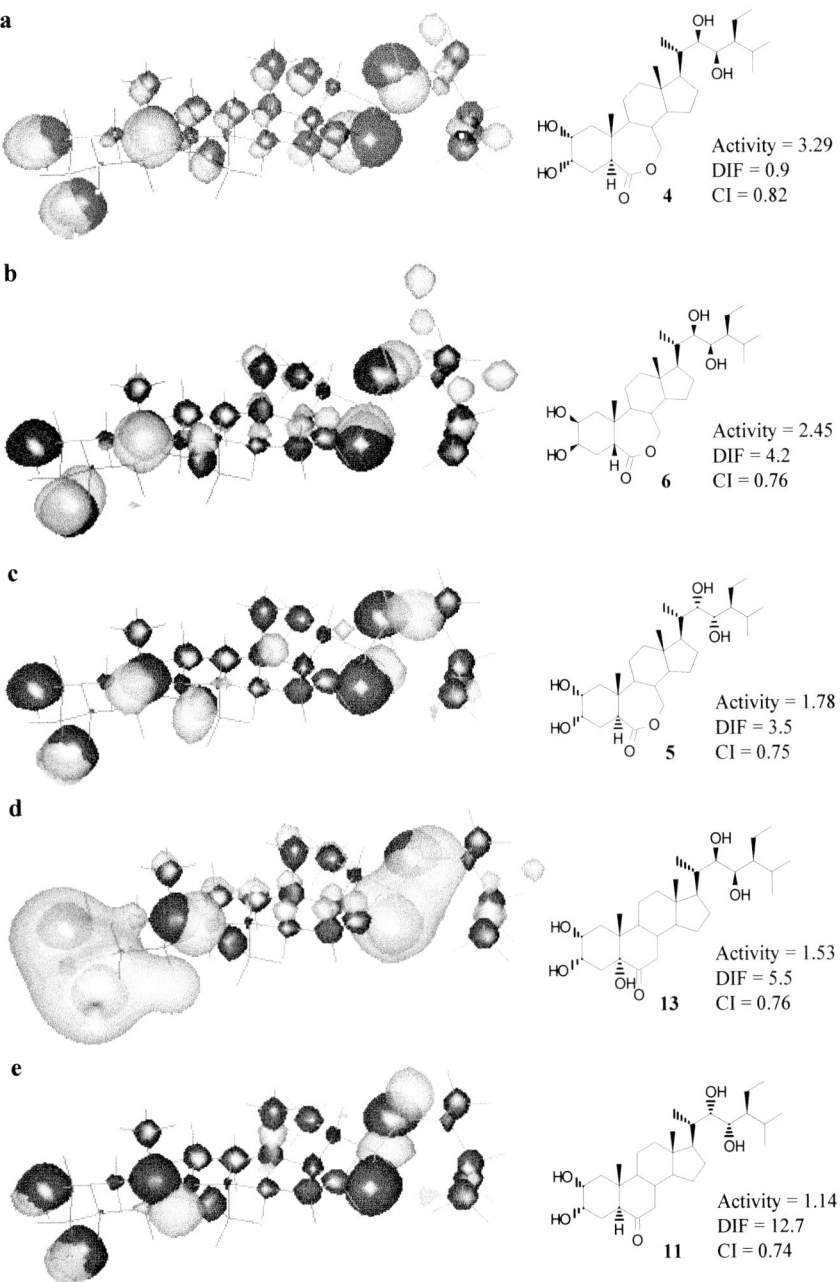

Fig. 10a-e. MEP maps overlapping between BL (1) (*in black*) and **4, 6, 5, 13,** and **11** (*in grey*). Activity [-log (dose)$_{45°}$] in rice lamina inclination test, DIF values, and CI values

with its very low CI value. For all the compounds shown in Fig. 9, the activity decreases with decreasing CI and increasing DIF values. Therefore, a good correlation is observed between activity and the DIF and CI similarity indices, but the CI seems to correlate better than the DIF index. It is important to remark here that **3** and **8** gave similar DIF values in spite of the very different spatial arrangement of the 22,23-glycol in both compounds, as a consequence of the loss of the three-dimensional information when the DIF is calculated as mentioned above.

Coming back to the discussion on the structural requirements postulated earlier (see section 2, this chapter), notice that all of the compounds here described fulfill those requirements despite their great differences in activity, which ranged from 100% for BL (**1**) to 3×10^{-4}% for **9**. Moreover, the comparison between Figs. 9b and 9d strengthen our finding about the close relationship between the functionalities present in a BR that are necessary to express activity. A single change in the configuration of the side-chain diol from *RR* to *SS* (**2** to **3**) not only modifies the spatial distribution of the side chain diols but also the overall shape of the molecule. The overlap of the lactone at the B ring and the $2\alpha,3\alpha$ diol between **1** and **2** is reduced in the case of **1** and **3** (Fig. 9d). More severe are the differences observed in the case of **8** (Fig. 9c) and **9** (Fig. 9e). So it seems that the CI similarity index could be a more precise way to establish the requirements.

On the contrary, in some cases, the CI does not explain the activity completely, and it is the DIF value that correlates better with it. An example of this is shown in Fig. 10b-e where the overlap MEP maps between BL (**1**) and four BR analogs, **5**, **6**, **11**, and **13** are shown, all of them having the same CI but eliciting different activity in accordance with their DIF values. In the case of **5**, the black area corresponding to the 2α-OH for BL (**1**) fully overlaps the grey area of **5**. On the other hand, it should be noted that the large areas observed for the two hydroxy groups regions of the hydroxyketone **13** are surrounding those of BL (**1**).

It seems that the differences observed in different parts of the molecules compensate each other, giving the same CI value. In these cases, only the DIF value allows us to discriminate between them because of the different spatial location of the oxygen atoms.

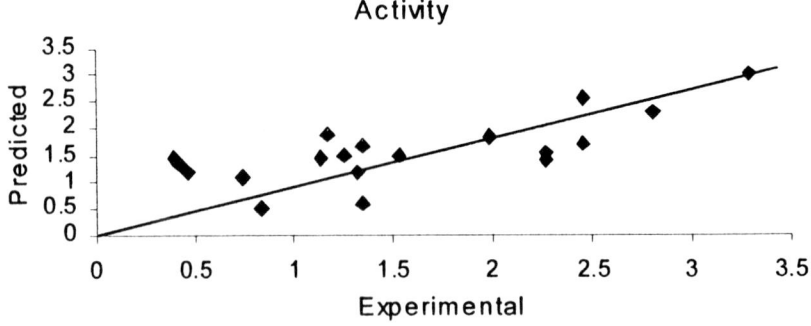

Fig. 11. Correlation between the activity predicted by means of a QSAR and the experimental one obtained in the rive lamina inclination test

One should take into account that different parts of the molecule could contribute in different degrees to the activity through different types of interaction with the receptor: electrostatic, hydrophobic, hydrogen bonding, and so on. Therefore, the changes introduced in specific parts of the molecule can produce a higher or lower activity modification depending on the contribution of this part of the molecule to the overall activity. Even in the case where only a minor structural modification is introduced without affecting the CI value, the higher the contribution of the modified zone of the molecule the greater is the change produced in the activity.

An important fact is observed in the case of lactone **6** where the activity, similar to that of EBL (**2**), is higher than that expected from its CI value. Notice that all the electrostatic potential sites of this compound overlay very well with those of BL (**1**) although having 2β,3β-diol and A/B *cis* junction, except the region near the 2β-OH, which is located very far from the 2α-OH for BL (**1**). These findings suggest that this group, or better this zone, is important for expressing high activity, as is the case of BL (**1**), but is not essential for eliciting high activity such as that in compound **6**.

3.5 Quantitative Structure-Activity Relationship

The last step shown in Fig. 3 is the establishment of a QSAR. At the moment, a good enough linear correlation has been obtained with the TSAR[*1] program using a multivariable regression method between the CI electrostatic similarity index and the activity for a set of BRs. From this result, it can be concluded that the higher the electrostatic similarity index (CI) of a compound with respect to BL (**1**), the more active it will be. Further work focused on improving this correlation is currently in progress. Figure 11 shows the correlation obtained between the activity predicted with the QSAR and the experimental one obtained in the RLIT.

4. Feasibility of Brassinosteroids Hydrogen Bonding with the Receptor

When considering that BR action may take place through a receptor-ligand complex, another interesting point to be determined concerns the type of interactions that can take place on binding and which of them contribute in a major way to the activity. In this sense, hydrogen bonding (H-bonding) interactions are being specifically analyzed. Therefore, if the region where the probability to form an H-bond is found, the interaction energy for these compounds is calculated, and some relationship is observed with the activity, it would indicate that this type of interaction may take place in the BR-receptor complex. Moreover, besides gaining more information about the mechanism of the BR-receptor interaction, it could help to discriminate between BRs and thus in the establishment of the SAR.

[*1] TSAR (1993) Oxford Molecular Ltd, Magdalen Center, Oxford Science Park, Sandford on Thames, Oxford, OX4 4GA, England

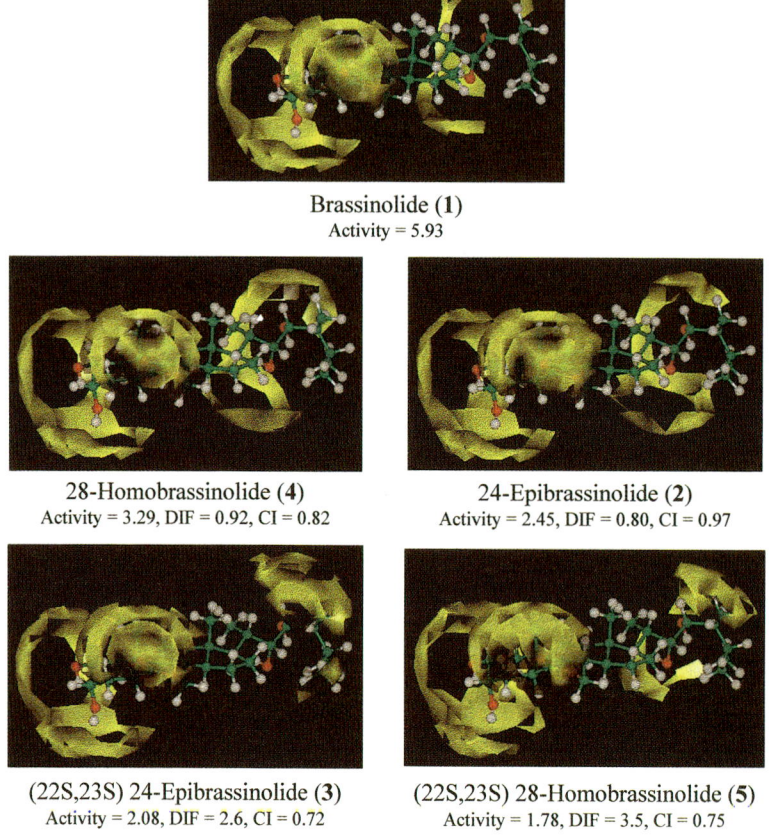

Fig. 12. GRID maps (areas with a higher probability to form H-bonds) using water probe at -3 kcal/mol, activity in rice lamina inclination test [-log $(dose)_{45}$], DIF value, and CI value for BRs **1** to **5**

The GRID force field (Goodford 1985, Wade et al 1993) has been used to calculate the interaction energy between the BR and different probes that simulate different types of interactions, e.g., H-bonding, steric, hydrophobic. Among the probes tested, water has been chosen to calculate the interaction energy of H-bonding owing to its capability to act as both acceptor and donor of H-bonds. The results currently obtained are analyzed in the following sections: side chain, A-ring, B-ring, and miscellaneous. It has been demonstrated throughout this chapter that small changes produced in one part of the molecule affect, in most cases, the overall spatial distribution of the entire molecule. To minimize this effect, the set of BRs to be compared in the following sections contain only structural modifications in the region studied.

9 Structure-Activity Relationship

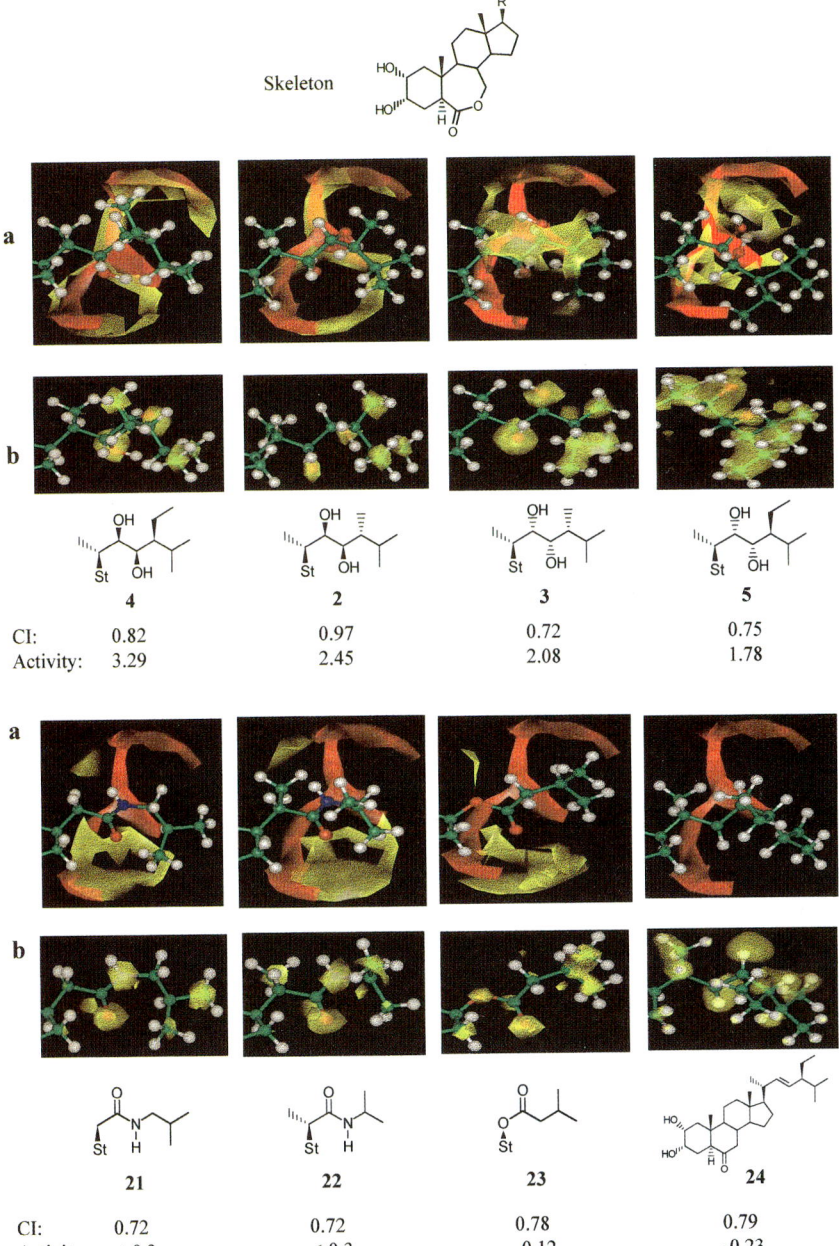

Fig. 13a,b. Side chain BRs analogs. **a** Overlapping of GRID maps for water probe at -3 KT between BL (**1**) (*in red*) and **2**, **3**, **4**, **5**, **21**, **22**, **23**, and **24** (*in yellow*). **b** Surface difference in MEP at -30 KT (*in yellow*) between BL (**1**) and **2**, **3**, **4**, **5**, **21**, **22**, **23**, and **24**. Activity [-log (dose)$_{45°}$] in rice lamina inclination test, and CI value

4.1 Side Chain Analogs

Figure 12 shows the global GRID map at a cutoff of -3 KT for **1** to **5** using water as probe. These compounds differ only in the alkyl substituent at C-24 and in the configuration of the diols at C-22,C23. The areas with a higher probability to form H-bonds are indicated in yellow. In the case of BL (**1**), the shape of the zones corresponding to the hydroxy groups on the A-ring and the side chain are similar. They are located on the same side facing the back. The zone corresponding to the lactone on the B-ring is pointing to the front.

By comparing these five BRs, one can observe that the larger difference is found in the side chain, in accordance with their structural features. To see better the spatial location of these areas, Fig. 13 only represents the side chain of BL (**1**) (in red) overlapped with each one of the other compounds, **2** to **5** (in yellow). Each picture has been rotated to show better the differences between side chains. For compounds **2** and **4**, the areas corresponding to the 23-OH are similar to that of BL (**1**). The great difference between **1** and 4 is a right shift in the area corresponding to the 22-OH group of **4**, and the activity decreases from 5.93 for **1** to 3.29 for **4**. This area is further shifted in the case of **2**, and the activity falls to 2.45. For **3**, the activity of which is 2.08, only a small overlap is observed between the areas corresponding to the 23-OH, and the area corresponding to the 22-OH is located in another plane. Finally there is **5**, with an activity of 1.78, for which no overlap is observed between the two hydroxy groups of the side chain, but the area corresponding to the high probability of H-bonding for the 23-OH of **5** is not so far from the one of **1**. So it seems that, even if the 22-OH were not perfectly located in the site, compounds having a zone with high probability of H-bonding near to the one of the 23-OH of BL (**1**) elicit important activity. The closer this zone is to the one of BL (**1**), the higher is the activity, e.g., compounds **2** to **5**.

This is in full agreement with the results obtained for the BR analogs **21**, **22**, and **23** (Fig. 13) synthesized by Kohout et al (1991), having amide or ester functions on the side chain but a $2\alpha,3\alpha$-diol at the A ring and a lactone at the B ring (Brosa et al (1998b). On them, only a zone close to the 22-OH is observed. In our RLIT, **21**, **22**, and **23** elicited only marginal activity, at least at doses lower than 2 µg per plant, although these compounds have elicited some activity in the bean second internode bioassay. As indicated in the introduction of this chapter, this is another example of the contradictions found when BRs are evaluated in different bioassays. Similarly, the marginal activity is shown by **24** where no oxygenate functions are present in the side chain.

So it seems that the location of the area corresponding to the 22-OH near that of BL (**1**) is important for high activity. However, the essential area with the highest probability of H-bonding to express activity must to be located in a specific zone and close to that of the 23-OH of BL (**1**)

Except for HBL (**4**) and EBL (**2**), all the BR analogs shown in Fig. 13 have similar CI values, indicating that the structural modifications among them are not enough to produce a change in the global electrostatic similarity index. On the con-

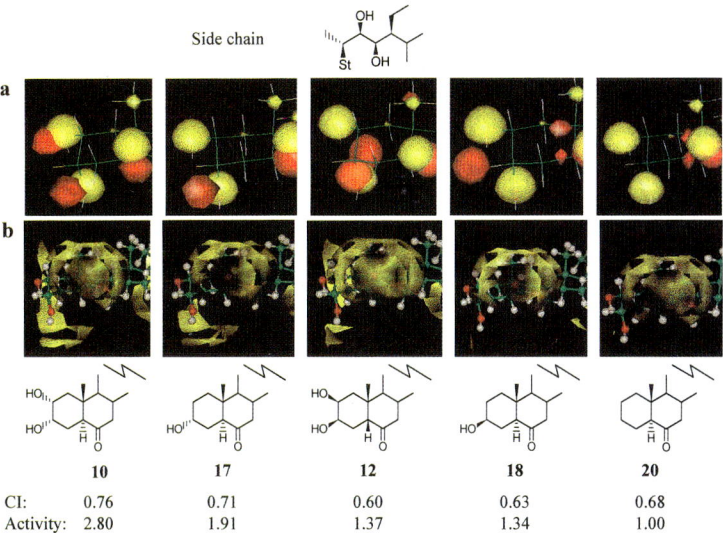

Fig. 14a,b. A ring BR analogs. **a** Overlapping of MEP maps at -30 KT between BL (**1**) (*in yellow*) and **10**, **12**, **17**, **18**, and **20** (*in red*). **b** Intersection of GRID maps (*in yellow*) for a water probe between BL (**1**) and **10**, **12**, **17**, **18**, and **20** at -6 KT. Activity [-log (dose)$_{45}$] in rice lamina inclination test, and CI value

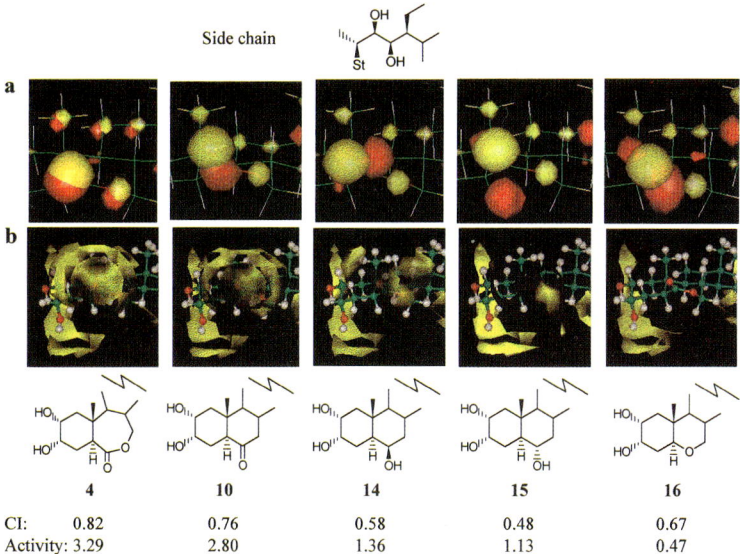

Fig. 15a,b. B ring BR analogs. **a** Overlapping of MEP maps at -30 KT between (**1**) (*in yellow*) and **4**, **10**, **14**, **15**, and **16** (*in red*). **b** Intersection of GRID maps (*in yellow*) for a water probe at -6KT between BL (**1**) and **4**, **10**, **14**, **15**, and **16**. Activity [-log (dose)$_{45}$] in rice lamina inclination test, and CI value

trary, a better explanation of the differences observed in the activity can been found by considering the feasibility of H-bonding,

The large difference in the activity observed when the 24S-methyl of BL (**1**) (5.93) is changed by a 24S-ethyl (**4**) (3.29) or a 24R-methyl (**2**) (2.45) is currently under study.

4.2 A-ring Analogs

Figure 14a shows the overlapped MEP maps at -30 KT for the A-ring and part of the B-ring between BL (**1**) (in yellow) and five BRs differing only in the-A ring, **10**, **12**, **17**, **18**, and **20** (in red). Figure 14b shows the intersection of GRID maps for a water probe of BL (**1**) and the same compounds (in yellow) at -6 KT. Thus, only the common areas between the two compounds to be compared are shown. Compounds **17** and **18** have been synthesized by L. Galagowsky (1997, unpublished results) and evaluated using our RLIT. The overlap of the rest of the molecule is similar for all the compounds.

Similar to the case of the side chains, no correlation between the CI and the activity has been detected for the series of A-ring analogs. Inspecting the MEP and GRID maps reveals some interesting findings. HCS (**10**) is the compound with the higher activity and has a 2α-OH. The rest of them without this functionality (**17**, **18** and **20**) or even having it with the opposite configuration (**12**) elicit some activity in a range of 1.91 for **17** to 1.00 for **20**. For all of them, the closer the MEP area for the 3-OH is to that of BL (**1**), the higher is the activity.

Another important observation is the different participation in activity of the side chain and the A-ring diols. By comparing **20** (activity = 1.00) (Fig. 14) lacking the A-ring diol with **24** (activity = -0.23) (Fig. 13) lacking the side chain diol, it was found that the H-bonding ability in the side chain is more important than that in the-A ring.

4.3 B-Ring Analogs

Figure 15a shows the overlapping MEP maps at -30 KT for the A- and B-rings, between BL (**1**) (in yellow) and five analogs differing only in the B-ring (in red). The difference between the MEP for BL (**1**) and each compound increases while the activity and CI value decreases except for the 6-oxabrassinosteroid **16**, whose CI value is higher than that expected. This is another example of the good correlation between CI and activity.

The lack of correlation for **16** can be explained when considering the inability of the ether function to form H-bonds. Figure 15b shows the intersection of GRID maps for a water probe (in yellow) between BL (**1**) and the B-ring analogs at -6 KT. The area on the B-ring where the probability to form an H-bond is higher progressively decreases in the order of **4**, **10**, **14**, **15**, and **16**, for which it completely disappears. Qualitatively there is a close relationship between this area and the activity.

9 Structure-Activity Relationship

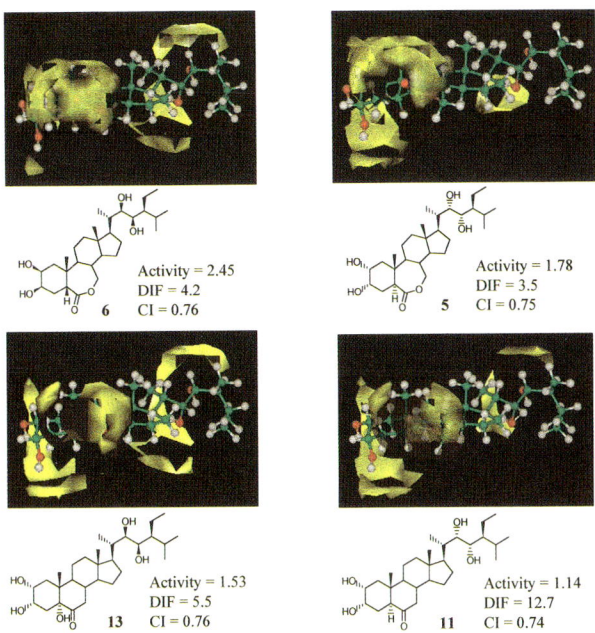

Fig. 16. Intersection of GRID maps (*in yellow*) for a water probe at -6 KT between BL (**1**) and **5**, **6**, **11**, and **13**. Activity, DIF, and CI

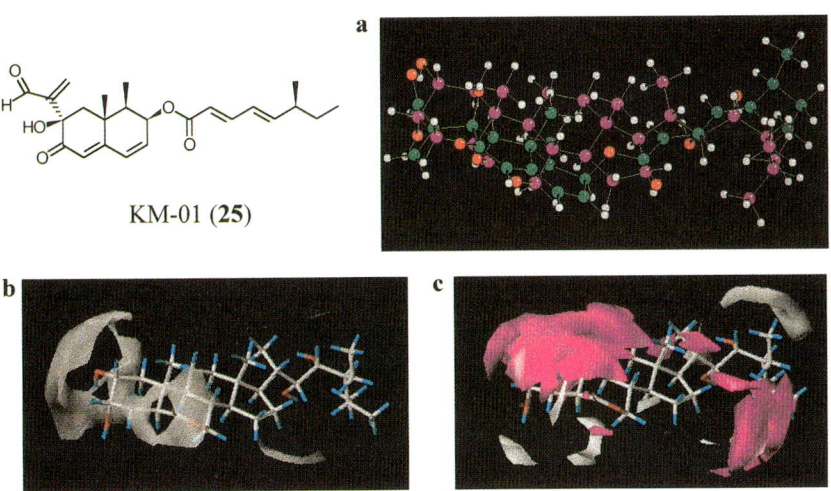

Fig. 17a-c. **a** Overlap between BL (**1**) (*in purple*) and KM-01 (**25**) (*in green*). **b** Nonspecific strong interactions for BL (**1**) and KM-01 (**25**) (*in white*). **c** Overlapping of specific strong interactions for BL (**1**) (*in white*) and KM-01 (**25**) (*in purple*)

Therefore, the inability to form an H-bond on the B ring of **16** is in agreement with its very low activity even with a CI value higher than that expected. Again, H-bonding analysis by a GRID search suggests that activity correlates somehow with H-bonding capability.

4.4 Miscellaneous

In this section, the same type of analysis is applied to H-bonding capability for the four compounds **5, 6, 11,** and **13**, that have the same CI value but elicit different activity. Figure 16 shows the intersection of GRID maps for a water probe (in yellow) between BL (**1**) and these four compounds at -6 KT. The effect of the electronegative functions at C-3 and C-23 are in agreement with the observations pointed out in the previous sections.

With respect to the contribution in the activity of the B-ring, it looks like the GRID surfaces wrap the upper part of the B-ring. A shift of this surface toward the lower region of the same ring or toward the C-ring produces a decrease in activity, in agreement with that observed in Fig. 15b. The hydroxyketone **13** has another factor that could decrease the ability of H-bonding with the receptor owing to its capability of forming an intramolecular H-bond between the 3α-OH and the 5α-OH (Brosa et al 1998).

5. Brassinosteroids Inhibitor

The first BL inhibitor, named KM-01 (**25**) (Fig. 20), was isolated from the fungus *Drechslera avenae* by Kim et al (1994). Its activity was first evaluated in the Raphanus test and RLIT. Whereas it was not active in both tests when applied alone, the co-application with BL (**1**) drastically reduced the hypocotyl elongation of radish and the bending angle of rice seedlings. Afterward it was evaluated in several bioassays (Kim et al 1995) indicating that no inhibitory effect on the activity of auxins, cytokinins, and ethylene were observed, but a synergistic effect with gibberellin (GA_3) and an inhibitory effect on abcisic acid were observed. Therefore it is the first selective BR inhibitor found in natural sources.

Compound **25** appears to be an antagonist of BR action and may compete with BRs for binding to the receptor at the same site. Therefore it should share some structural and electrostatic features. Accordingly, the same model used to establish the SAR is currently been applied to **25**, and our preliminary results are discussed below.

To find the conformation of **25** most similar to that of BL (**1**) a simultaneous optimization of the CI with respect to the BL (**1**) and of the energy was performed with the ASP program. Two similarity indices were calculated: the electrostatic CI and a shape similarity index. The high CI value obtained (0.729) indicates the great similarity in the electrostatic charge distribution. On the other hand, the low shape similarity found (0.484) indicates geometrical differences.

Although the structures of KM-01 (**25**) and BL (**1**) are very different, their overall three-dimensional shape overlap well as shown in Fig. 17a. In this figure, where **1** is represented in green and **25** in purple, one can observe that the aldehyde group of **25** is located close to the 2α-OH of **1**, the ketone of **25** is near the C-6 lactone carbonyl of **1**, the carbonyl ester of **25** is oriented to the 22R-OH of **1**, and part of the side chain of **25** is close to that of **1**.

The methodology developed by Cruciani and Goodford (1994) has been used to determine the essential zones for binding and for eliciting activity. The procedure consists of evaluating the interaction between a molecule and different probes, followed by an analysis of principal components. Several probes have been used in this study: hydrophobics, sterics, electrostatics, and H-bonding.

As has been indicated, if the BR and the inhibitor bind to the same receptor in a similar way, they should have minimal structural similarity, which will be responsible for binding to the receptor. Figure 17b shows the common *nonspecific* areas (in white) for **1** and **25**, corresponding to the zones of the 2α-OH and C-6 carbonyl for **1**, and most likely responsible for binding. Figure 17c shows the overlap of the strong and *specific* interactions for **1** (in white) and **25** (in purple). The white areas of **1** correspond to the 3α-OH, 23R-OH, and C-6 carbonyl, which could be mainly responsible for eliciting activity. Notice that these areas, although smaller, fit with the GRID map found for BL (**1**) (Fig. 12). The areas corresponding to the B ring are complementary in both Figs. 17b and 17c. Therefore, while the areas of Fig. 17b are those responsible for binding, the white areas of Fig. 17c are those responsible for activity.

6. Conclusions

This chapter has presented the state of the art in structure-activity relationships for BRs, and it is the first reported evidence of the limited scope of applicability and weakness of the structural requirements proposed in the literature for eliciting high BR activity. With the aim of looking for a new strategy to define these requirements, a model based on brassinosteroid-receptor interaction that provides a useful way to explain the activity of different BRs from the structural point of view has been described.

By means of molecular modeling techniques it has been demonstrated that the electrostatic charges play an important role in the description of the activity as evidenced by its good correlation with the DIF value or with the Carbó electrostatic potential similarity index (CI) in most of the cases studied. However, due to the loss of 3D information when the DIF value is calculated, and the limited structural differences between the set of BRs examined, the activity cannot be completely explained with these two similarity index in some cases.

A better explanation of the differences observed is found by considering the putative H-bonding interactions in the BR-receptor complex. Based on the GRID methodology over the set of BRs studied, the results obtained at present have allowed us to provide defined information about the areas of the molecule responsible for binding and for eliciting activity. The electronegative part of the side chain seems to be more important than that of the A-ring diol, in accordance with a Free Wilson analysis. Moreover, our findings suggest that, whereas the areas located near the C-22 and C-2 hydroxy groups of BL (1) are responsible for binding, the ones near the C-23 and C-3 hydroxy groups seem to be responsible for eliciting activity. With respect to the B-ring, the contribution to binding should be located in the lower region of the ring and that responsible for eliciting activity in its upper part.

These results, fit perfectly well with those obtained with the BR inhibitor KM-01, although being the last ones preliminary. More precise knowledge about the putative receptor structure has been acquired when including the inhibitor in our studies.

In summary, the methodology developed to find a relationship between structure and activity opens a new way to explain the activity of different BRs from the structural point of view.

References

Brosa C (1997) Biological effects of brassinosteroids. In: Parish EJ, Nes D (Eds) Biochemistry and Functions of Sterols. CRC Press, Boca Raton, pp 201-220
Brosa C, Peracaula R, Puig R, et al (1992) Use of dihydroquinidine 9-O-(9'-phenantryl) ether in osmium-catalized asymmetric dihydroxylation in the synthesis of brassinosteroids. Tetrahedron Lett 33: 7057-7060
Brosa C, Nusimovich S, Peracaula R (1994) Synthesis of new brassinosteroids with potential activity as antiecdysteroids. Steroids 59: 463-467
Brosa C, Soca L, Terricabras E, et al (1996a) Brassinosteroids: Looking for a practical solution. Proc Plant Growth Regul Soc Am 23: 21-26
Brosa C, Zamora I, Capdevila JM (1996b) Brassinosteroids: A new way to define the structural requirements. Tetrahedron 52: 2435-2448
Brosa C, Puig R, Comas X, et al (1996c) New synthetic strategy for the synthesis of 24-epibrassinolide. Steroids 61: 540-543
Brosa C, Zamora I, Terricabras E, et al (1997) Synthesis and molecular modeling: related approaches to the progress in brassinosteroid research. Lipids 32: 1341-1347
Brosa C, Soca L, Terricabras E, et al (1998a) New synthetic brassinosteroids: a 5α-hydroxy-6-ketone analog with strong plant growth promoting activity. Tetrahedron 54: 12337-12348
Brosa C, Zamora I, Kohout L, et al (1998b) The effect of electrostatic properties and feasibility to form H-bonds on the activity of brassinosteroid side chain analogs. Collect Czech Chem Commun 63: 1635-1645
Carbó R, Leyda L, Arnau M (1980) An electron density measure of the similarity between two compounds. Int. J Quantum Chem 17: 1185.
Cerana R, Lado P, Anastasia M, et al (1984) Regulating effects of brassinosteroids and of sterols on growth and H^+ secretion in maize roots. Z Pflanzenphysiol 114: 221-225
Clouse SD (1996) Plant hormones: brassinosteroids in the spotlight. Curr Biol 6: 658-661
Clouse SD, Zurek DM (1991) Molecular analysis of brassinolide action in plant growth and

development. In: Cutler HG, Yokota T, Adam G (Eds) Brassinosteroids: Chemistry, Bioactivity and Applications. ACS Symp Ser 474. Amer Chem Soc, Washington, DC, pp 122-140

Cruciani G, Goodford PJ (1994) A search for specificity in DNA-drug interactions. J Mol Graphics 12: 116-129

Cutler HG, Yokota T, Adam G (1991) Brassinosteroids: Chemistry, Bioactivity and Applications. ACS Symp Ser 474, Amer Chem Soc, Washington, DC

Ferrer JC, Lalueza R, Brosa C, et al (1990) Short step synthesis of (22E,24R)-5α-ergosta-2,22-dien-6-one, a key intermediate for the preparation of 24-epibrassinolide. Tetrahedron Lett 27: 3941-3942

Fujioka S, Inoue T, Takatsuto S, et al (1995) Biological activities of biosynthetically-related congeners of brassinolide. Biosci Biotechnol Biochem 59: 1973-1975

Geuns JMC (1983) Plant steroid hormones. Biochem Soc Trans 11: 543-548

Good AC, Hodgkin EE, Richards WG (1992) J Chem Inf Comput Sci 32: 188

Good AC, Utilization of gaussian function for the rapid evaluation of molecular similarity

Goodford PJ (1985) A computational procedure for determining energically favorable binding sites on biologically important macromolecules. J Med Chem 28: 849-857

Grove MD, Spencer GF, Rohwedder WK, et al (1979) Brassinolide, a plant growth-promoting steroid isolated from *Brassica napus* pollen. Nature 281: 216-217

Kim S-K, Mizuno K, Hatori M, et al (1994) A brassinolide-inhibitor KM-01, its isolation and structure elucidation from a fungus *Drechslera avenae*. Tetrahedron Lett 35: 1731-1743

Kim S-K, Asano T, Marumo S (1995) Biological activity of brassinosteroid inhibitor KM01 produced by a fungus *Drechslera avenae*. Biosci Biotech Biochem 59: 1394-1397

Klebe G, Mietzner T, Weber F (1994) Different approaches toward an automatic structural alignment of drug molecules: applications to sterol mimics, thrombine and thermolysin inhibitors. J Comput Aided Mol Design 8: 751-778

Kohout L Strnad M (1989) Brassinolide analogs without a side chain. Collect Czech Chem Commun 54: 1019-1027

Kohout L Strnad M (1992) Brassinosteroids with ester function with five carbon atoms at the 20 position. Collect Czech Chem Commun 57: 1731-1738

Kohout L, Velgova H, Strnad M, et al (1987) Steroids. Part CCCXXVIII. Brassinosteroids with androstane and pregnane skeleton. Collect Czech Chem Commun 52: 476

Kohout L, Strnad M, Kamìnek M (1991) Types of brassinosteroids and their bioassays. In: Cutler, HG, Yokota, T, Adam, G (Eds) Brassinosteroids: Chemistry, Bioactivity and Applications. ACS Symp Ser 474. Amer Chem Soc, Washington, DC, pp 56-73

Kohout L, Kasal A, Strnad M (1996) Steroids. Part CCCLXXIII. Pregnane-type brassinosteroids with a four-carbon ester functionality in position 20. Collect Czech Chem Commun 61: 930-940

Kondo H Mori K (1983) Synthesis of brassinolide analogs with and without the steroidal side chain. Agric Biol Chem 47: 97-102

Kubinyi H (1993) QSAR:Hansch analysis and related approaches. In: Mannhold R, Krogsgaard-Larsen P, Timmerman H (Eds) Methods and Principles in Medicinal Chemistry. VCH Publishers, New York

Kubinyi H, Kehrhahn O (1976) Quantitative structure-activity relationships. 1. The modified Free-Wilson approach. J Med Chem 19: 578-586

Kutschabsky L, Adam G, Vorbrodt H-H (1990) Molecular and crystal structure of (22S,23S) homobrassinolide. Z Chem 30: 136

Li J, Nagpal P, Vitart V, et al (1996) A role for brassinosteroids in light-dependent development of *Arabidopsis*. Science 272: 398-401

Li J, Chory J (1997) A putative leucine-rich repent receptor kinase involved in brassinosteroid signal transduction. Cell 90: 929-938

McMorris TC, Patil PA, Chavez RG, et al (1994) Synthesis and biological activity of 28-homobrassinolide and analogues. Phytochemistry 36: 585-589

Stoldt M, Porzel A, Adam G, et al (1997) Side chain conformation of the growth-promoting phytohormones brassinolide and 24-epibrassinolide. Magn Reson Chem 35: 629-636

Takatsuto S, Ikekawa N (1984) Synthesis and activity of plant growth-promoting steroids, (22R, 23R, 24S)-28-homobrassinosteroids, with modifications in rings A and B. J Chem Soc Perkin Trans I 1984: 439-447

Takatsuto S, Yazawa N, Ikekawa N (1984) Synthesis and biological activity of brassinolide analogues, 26,27-bisnorbrassinolide and its 6-oxo analogues. Phytochemistry 23: 525-528

Takatsuto S, Yazawa N, Ikekawa N, et al (1983a) Synthesis of (24R)-28-homobrassinolide analogues and structure-activity relationships of brassinosteroids in the rice lamina inclination test. Phytochemistry 22: 1393-1397

Takatsuto S, Yazawa N, Ikekawa N, et al (1983b) Structure-activity relationship of brassinosteroids. Phytochemistry 22: 2437-2441

Takatsuto S, Ikekawa N, Morishita T, et al (1987) Structure-activity relationship of brassinosteroids with respect to the A/B-ring functional groups. Chem Pharm Bull 35: 211-216

Takeno K, Pharis RP (1982) Brassinosteroid-induced bending of the leaf lamina of dwarf rice seedlings: an auxin-mediated phenomenon. Plant Cell Physiol 23: 1275-1281

Thompson MJ, Meudt WJ, Mandava NB, et al (1982) Synthesis of brassinosteroids and relationship of structure to plant growth-promoting effects. Steroids 39: 89-105

Wada K, Marumo S (1981) Synthesis and plant growth-promoting activity of brassinolide analogues. Agric Biol Chem 45: 2579-2585

Wade RC, Clark KJ, Goodford PJ (1993) Further development of hydrogen bond functions for use in determining energically favorable binding sites on molecules of known structure. 1. Ligand probe groups with the ability to form two hydrogen bonds. J Med Chem 36: 140-147

Yokota T, Mori K (1992) Molecular structure and biological activity of brassinolide and related brassinosteroids. In: Bohl M, Duax WL (Eds) Molecular Structure and Biological Activity of Steroids. CRC, Boca Raton, pp 317-340

Yokota T, Morita M, Takahashi N (1983) 6-Deoxycastasterone and 6-deoxodolichosterone: putative precursors for brassinolide-related steroids from *Phaseolus vulgaris*. Agric Biol Chem 47: 2149-2151

10
Practical Application of Brassinosteroids in Agricultural Fields

YASUO KAMURO[1] AND SUGURU TAKATSUTO[2]

[1] BAL Planning Co., Ltd., 2-15-16, Hanaike, Ichinomiya, Aichi 491, Japan
[2] Department of Chemistry, Joetsu University of Education, Joetsu, Niigata 943, Japan

1. Introduction

In 1979, brassinolide was isolated from the pollen of rape and its structure determined (Grove et al 1979). Because brassinolide is the first bioactive steroid with high plant growth-promoting activity (Mandava et al 1981, Mandava 1988, Yopp et al 1981, Adam et al 1991), much effort was devoted to the synthesis of brassinolide, naturally occurring analogs, and several related bioactive steroids, which are collectively called brassinosteroids (BRs). Having these steroids in hand, extensive research of BRs started simultaneously in the fields of phytophysiological activity, mode of action, chemical synthetic methodology, structure-activity relationship, microanalysis, distribution in the plant kingdom, search for new related compounds, biosynthesis, metabolism, and application in agricultural fields (see each preceding chapter, this book).

With respect to the plant growth-regulating effect of BRs on field crops, many research results were reported within 10 years after the discovery of brassinolide. In the latter half of the 1980s, three bioactive BRs, brassinolide, 24-epibrassinolide, and (22S,23S)-28-homobrassinolide (Fig. 1), were selected as practical candidates for agricultural uses, and their field tests were extensively and intensively carried out in Japan, China, and many other countries. These three BRs showed almost equally high biological activity in a wide variety of bioassay systems. The results strongly suggested the possibility that the plant growth-regulating effects elicited by BRs could be applied to a wide variety of agricultural fields. However, as the plant materials changed from explants to young plants and then to adult plants, and as the

Key words: brassinolide, 24-epibrassinolide, (22S,23S)-28-homobrassinolide, TS303, long-lasting activity, practical test, germination, vegetative growth, rooting, reproductive development, seed-setting, fruit-setting, environmental stress, combination effect, TNZ303

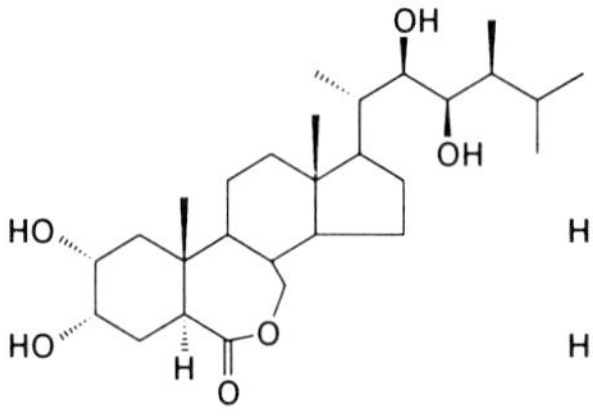

Brassinolide

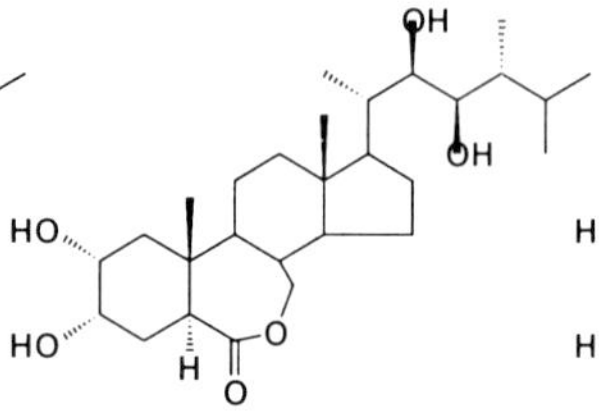

24-Epibrassinolide

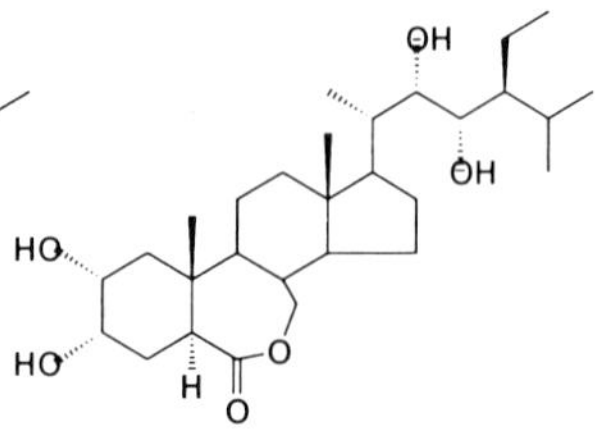

(22S,23S)-28-Homobrassinolide

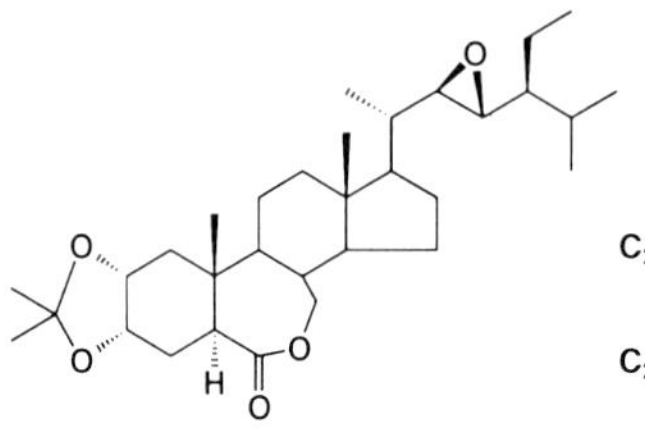

Compound No. 6

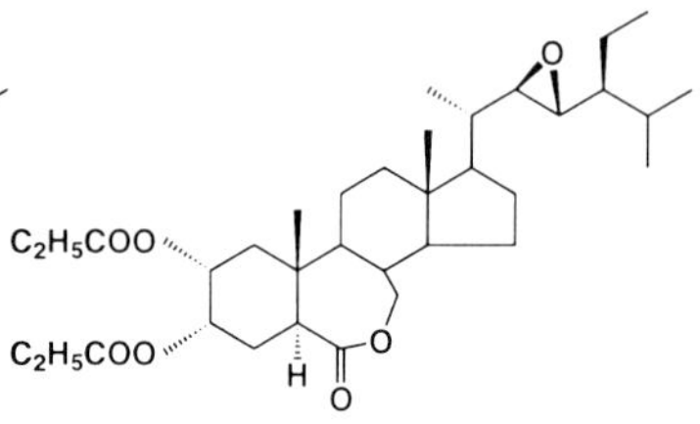

TS303

Fig. 1. Structures of brassinolide and related steroids.

growth conditions changed from growth chambers to greenhouses and then fields, the plant growth-regulating effects gradually decreased and stable, significant practical effects on crops were not obtained. Therefore, field trials with these three BRs are now suspended in Japan and European countries. There is still the possibility that the field trials for practical use will be resumed if the formulation and treatment method of these BRs can be improved. On the other hand, in China (22*S*,23*S*)-28-homobrassinolide has been registered as a plant growth regulator to increase the yields of tobacco, sugar cane, rape seeds, tea, and some fruits (registration nos. LS94572 and LS94573), while in Russia and Belarus 24-epibrassinolide has been registered as a regulator of potato, tomato, cucumber, pepper, and barley (Khripach et al 1997).

Nonetheless, searching for structurally related compounds with greater stability under field test conditions is the most basic and important solution to the problem. From this point of view, we have made an effort to find such a practical compound for many years. We have screened a large number of BR-related compounds, not with high biological activity at the bioassay level but with stable biological effect at the field crop level. As a consequence, we have selected a BR analog TS303 (see Fig. 1) as a practical candidate for agricultural application; it has been shown to be superior to brassinolide, 24-epibrassinolide, and (22*S*,23*S*)-28-homobrassinolide under a wide variety of field test conditions. Official tests of TS303 are being carried out in many countries so it can be registered as a plant growth regulator. There is no other compound now being tested officially for registration. We describe in this chapter the plant growth-regulating effects of BRs and TS303 in comparison with those of brassinolide.

2. Application of Brassinosteroids to Crops

Brassinolide has unique multiple effects on a wide variety of growth phenomena in the life cycle of plants at a very low concentration, 1×10^{-10} M (see the chapter by J Sasse, this volume). Development of large-scale synthesis of BRs allowed extensive application studies of BRs in agricultural fields, especially in Japan. In the latter half of the 1980s, there were many reports suggesting that BRs will be put into practical use in the near future. Several reviews summarizing these details are available (Hamada 1986, Mandava 1988, Ikekawa and Zhao 1991, Takeuchi et al 1992, Sasse 1997, Khripach et al 1997).

Many pesticide companies have undertaken practical experiments using brassinolide, 24-epibrassinolide, and (22S,23S)-28-homobrassinolide. To our disappointment, as the tests have extended to the farmer's level, the extent of effectiveness and the stability of the results have decreased as described above. However, many scientists have expressed consistent interest in the specific plant growth-regulating effects elicited by BRs. In this section, we summarize the practical results by testing the three BRs on field crops, mainly by referring to the papers published in the 1990s.

2.1 Promoting Effect on Germination

From the beginning of the practical application of BRs, it has been well known that germination has been promoted by treatment of seeds with BRs before sowing. The effects on seeds were expressed as stimulation of germination, increased germination percentage, promotion of rooting and root growth, increased plant growth of the seedlings, and increased tillers and vegetative buds. As a consequence, these effects bring about yield increases as a whole (Takematsu and Takeuchi 1989, Takahashi et al 1994). In Russia and Belarus, 24-epibrassinolide has been officially registered as a seed treatment agent for various crops (Khripach et al 1997). Furthermore, it has been reported that BRs have promoted germination of not only crop seeds but also the seeds of weeds and parasitic plants (Takeuchi et al 1991, 1995, 1996). They suggested that BRs will be used practically for germination control of weeds.

With respect to seed treatment, soaking seeds in an aqueous solution of test compounds is good for rice seeds, for which water absorption presents no practical problem. On the other hand, the seeds of leguminous crops should be sprayed, coated, or encrusted with test compounds. In the case of the soaking treatment of the seeds of leguminous crops, absorption of water by the seeds before sowing has had a bad effect on their germination (Y. Kamuro, unpublished work, 1994).

2.2 Promotion of Vegetative Growth and Rooting

Among the reports of the practical application of BRs, those dealing with the promotive effect of BRs on vegetative growth (Brawn and Wild 1984, Takematsu and Takeuchi 1989, Takahashi et al 1994, Pipattanawong et al 1996) and rooting of cuttings (Iwahori et al 1990, Yoshioka et al 1990, Roddick and Guan 1991, Roddick and Ikekawa 1992, Sasse and Sands 1992, Ronsch et al 1993, Sasse and Sasse 1994) are the most predominant. Much attention has been paid to the mode of action of BRs that affects qualitatively the morphogenesis of plants, such as the increases of the number of leaves, leaf area, fresh weight and dry weight of foliage and roots, foliage age, and the number and growth of productive branches and tillers. As a consequence, it is thought that BRs promote the number of ears in gramineous crops, pods in leguminous crops, fruits, and tubers, resulting in yield increases of these crops. BRs are also reported to promote of root nodule development (Takahashi et al 1994) and absorption of effective components of manure or fertilizer and of minerals in soils (Pirogovskaya et al 1996, Ronsch et al 1996). These multiple modes of action are characteristic of BRs.

2.3 Effect on Reproductive Growth

Brassinolides had a positive effect on the flowering of a long-day plant, *Raphanus sativus*, but no such effect on that of a short-day plant, *Perilla frutescence* (Suge 1986). Brassinolide induced sex modification in the staminate inflorescence in *Luffa cylindrica* (Suge 1986). By spraying satsuma mandarin with BRs in mid-winter, the number of its flowers decreased (Sugiyama 1994). In grapefruits, spraying with BRs in autumn increased the number of its flowers, while such treatment in late winter decreased the number of its flowers (Yoshioka et al 1990). In strawberry, BRs have a promotive effect on the number of inflorescence and flowers when applied by spray treatment (Pipattanawong et al 1996). Yield increases are confirmed by many field tests using cereals and leguminous crops. Thus, BRs may have a positive influence on reproductive growth. It is unknown how BRs affect reproductive growth. It is desirable that further study should be undertaken on the practical effect and mode of action of BRs from these viewpoints.

2.4 Effects on Seed- and Fruit-Setting

Many reports have been published on the promotive effects of BRs on the yield increases of field crops such as cereals (Takematsu and Takeuchi 1989, Hirai et al 1991, Fujii et al 1991, Fujii and Saka 1992, Yand et al 1992, Kamuro and Takatsuto 1992), legumes (Takematsu and Takeuchi 1989, Kamuro and Takatsuto 1992), fruits (Sugiyama and Kuraishi 1989, Maotani et al 1989, Iwahori et al 1990, Kamuro and Takatsuto 1992, Wang et al 1994, Xu et al 1994a), and rapeseed (Takematsu and Takeuchi 1989). These effects seems to be mediated through the promotion of fertilization and fructification. It is known that BR promotes of germination and growth of pollen tubes (Hewitt et al 1985) and helps in overcoming the effect of self-incompatibility (Lee et al 1990). Meanwhile, the effects of BRs on seed- and fruit-setting may also be derived from a combination of preventive effects on flower and fruit drop (Maotani et al 1989, Iwahori et al 1990, Xu et al 1994a), enhancement of photosynthetic capacity (Brawn and Wild 1984), and translocation of photosynthates (Fujii and Saka 1992).

The major difficulty in putting brassinolide, 24-epibrassinolide, and (22S,23S)-28-homobrassinolide into practical use is the low stability of their effects in field trials. Kamuro and Takatsuto (1991) have pointed out that these BRs have an effect only for a short time. In the pot test or small scale field test, it is possible to treat the plants with BRs at the exact time of flowering to show a constant promotive effect on fertilization. However, in field crops, the period of flowering ranges from one week to several weeks or months. At the field level, each plant in the community does not come to flower on the same day but, rather, flowers more often on different days. As a consequence, it is thought that effective results have not been achieved by use of BRs with low persistence of biological effect. Although the crop yield can be increased by repeated treatment with these BRs, this multiple treatment is apparently an impractical solution. Selection of the compound with long-lasting biological activity and its field test results will be described in the latter section of this chapter.

2.5 Resistance Against Environmental Stress

Among environmental factors affecting the growth of crops, temperature conditions are especially important. The promotive effects by BRs were obtained especially under unfavorable low temperature conditions (Hamada 1986, Takematsu and Takeuchi 1989, Takeuchi et al 1996). The yield of field crops is directly influenced by germination percentage, rooting percentage, and seed- or fruit-set percentage. These yield components greatly depend on the temperature at the time of seed sowing, transplanting, and flowering. It is hoped that stability and improvement of agricultural production can be achieved by application of BRs, which may positively influence these components.

Brassinosteroids strengthen drought resistance and show increasing effect on plant growth and yield, even under soil water deficit conditions (Shen et al 1990, Singh et al 1993, Sairam 1994, Xu et al 1994b). Furthermore, BRs exhibit additional effects such as salt tolerance (Takematsu and Takeuchi 1989, Sasse et al 1995), herbicide safening (Takematsu and Takeuchi 1989, Kim et al 1993), and disease resistance (Khripach et al 1997, Volynets et al 1997). BRs were also suggested to be involved in fungal control (Adam et al 1991) and insect control (Richter and Koolman 1991).

Thus, BRs increase the ability of resistance against a wide variety of environmental stresses under field conditions. Further insight into the physiological mechanism of action of these effects is described in a preceding chapter (see the chapter by J Sasse, this volume). There have been few reports on the mode of action of these effects in field crops. The role of BRs in protecting plants against environmental stresses will be an important research theme for clarifying the mode of action of BRs and may contribute greatly to the usage of BRs in agricultural production.

2.6 Environmental Effects on the Results of Field Trials

Since environmental factors greatly affect the test results of BRs, cultivation conditions of temperature, season, day length, and soil property must be taken into account. In general, most plant growth regulators exhibit their effects under optimal temperature conditions for plants. BRs have merit in that their useful effects are exhibited under suboptimal temperature conditions for plants. Growth phenomena such as germination, vegetative growth, rooting, fruit-setting, fruit thickening, and fruit ripening are greatly affected by temperature, consequently bringing about a

Table 1. Effect of brassinolide (BL) on dry weight in young rice plants under different temperature conditions

Conc. of BL	% D. W. weight against control	
ppm	15-18°C	25-35°C
1.0	116.5	94.2
0.01	112.0	105.9
control	100.0 (1.2 g)	100.0 (1.7 g)

Rice plants were sprayed at the fifth true leaf stage with 1.0 or 0.01 ppm of BL and grown under 15°C night/18°C day or 25°C night/35°C day. The average dry weight (D. W.) per plant was measured 1 month after the treatment.

Table 2. Fruit setting effect of brassinolide (BL) on tomato in winter or in summer

Season	% Fruit weight against control	
	BL 0.1 ppm	control
Winter	256.2	100.0 (57.5 g)
Summer	119.8	100.0 (144.7 g)

Tomato plants were cultivated in a greenhouse during winter or field conditions during summer. BL (0.1 ppm) was sprayed at blooming time on each flower and average fruit weight was measured 30 days after the treatment.

great fluctuation of crop yields.

There are many papers pointing out that BRs tend to show their effects more under low temperature conditions. This means that the test results depend on whether the field test is carried out under optimum temperature condition for the test plants. For example, Table 1 (Y. Kamuro, unpublished work, 1991) shows that the effect of brassinolide on the growth of rice differed greatly at different cultivation temperatures (Kamuro and Takatsuto 1991). As shown in Table 2, in the winter season brassinolide showed a distinguished promotive effect on fruit-setting and thickening of tomato, while in the summer season brassinolide showed the promotive effect to a much lesser extent (Kamuro and Takatsuto 1991). Thus, it may be difficult for us to evaluate the effects of BRs when tests are conducted under optimal growth conditions. Therefore, a potential market for BRs is expected to be in the area of cultivated crops, which are easily affected by low temperature.

In addition to the temperature, the effects of BRs are also greatly affected by light conditions such as day length (Kamuro and Inada 1987, Kamuro and Takatsuto 1991) (see Fig. 2), light or dark (Krizek and Worley 1981, Mandava et al 1981), and spectral quality (Krizek and Worley 1981, Krizek and Mandava 1983, Meudt and Porath 1986, Kamuro and Inada 1991, Kalituho et al 1997), indicating relationships between phytochromes and BR action. Therefore, it is to be expected that the test results in different seasons are not always the same, even employing the same crop. As shown in Table 2, the thickening effect of brassinolide on tomato varied between summer and winter. Therefore, it is important that the localities and seasons should be taken into account to evaluate the test results.

2.7 Interaction of BRs with Other PGRs in Agricultural Use

There are several reviews on the interactions of BRs with auxin, gibberellin, cytokinin, ethylene, and abscisic acid (Meudt 1987, Mandava 1988, Katsumi 1991, Roddick and Guan 1991, Sasse 1991, Sakurai and Fujioka 1993). BRs show multiple interactions with these plant hormones, which are characteristic of BRs. Therefore, the mixed-treatment effects of BRs and other plant hormones are expected to be important for the purpose of the growth control of field crops. However, to date, most of the papers that have been published on these effects were obtained by bioassays or tests using young plants.

There are a few practical studies concerning the interaction effects between BRs and gibberellin on the germination of weed seeds (Takeuchi et al 1991, 1995, 1996) and the production of seedless loquat fruits (Tani et al 1989) and those between BRs and abscisic acid on drought resistance of sorghum (Xu et al 1994b). Nevertheless, it seems difficult to obtain stable and significant results because brassinolide, 24-epibrassinolide, and (22S,23S)-28-homobrassinolide have been used in the tests.

At present, the mixed treatments of a BR analog (TS303) with the other plant hormones are being exploited for the purpose of controlling field crops. The details of these effects will be described in section 3, this chapter.

Fig. 2. Effect of brassinolide on radish growth under long-day (*left*) or short-day (*right*) conditions. Radish plants were grown at 25°C under long-day (L. D.: 16 hrs) or short-day (S. D.: 8 hrs). Brassinolide (BL) was sprayed twice at the fourth true leaf stage and 7 days after the first treatment. Photos were taken 11 days after the first treatment.

Fig. 3. The mode of action of brassinolide (BL) and Compound No. 6 on fruit-setting of tomato. **a**: Nontreatment. **b**: BL, 0.1 ppm. **c**: Compound No. 6, 0.1 ppm.

Fig. 4. Effects of seed-treatment of TNZ303 on the growth of rice plants.

10 Application of Brassinosteroids   231

Fig. 5. Fruit-setting effects of TS303 on grapevine (var. Kyoho) (*left*) and mango (*right*).

Fig. 6. Effects of mixed treatment of TS303 and PDJ on preventing cold damage in benjamin (*left*) and pear (*right*) trees.

3. Practical Application of Brassinosteroids with Long-Lasting Activity

There were questions about the practicability of brassinolide, 24-epibrassinolide, and (22S,23S)-28-homobrassinolide because these BRs are thought to have a short period (2 or 3 days at most) of active persistence (Kamuro and Takatsuto 1991). To increase the crop yield by improving fertilization and fructification, it is desirable for BRs to be active throughout the whole flowering time. To meet this requirement, we selected a BR analog TS303 (see Fig. 1) as the most promising compound with long-lasting activity. It is now generally accepted that the practical effects of brassinolide, 24-epibrassinolide, and (22S,23S)-28-homobrassinolide are inadequate for agricultural use. It is curious why there are no other reports with special emphasis on the relationship between the structure of BRs and their persistent biological activity.

3.1 Selection of Compound TS303

We began the selection of compounds with long-lasting biological activity in 1986 and have aimed for compounds with persistent activity of more than 15-20 days. The normal bioassay systems easily evaluate the biological activity of test compounds within a few days but cannot determine the length of persistence of the biological activity. The first problem is how to evaluate the persistence of the biological activity of a number of BR-related compounds. Ultimately, we compared the effects of each compound on the fruit-setting of tomato cultivated in a greenhouse during winter. The temperature in a greenhouse was lowered to the extent that the tomato plant does not give fruit-setting during the blooming period. The tomato plant has several flower buds in each cluster. Under the low temperature condition, each bud comes into bloom every few days, and it requires about 20 days for the blooming period of one cluster. Each test compound was sprayed specifically on a cluster on the day when the first flower of each cluster came into bloom. As shown in Fig. 3, fruit-setting was not observed in untreated clusters. When sprayed with brassinolide, only the first flower gave fruit-setting, but the fruit-setting effect was not observed in the second flower, which came into bloom several days after spraying. On the other hand, Compound No. 6 (see Fig. 1) did not show any fruit-setting effect on the first flower, but exhibited such effect on later flowers, which came into bloom several days after spraying. These results suggest that Compound No. 6 itself does not show any biological activity but is transformed into a biologically active form in plant tissues (Kamuro and Takatsuto 1991). The strength of the persistent activity elicited by test compounds was estimated by the number of fruit-set per cluster. In a bioassay system such as a lamina joint test, brassinolide showed very strong biological activity, while Compound No. 6 was almost inactive. Many compounds were synthesized that showed a growth effect similar to that of Compound No. 6 (Kamuro and Takatsuto 1991). After screening a large number of synthetic compounds related to

Compound No. 6, we have ultimately selected a compound termed TS303 as the most promising candidate for practical use.

3.2 Present Situation of the Practical Tests of TS303

The comparison of results between the effects of Compound No. 6, TS303, and brassinolide on field crops have been reported (Kamuro and Takatsuto 1991, 1992, Takatsuto et al 1996). TS303 has been officially tested for registration in many countries after 1996. At present, no other BR-related compound except TS303 has been developed as a candidate for practical use. The present situation of the field trials of TS303 is described.

3.2.1 Effects of Seed Treatment on the Growth and Yields of Crops

The most extensively investigated practical application of TS303 is seed treatment. This treatment was conducted in combination with a jasmonic acid analog. The mixed formulation (test agent name: TNZ303-liquid) consists of 30 mg each of TS303 and *n*-propyl dihydrojasmonate (PDJ) per liter. PDJ itself promotes germination at a very low concentration and showed a synergistic effect with TS303 (Kamuro et al 1996). PDJ has been also officially tested for registration in many countries (Fujisawa et al 1996, 1997). The standard dilution rate of TNZ303 with water for practical use is 3000 times. Thus, the effective concentrations for seed treatment are as low as 0.01 ppm for each component. When seeds are soaked in an aqueous solution of TNZ303, the appropriate soaking period depends on the kinds of seeds, ranging from several hours to overnight. TNZ303 was also found to be effective in germination when a mixture with a coating agent was sprayed on seeds. In this case, a 1:3000 mixture of TNZ303 and a seed coating agent is most suitable.

A great effect on rice was confirmed by seed treatment with TNZ303 before direct sowing. Table 3 shows the results of plant growth and yield increase in the official tests for registration, in which seeds were dipped overnight in a 3000 times diluted aqueous solution of TNZ303. The percentage of establishment of seedlings, root growth, and tillering were significantly promoted (Fig. 4), and consequently, the number of ears per area and yield were stably increased. These promoting effects were observed even more when environmental conditions such as temperature and soil conditions (drought, salt, herbicide, soil disease) were unfavorable for rice plants (the official test results for registration in Japan, Korea, and China, 1996, 1997).

In Russia and Poland, the spray treatment of seeds with TNZ303 is now being applied to wheat, potato, rape seed, and sugar beet. As shown in Table 4, significant yield increasing effects were observed in the official tests. It is interesting that when treated with TNZ303 the rate of plants infected with soil disease decreased, indicating that TNZ303 exhibited a disease resistance effect. In these test results, TNZ303 was already registered on rice and cotton in China (registration no. LS 98034). TNZ303 also passed the efficacy tests on rice in Korea and those on wheat and potato in Russia. The seed treatment with TNZ303 in important crops is expected in many more countries in the future.

Table 3. Germination promoting and yield increasing effects of TNZ303 and brassinolide (BL) on rice plants in official test for registration

Country		% Establishment		% Growth against control			
		Control	TNZ303	Height	Foliage D.W.	Tillers	Root D.W.
Japan	(1)	17.5	42.5	101.2	107.9	104.8	155.8
(1996)	(2)	37.1	53.2	88.1	97.4	115.3	114.0
	(3)	63.9	72.0	98.4	96.7	-	83.0
	(4)	78.2	87.7	101.3	105.6	108.3	-
	(5)	75.0	93.6	99.2	85.7	103.0	127.3
	(6)	61.2	81.6	103.7	113.5	-	-
	Ave.	55.5	71.8	98.7	101.1	107.9	120.0

		% Yield increase against control						
		1996			1997			
		(1)	(2)	(3)	(1)	(2)	(3)	Ave.
China	TNZ303	12.0	7.4	5.4	13.7	8.5	5.1	8.7
	BL	8.0	4.5	2.8	7.0	0.8	3.7	4.5
Korea	TNZ303				8.1	6.0	5.0	6.4

Treatment dose: 0.01 ppm, seed dipping treatment.

Table 4. Effects of TNZ303 on yield increase and disease resistance in wheat, potato, rapeseed, and beet in official test for registration

Country : Crop	% Yield increase against control					
	1996		1997			
	(1)	(2)	(1)	(2)	(3)	Ave.
Russia : wheat	23.5		46.5	22.4	8.5	25.2
: potato	16.9		28.3	27.7	15.5	22.1
Poland : wheat	13.9		29.6	10.9	9.3	15.9
: potato	27.0	11.1	8.9			15.7
: rapeseed			15.9	6.7		11.2
: beet	38.5	19.8	14.6	6.1		19.8

% Decrease of diseased plants against control
Country : Crop (1996): Disease

Russia : wheat : Fusarium	-21.7%	Poland : potato : Rhyzoctonia	-54.1%
: Ophiobolus	-61.9%	: Phytophtora	-33.7%
: Septoria	-21.7%		

Treatment dose: 0.01 ppm, spray treatment to seeds.

Table 5. Growth promoting and yield increasing effects of TNZ303 and brassinolide (BL) on rice plants in official test for registration

Country		% Growth against control (30-40 days after treatment)			
		Height	Tillers	Leaves D.W.	Roots D.W.
Japan	1996(1)	104	163	108	110
	(2)	103	119	134	129
	1997(1)	108	-	107	116
	(2)	108	-	116	122
	(3)	104	104	106	-
	Ave.	105	129	114	119

		% Yield increase against control			
		(1)	(2)	(3)	Ave.
China	TNZ303	13.2	11.8	10.1	11.7
(1997)	BL	4.5	3.3	6.2	4.7

Treatment dose: 0.01 ppm, spray treatment to seedlings.

3.2.2 Effects on Vegetative Growth and Rooting

In the cultivation of rice in Japan, Korea, and China, a transplanting technique is employed more frequently than direct sowing. It is very important to promote rooting and vegetative growth after transplanting. A significant promotive effect was observed by spray treatment with TNZ303 (3000 times dilution) 3-5 days before transplanting, as shown in Table 5. Auxin agents are generally used for the promotion of rooting of cuttings, and favorable combination effects were observed by mixtures of TS303 or TNZ303 with auxin agents (Kamuro et al 1996, Watanabe et al 1997). Since the practical tests of TS303 in this application area have just started, it is now hoped that TS303 will be applied to many kinds of crops.

3.2.3 Effects on Reproductive Growth, Seed-Setting, and Fruit-Setting

The most economically important action of BRs is to promote fertilization resulting in the promotion of seed-setting and fruit-setting in a wide variety of crops. With the aim of the stability of yield and/or yield increase, a large potential market of BRs is expected for crops in which seeds or fruits are harvested.

In Japan, official tests for registration of TS303 are now being carried out with grapevine (var. Kyoho) and cherry (var. Satonishiki). These are important economic crops, and their fruit-settings are known to be very unstable. In grapevine, the promotive effects were obtained by spraying a 0.01 ppm solution of TS303 (Table 6, Fig. 5). In these experiments, spraying just before flowering was not particularly effective, while a highly promotive effect was observed by spraying 6-10 days before the start of flowering. The results show that the characteristic action of TS303 takes about 5 days to exhibit plant growth-regulating activity (Fig. 3). This demon-

Table 6. Fruit-setting effect of TS303 on grapevine (var. Kyoho) in official test for registration

Country	Application day*	Place	Year	No. of berries/cluster Treated	No. of berries/cluster Control
Japan	-10	Nagano	'95	42.4	28.3
	-10	Aichi	'97	34.8	26.2
	-9	Yamagata	'97	36.7	30.9
	-8	Yamanashi	'94	45.1	26.9
	-8	Nagano	'97	20.7	18.8
	-7	Chiba	'94	25.9	7.9
	-6	Yamagata	'95	48.9	37.7
	-6	Yamanashi	'95	39.8	38.4
	-6	Hiroshima	'95	19.3	14.0
	-5	Nagano	'96	26.8	28.3
	-5	Nagano	'96	20.0	19.0
	-5	Hiroshima	'96	9.8	13.0
	-4	Yamanashi	'97	29.0	24.5
	-4	Yamagata	'96	17.2	13.2
	-4	Yamagata	'96	15.4	7.9
Korea	-10	(1)	'97	34.7	22.9
	-10	(2)	'97	26.3	4.6
	-10	(3)	'97	17.0	11.8

Spray dose: 0.01 ppm.
*Days before the start of blooming.

strates that in the case of crops with a short flowering period the timing of the spray treatment is an important factor for TS303 to exhibit its promotive effect to the full extent. In Korea and China, good fruit-setting effects of TS303 were also confirmed in Kyoho (Table 6). Even in the case of mango, in which fruit-setting is usually unstable and thickening of fruit is also not uniform, TS303- sprayed plants showed a uniform fruit-setting effect (Fig. 5).

Crops whose flowering period ranges over several months require repeated spraying with TS303 over an interval of 3-4 weeks, because the biological persistence of TS303 is about 3 weeks. In a practical test for strawberry conducted in Japan in 1997, once per month spraying for a total of 5 months resulted in 117-124% increase in the total yield (Pipattanawong et al 1996). For red pepper in Korea, four times spraying (in a interval of 20 days) with a 0.01 ppm solution of TS303 resulted in 131.2% yield increase (the result of a demonstration test by Bi Wang Industrial Co., 1997). These yield-increasing effects obtained by repetitive treatments over a long period may be attributable to a combined effect on increased numbers of flower-buds and fruit-setting. This view is supported by the results (Pipattanawong et al 1996). In addition, the repetitive sprayings over an interval of 3-4 weeks seems to be no problem for practical application, since TS303 can be mixed in the spray tank with fungicides or insecticides.

3.2.4 Resistance Against Environmental Stress

Under optimal cultivation conditions for crops, the effects of TS303 and BRs are generally obscure. It can be said that the above-mentioned effects observed in germination, vegetative growth, rooting, reproductive development, and seed- or fruit-setting are based on a kind of resistance against environmental stress. TS303 exhibits its effect even under apparently unfavorable cultivation conditions that bring injury to crops. For example, as a response to cold injury during winter, fruit and leaf drop are crucial problems. In a test employing benjamin plants (Fig. 6), TS303 showed a significant preventive effect against cold damage and, furthermore, had a synergistic effect with PDJ, a jasmonic acid analog (Kamuro et al 1996). In a test for preventing cold damage in citrus during winter, a spray treatment with a mixture of TS303 (0.01 ppm) and PDJ (10 ppm) also exhibited a significant effect (Kamuro et al 1997). The treatments with three times repeatedly at the beginning of December, January, and February prevented defoliation and fruit-drop during winter before the March harvest time (Table 7). Figure 6 also shows the result with pear trees conducted in a field trial. Pear trees were sprayed with a mixture of TS303 (0.01 ppm) and PDJ (10 ppm) 1 month before the first frost. The leaves of nontreated trees were almost completely dropped, while those of the trees sprayed with the mixture showed little leaf drop. In the rice crop regions of Japan, Korea, and China, cool summer is a major and crucial problem, causing two types of cool-weather damage on rice plant: delayed growth and floral impotence. It was found that TS303 exhibited preventive effects in both cases (Kamuro et al 1997). When a period of cold damage is 2-3 weeks, a single treatment with TS303 alone or a mixture with PDJ is recommended. However, when more prolonged periods of cold damage are expected for crops, it is necessary to spray repeatedly at intervals of 3-4 weeks.

The effects of TS303 in combination with PDJ on environment stresses such as drought, salt, herbicide, and diseases are described in section 3.2.1, this chapter. However, the practical application in these area awaits further investigation.

Table 7. Effects of TS303, PDJ and mixed-treatments for preventing defoliation and fruit-drop from cold damages of citrus

Compounds: concentration (ppm)		% Defoliation	% Fruit-drop
TS303	0.01	21.2	7.0
PDJ	200	43.1	26.2
	50	28.6	12.9
	10	16.4	8.0
TS303 0.01 + PDJ10		7.0	3.7
Untreated control		32.7	14.6

4. Conclusions

Brassinosteroids show a wide variety of growth-regulating effects on plants that are useful to agricultural production. In the latter half of the 1980s, these effects were extensively tested under field cultivation conditions. However, their results turned out to be disappointing in many cases. The major reason for this is the rapid metabolism (inactivation) of BRs. In the search for a compound with long-lasting biological activity, a large number of compounds have been evaluated for their persistent activity by employing the tomato fruit-setting assay. Consequently, TS303 has been selected as the most promising developmental candidate for agricultural use. This compound and TNZ303 (a mixture of TS303 and PDJ) have now been officially tested for registration in many countries. Significant and stable results have been obtained for promotion of germination, vegetative growth, and rooting. Furthermore, remarkable promotive effects have been observed on reproductive growth, seed- and fruit-settings, and resistance against environmental stress. Research on BR applications such as developing TS303 will probably bring great benefits to human beings in the near future.

References

Adam G, Marquardt V, Vorbrodt HM, et al (1991) Aspects of synthesis and bioactivity of brassinosteroids. In: Cutler GH, Yokota T, Adam G (Eds) Brassinosteroids: Chemistry, Bioactivity, and Application. ACS Symp Ser 474. Amer Chem Soc, Washington, DC, pp 74-85

Brawn P, Wild A (1984) The influence of brassinosteroid on growth and parameters of photosynthesis of wheat and mustard plants. J Plant Physiol 116:189-196

Fujii S, Saka H (1992) Growth regulation action of brassinolide on plants. II. Effect of brassinolide on the translocation of assimilate in rice plants during the ripening stage. Jpn J Crop Sci 61:193-199

Fujii S, Hirai K, Saka H (1991) Growth regulating action of brassinolide in rice plants. In: Cutler GH, Yokota T, Adam G (Eds) Brassinosteroids: Chemistry, Bioactivity, and Application. ACS Symp Ser 474. Amer Chem Soc, Washington, DC, pp 306-311

Fujisawa H, Seto H, Yoshida S, et al (1996) Promoting effects of jasmonic acid analog, n-propyl dihydrojasmonate (PDJ), on plant growth. Proc Plant Growth Regul Soc Amer 23: 111-116

Fujisawa H, Koshiyama M, Seto H, et al (1997) Effects of jasmonic acid compound on fruit setting, fruit growth, ripening and cold-resistance. Proceedings of the 8th International Symposium on Plant Bioregulators. Acta Hort 463:261-266

Grove MD, Spencer GF, Rohwedder WK, et al (1979) Brassinolide; a plant growth promoting steroid isolated from *Brassica napus* pollen. Nature 281:216-217

Hamada K (1986) Brassinolide in crop cultivation. In: Plant Growth Regulators in Agriculture FFTC Book Ser 34, Food and Fertilizer Technology Center for the Asian and Pacific Region, Taipei, pp 188-197

Hewitt FR, Hough T, O'Neill P, et al (1985) Effects of brassinolide and other growth regulators on germination and growth of pollen tubes of *Prunus avium* using a multiple hanging-drop assay. Aust J Plant Physiol 12:201-211

Hirai K, Fujii S, Honjo K (1991) Plant growth regulating action of brassinolide. I. The effect of brassinolide on the ripening rice plants under low temperature condition. Jpn J Crop Sci 60:29-35

Ikekawa N, Zhao YJ (1991) Application of 24-epibrassinolide in agriculture. In: Cutler GH, Yokota T, Adam G (Eds) Brassinosteroids: Chemistry, Bioactivity, and Application. ACS Symp Ser 474. Amer Chem Soc, Washington, DC, pp 280-291

Iwahori S, Tominaga S, Higuchi S (1990) Retardation of abscission of citrus leaf and fruitlet explants by brassinolide. Plant Growth Regul 9:119-125

Kalituho LN, Chaica MT, Kabashnikova LF, et al (1997) On the phytochrome mediated action of brassinosteroids. Proc Plant Growth Regul Soc Amer 24:140-145

Kamuro Y, Inada K (1987) Effect of light conditions on brassinolide induced mung bean epicotyl elongation and radish growth. Proc Plant Growth Regul Soc Amer 14: 221-224

Kamuro Y, Inada K (1991) The effect of brassinolide on the light induced growth inhibition in mung bean epicotyl. Plant Growth Regul 10:37-43

Kamuro Y, Takatsuto S (1991) Capability for and problems of practical uses of brassinosteroids. In: Cutler GH, Yokota T, Adam G (Eds) Brassinosteroids: Chemistry, Bioactivity, and Application. ACS Symp Ser 474. Amer Chem Soc, Washington, DC, pp 292-297

Kamuro Y, Takatsuto S (1992) Practical uses of brassinolide derivative: Promoting effects on fertilization and seed-setting. Proc Plant Growth Regul Soc Amer 19:275-277

Kamuro Y, Takatsuto S, Noguchi T, et al (1996) Application of a long-lasting brassinosteroid TS303 in combination with other plant growth regulators. Proc Plant Growth Regul Soc Amer 23:27-31

Kamuro Y, Takatsuto S, Noguchi T, et al (1997) Practical effects of brassinosteroid compound: TS303. Proc Plant Growth Regul Soc Amer 24:111-116

Katsumi M (1991) Physiological modes of brassinolide action in cucumber hypocotyl growth. In: Cutler GH, Yokota T, Adam G (Eds) Brassinosteroids: Chemistry, Bioactivity, and Application. ACS Symp Ser 474. Amer Chem Soc, Washington, DC, pp 246-254

Khripach VA, Zhabinskii VN, Malevannaya NN (1997) Recent advances in brassinosteroid study and application. Proc Plant Growth Regul Soc Amer 24: 101-106

Kim KU, Kwon ST, Shim DH, et al (1993) Effects of herbicide safener on rice sprouted seedlings for machine transplanting in Korea. Acta Phytopathol Entomol Hung 28:2-4

Krizek DT, Worley JF (1981) The influence of spectral quality on the internodal response of intact bean plants to brassins. Physiol Plant 51:259-264

Krizek DT, Mandava NB (1983) Influence of spectral quality on the growth response of intact bean plants to brassinolide, a growth promoting steroidal lactone. Physiol Plant 57:317-329

Lee JM, Baek NK, Lee CW, et al (1990) Interaction between CO_2 and chemical treatment on overcoming self-incompatibility in cruciferous crops. Proc Intr Horticultural Cong, pp 447

Mandava NB (1988) Plant growth promoting brassinosteroids. Annu Rev Plant Physiol Plant Mol Biol 39:23-52

Mandava NB, Sasse JM, Yopp JH (1981) Brassinolide, a growth promoting steroidal lactone. II. Activity in selected gibberellin and cytokinin bioassays. Physiol Plant 53:453-461

Maotani T, Suzuki A, Nishimura T, et al (1989) Control of physiological fruit drop of Japanese persimmon 'Hiratanenashi'. J Japan Soc Hort Sci 58:557-562

Meudt WJ (1987) Chemical and biological aspects of brassinolide. In: Fuller G, Nes WD (Eds) Ecology and Metabolism of Plant Lipids. ACS Symp Ser 325. Amer Chem Soc, Washington, DC, pp 53-75

Meudt WJ, Porath D (1986) Brassinosteroid effect on red light controlled bean plumular hook opening. Proc Plant Growth Regul Soc Amer 13:129

Pipattanawong N, Fujishige N, Yamane K, et al (1996) Effects of brassinosteroid on vegetative and reproductive growth in two day-neutral strawberries. J Japan Soc Hort Sci 65:651-654

Pirogovskaya GV, Bogdevitch IM, Nanmova GV, et al (1996) New forms of mineral fertilizers with additives of plant growth regulators. Proc Plant Growth Regul Soc Amer 23:146-151

Richter K, Koolman J (1991) Antiecdysteroid effects of brassinosteroids in insects. In: Cutler GH, Yokota T, Adam G (Eds) Brassinosteroids: Chemistry, Bioactivity, and Application. ACS Symp Ser 474. Amer Chem Soc, Washington, DC, pp 265-278

Roddick JG, Guan M (1991) Brassinosteroids and root development. In: Cutler GH, Yokota T, Adam G (Eds) Brassinosteroids: Chemistry, Bioactivity, and Application. ACS Symp Ser 474. Amer Chem Soc, Washington, DC, pp 231-245

Roddick JG, Ikekawa N (1992) Modification of root and shoot development in monocotyledon and dicotyledon seedlings by 24-epibrassinolide. J Plant Physiol 140:70-74

Ronsch H, Adam G, Matshe J, et al (1993) Influence of (22S,23S)-homobrassinolide on rooting capacity and survival of adult Norway spruce cuttings. Tree Physiol 12:71-80

Ronsch H, Adam G, Voigt B (1996) Retardation of needle chlorosis by brassinosteroids in magnesium-deficient seedlings of Norway spruce. Proc Plant Growth Regul Soc Amer 23:62

Sairam RK (1994) Effects of homobrassinolide application on plant metabolism and grain yield under irrigated and moisture-stress conditions of two wheat varieties. Plant Growth Regul 13:147-159

Sakurai A, Fujioka S (1993) The current status of physiology and biochemistry of brassinosteroids. Plant Growth Regul 13:147-159

Sasse JM (1991) Brassinolide induced elongation. In: Cutler GH, Yokota T, Adam G (Eds) Brassinosteroids: Chemistry, Bioactivity, and Application. ACS Symp Ser 474. Amer Chem Soc, Washington, DC, pp 255-264

Sasse JM (1997) Recent progress in brassinosteroid research. Physiol Plantarum 100:696-701

Sasse JM, Sands R (1992) Brassinosteroids and transplantation stress. Proc Plant Growth Regul Soc Amer 19:135-138

Sasse JM, Sasse JM (1994) Brassinosteroids and roots. Proc Plant Growth Regul Soc Amer 21:228-232

Sasse JM, Smith R, Hudson I (1995) Effects of 24-epibrassinolide on germination of seed of *Eucalyptus camaldulensis* in saline conditions. Proc Plant Growth Regul Soc Amer 22:136-141

Shen XY, Dai JY, Hu AC, et al (1990) Studies on physiological effects of brassinolide on drought resistance in maize. J Shenyang Agric Univ 21:191-195

Singh J, Nakamura S, Ota Y (1993) Effect of epibrassinolide on gram (*Cicer arietinum*) plants grown under water stress in the gurenile stage. Indian J Agric Sci 63:395-397

Suge H (1986) Reproductive development of higher plants as influenced by brassinolide. Plant Cell Physiol 27:199-205

Sugiyama K (1994) Reducing flowering of satsuma mandarin with GA and brassinosteroids. Proc Intr Horticultural Cong no.26-10

Sugiyama K, Kuraishi S (1989) Stimulation of fruit set of Morita navel orange with brassinolide. Acta Hort 239:345-348

Takahashi H, Singh J, Koshioka, et al (1994) Effects of epibrassinolide application on plant growth, yield components and yield of green gram (*Vigna radiata* (L.) Wilczek). Jpn J

Trop Agric 38:227-231
Takematsu T, Takeuchi Y (1989) Effects of brassinosteroids on growth and yields of crops. Proc Japan Acad 65 (Ser B):149-152
Takatsuto S, Kamuro Y, Watanabe T, et al (1996) Synthesis and plant growth promoting effects of brassinosteroid compound TS303. Proc Plant Growth Regul Soc Amer 23: 15-20
Takeuchi Y, Worsham AD, Awad AE (1991) Effects of brassinolide on conditioning and germination of witchweed (*Striga asiantica*) seeds. In: Cutler GH, Yokota T, Adam G (Eds) Brassinosteroids: Chemistry, Bioactivity, and Application. ACS Symp Ser 474. Amer Chem Soc, Washington, DC, pp 298-305
Takeuchi Y, Ogasawara M, Konnai M, et al (1992) Application of brassinosteroids in agriculture in Japan. Proc Plant Growth Regul Soc Amer 19:343-352
Takeuchi Y, Yoneyama K, Ogasawara Y, et al (1995) Effects of brassinosteroids on seed germination of weeds. Proc Plant Growth Regul Soc Amer 22:781-786
Takeuchi Y, Omigawa Y, Ogasawara, et al (1996) Effects of brassinosteroids on conditioning and germination of clover broomrape (*Orobanche miror*) seeds. Plant Growth Regul 16:153-160
Tani H, Yuda E, Doi S, et al (1989) Studies on production of seedless loquat fruits. J Jpn Soc Hort Sci 58:126-127
Volynets AP, Pshenichnaya LA, Manzhelesova NE, et al (1997) The nature of protective action of 24-epibrassinolide on barley plants. Proc Plant Growth Regul Soc Amer 24:133-137
Wang Y, Luo W, Xu R, et al (1994) Effect of epibrassinolide on growth and fruit quality of water melon. Zhiwu Shengh Tongxun 30:423-425
Watanabe T, Noguchi T, Kuriyama H, et al (1997) Effects of brassinosteroid compound TS303 on fruit setting, fruit growth, taking roots and cold-resistance. Proc 8th International Symp on Plant Bioregulators. Acta Hort 463:267-270
Xu RJ, Li XD, He YJ, et al (1994a) Effects of treatment with epibrassinolide and chololic lactone on the fruit-set and ripening in some grape cultivation. J Shanghai Agric Coll 12:90-95
Xu HL, Shida A, Futatsuya F, et al (1994b) Effects of epibrassinolide and abscisic acid on sorghum plants growing under soil water deficit. Jpn J Crop Sci 63:671-681
Yand ZS, Shi GA, Jin JH (1992) Effects of epibrassinolide and triadimefon on winter wheat yield and its physiological response. Acta Univ Agric Boreali Occidentials 20:47-50
Yopp JH, Mandava NB, Sasse JM (1981) Brassinolide, a growth promoting steroidal lactone. I. Activity in selected auxin bioassay. Physiol Plant 53:445-452
Yoshioka T, Nesumi H, Ito Y, et al (1990) Selection of plant growth regulators affecting citrus precocious flowering. J Jpn Soc Hort Sci 59:44-45

INDEX

A

abscisic acid 145, 152
ACC synthase 164
accumulation
    malondialdehyde 150
acropetal transport 115
active conformation (AC) 196
activity
    catalase 150
    peroxidase 150
acyl conjugates 32, 118
alcohol dehydrogenase 153
alfalfa 140
alga 29
Alnus glutinosa L. 35
Amblyomma hebraeum 153
m-aminophenylboronates 62, 63
angiosperms 29
animal receptor kinases 179
antagonist of BR action 218
anther
    Cryptomeria japonica 39
    Erythronium japonicum Decne 32
    Japanese cedar 39
    Lilium longiflorum 105
    lily 106
anti-stress effects 9
antiserum against castasterone (antiserum
    against CS) 50
aphid-infested leaves 5
Apium graveolens 8, 38
Apocynaceae 32
aquaporin synthesis 146
Arabidopsis (*Arabidopsis thaliana*)
    35, 97, 106, 141, 149, 151, 163, 164, 169
Arizona cypress 38
asymmetric growth 141
atmospheric pressure chemical ionization
    (APCI) 63
AUUUUA motif 166

auxin 145, 150, 166, 168
auxin-induced elongation 145
Avena curvature test 8
Avena straight growth test 8
azuki bean 143

B

Baeyer-Villiger oxidation 72
barley 152
barley ES43 protein 181
basipetal transport 115
bean 140, 150
bean first and second internode bioassays 194
bean second internode bioassay 2, 8, 48
benjamin plants 237
Beta vulgaris L. 35
betacyanin synthesis 9
Betulaceae 35
bile acid 70
biomass production 140
bismethaneboronates (BMBs) 54
Brassica campestris 35
Brassica chinensis 167
Brassica juncea 149
Brassica napus 21, 35
brassicasterol 87
brassinolide (BL) 3, 21, 30, 70, 106, 223
    [24,28-^{3}H]- 127
    [24,28-^{3}H$_2$]- 56, 87
    [26,28-^{2}H$_6$]- 57
    [^{3}H]- 114, 116
    -induced elongation 145
    inhibitor 218
brassinosteroids (BRs) 2
    [5,7,7-^{3}H]- 127
    conjugates 36
    dwarfs 169
    signal transduction pathway 180, 182
    -deficient 164
    -deficient mutants 13

-insensitive 164
-insensitive mutants 13, 178
-receptor interaction 211
-regulated gene 180
^{2}H-labeled- 57
brassins 2, 4, 137
*BRI*1 13, 178, 180
*bri*1 13, 169, 178
*BRI*1 gene 146
broad bean 37
bromegrass 152, 165
*BRU*1 164, 165
buckwheat 38

C

C-22 aldehyde 70, 75, 77, 78, 80, 81, 83
C_{27}-BRs 22, 107
C_{28}-BRs 22, 93, 107
C_{29}-BRs 22, 107
Calliphora vicina 153
calmodulin 168
cambial region
   Pinus silverstris 39
   Scots pine 39
cambial scrapings
   Pinus silvestris 169
cambium 148
campestanol 101, 172
   [$^{13}C_5$]- 59
   [^{14}C]- 59
   [26,28-$^2H_6$]- 59
campesterol 95, 100, 172, 173
   [26,28-$^2H_6$]- 59
^{13}C-/^{14}C-labeled campesterol 95
Cannabis sativa 37, 63
Carbó similarity index (CI) 208
carrot cell cultures 122
Caryophyllaceae 35
Cassia tora 37
Castanea crenata 36
castasterone (CS) 6, 7, 21, 30, 74, 85, 106
   [24,28-^3H]- 127
   [24,28-$^3H_2$]- 56
   [26,28-$^2H_6$]- 57

[$^2H_6$]- 96
[^3H]- 114, 116
cat-tail 7
catalase 150
Catharanthus roseus
   32, 93, 103, 105, 106, 118
Catharanthus roseus crown gall cells 12
cathasterone (CT) 32, 103, 175
cattail 32
*cbb*1-*cbb*3 (*cabbage*) 142, 169
celery 38
cell cultures
   bromegrass 152, 165
   serradella 124
   tomato 128
cell division 146
cell expansion 167
cellulose microfibrils 167
cellulose synthase 167
cereals 226
chalcone synthase 153
charcoal chromatography 52
Chenopodiaceae 35
cherry 235
chestnut 6, 36
chilled rice, maize, and cucumber 152
Chinese cabbage 7, 35, 147
Chironomus tentans 153
Chlorella vulgaris 140
chlorophyll biosynthesis 152
chlorophyll levels 140
5α-cholestane skeleton 26
cinnamate-4-hydroxylase 165, 168
citrus 237
Citrus sinensis 38
Citrus unshiu 38
Claisen rearrangement 74
clover broomrape 147
CO_2 fixation 140
cockroach 153
cold injury 237
compactin 95
Compositae 35
compound No. 6 232
Convolvulaceae 35

cool-weather damage 237
cotton 233
cotyledons
  cucumber 150
cpd 13, 98, 103, 148, 149, 153, 169, 175
CPD gene 13, 98, 107, 174, 175
CPD promoter-*uidA* reporter gene 107
crinosterol 81, 85
  [26,28-^{2}H$_6$]- 57, 59
crown gall cells
  Catharanthus roseus 93
Cruciferae 35
Cryptomeria japonica 39
cu-3 14
cucumber (*Cucumis sativus*)
  115, 118, 142, 150
  seedlings 141
Cucurbita maxima 167
cultured cells 12, 103
  L. longiflorum 105
  lily 106
  Catharanthus roseus 32, 59, 103
Cupressaceae 38
Cupressus arizonica 12, 38, 62, 105
1-cyanoisoindole-2-*m*-phenylboronic acid 61
cycloartenol 100
CYP85 175
CYP88 175
CYP90 174, 175
cysteine protease 165
cytochrome P450 98, 103, 174
cytochrome P450 inducers 132
cytochrome P450 inhibitors 132
cytokinin 145, 168

D

d 14
dansylaminophenylboronates 63
dansylaminophenylboronic acid 61
de-etiolated dwarfs
  Arabidopsis thaliana 97
de-etiolation 151
23-dehydrobrassinolide (23-dehydroBL)
  39, 40

3-dehydro-24-epicastasterone (3-dehydro-24-
  epiCS) 129
3-dehydro-6-deoxoteasterone (6-deoxo3DT)
  7, 38, 105, 106
3-dehydro-Δ^4-campesterol 12
3-dehydrocampestanol 12
3-dehydroteasterone 12, 105
dent corn 31
6-deoxobrassinosteroids (6-deoxoBRs)
  12, 108
6-deoxo-24-epicastasterone 8
6-deoxo-25-methyldolichosterone (6-deoxo-25-
  Me-DS) 36
6-deoxo-28-homodolichosterone (6-deoxo-28-
  homoDS) 36
6-deoxo-28-norcastasterone 8
6-deoxocastasterone (6-deoxoCS) 12, 35, 105
6-deoxocathasterone (6-deoxoCT) 106
  [26,28-^{2}H$_6$]- 60
6-deoxodolichosterone (6-deoxoDS) 36
6-deoxoteasterone (6-deoxoTE) 32, 35, 105
6-deoxotyphasterol (6-deoxoTY)
  7, 35, 38, 105, 106
2-deoxy-25-methyldolichosterone (2-deoxy-25-
  Me-DS) 36
2-deoxybrassinolide (2-deoxyBL) 8, 37
6-deoxy brassinosteroids 85
det2 13, 97, 101, 150, 169, 173
DET2 gene 13, 97, 101, 173
deuterio-labeled campesterol 101
dicots 29
2,24-diepibrassinolide (2,24-diepiBL) 118
3,24-diepibrassinolide (3,24-diepiBL)
  118, 124
2,3-diepi-25-methyldolichosterone (2,3-diepi-
  25-Me-DS) 36
2,3-diepicastasterone (2,3-diepiCS) 36
3,24-diepicastasterone (3,24-diepiCS) 36, 124
differentiation of xylem 148
2α,3β-dihydroxy-B-homo-6a-oxa-5α-
  pregnane-6,20-dione 124
22α,23α-dihydroxycampesterol 177
DIM1 173
diminuto (*dim, dim1*)
  13, 59, 98, 101, 169, 173

1,3-dipolar cycloaddition 78
disease resistance 227
dismutase
    superoxide 150
Distylium factors 5
Distylium racemosum 5, 36, 105
dolicholide (DL) 7, 22, 37, 81, 87
Dolichos lablab 7, 37
dolichosterone (DS) 7, 22, 37
double mutant 176
DPY 175
dpy 14, 175
    tomato 175
drought resistance 227
drought stress 152
Dryocosmus kuriphilus 6
dwarf rice lamina inclination assay 49
dwf1 172
DWF1 gene 173
dwf4 13, 98, 103, 174
DWF4 genes 98, 103, 174
DWF5 172
dwf5 172
dwf7 172
dwf8 176

E

early C6-oxidation pathway 12, 103
ecdysteroid 153
electrochemical detection 61
electrospray ionization (ESI) 63
elongation and radial growth 140
elongation of maize root segments 192
enzyme-linked immunosorbent assay (ELISA) 49
2-epi-25-methyldolichosterone (2-epi-25-Me-DS) 36
epibrassinolide
    [24,28-^{3}H$_2$]- 87
24-epibrassinolide (24-epiBL)
    4, 10, 11, 35, 37, 85, 114, 123, 128, 152, 153, 173, 204, 223, 225
    [^{14}C]24- 118
    (22R, 23R)- 85

(22S, 23S)- 85
[^{3}H]-24- 124
[5,7,7-^{2}H$_3$]-24- 58
[5,7,7-^{3}H$_3$]-24- 88
^{14}C-labeled- 57
epibrassinosteroid (epiBR)
    [^{14}C]- 115
24-epicampesterol 95
2-epicastasterone (2-epiCS) 36
24-epicastasterone (24-epiCS) 35, 123, 128, 173, 208
    (22S,23S)- 208
    [^{3}H]- 124
    ^{3}H- and ^{2}H-labeled- 57
3-epicastasterone (3-epiCS) 36, 106, 118
3-epi-1α-hydroxycastasteorne (3-epi-1α-OH-CS) 36
3-epi-2-deoxy-25-methyldolichosterone (3-epi-2-deo 36
3-epi-6-deoxocastatseorne (3-epi-6-deoxoCS) 36
epicotyls
    soybean 164, 167
epinasty 141
episterol 171
24-epiTE 128
24-epiTE glycosides 129
Equisetum arvense L. 39
ergosterol 85
Erythronium japonicum Decne 32
estrogen 166
ethephon 145
ethylene biosynthesis 141
ethylene production 138
Eucalyptus 169
Eucalyptus camaldulensis 147, 152
European alder 35
expansins 167
expansion of cotyledon 142
expansion of leaf blades 140
explants
    mung bean 116
expression of *BRU1* 142
extraction 50

F

FAB-MS 116
Fagaceae 36
Fagopyrum esculentum 38
fatty acid conjugates 127
fertilization 226, 235
field trials 10
flower 36
flower and fruit drop 227
flower induction 149
flowering
   Perilla frutescence 226
   Raphanus sativus 226
fluorimetric detection 61
freezing tolerance 152
fructification 227
fruit ripening 150
fruit set 149
fruit-setting 235
fruit-setting of tomato 232
fruiting bodies 154
fruits
   Chinese cabbage 7
fungal control 227
fungal development 154
fungal infections 153
fusicoccin (FC) 143

G

GA dwarfs 169
garden pea 37, 99
gas chromatography-mass spectrometry (GC/MS) 54, 96, 115
GC-EI-SIM 55
GC-MS-selected ion monitoring (SIM) 54
geometry similarity index [DIF] 206
geotropic response 150
germination 225
   clover broomrape 147
   Eucalyptus camaldulensis 152
   moth bean 152
   witchweed 147

gibberellin bioassays 8
glucanases 167
23-glucosylbrassinolide 40
2-O-β-D-glucopyranosyl-3,24-diepicastasterone (2-O-β-D-glucopyranosyl-3,24-diepiCS) 129
3-O-β-D-glucopyranosyl-3,24-diepicastasterone (3-O-β-D-glucopyranosyl-3,24-diepiCS) 129
23-O-β-D-glucopyranosyl-2-epi-25-methydolichosterone 36
23-O-β-D-glucopyranosyl-2-epi-25-methyldolichosterone 116
23-O-β-D-glucopyranosyl-25-methyldolichosterone 36, 116
23-O-β-D-glucopyranosyl-brassinolide (23-O-β-D-glucopyranosyl-BL) 116
25-β-D-glucopyranosyloxy-24-epibrassinolide (25-β-D-glucopyranosyloxy-24-epiBL) 131, 132
26-β-D-glucopyranosyloxy-24-epibrassinolide (26-β-D-glucopyranosyloxy-24-epiBL) 131, 134
β-D-glucopyranosyl-(1→6)-β-D-glucopyranoside (gentiobiose), β-D-glucopyranosyl-(1→4)-β-D-galactopyranoside (lycobiose) 129
glycosidic metabolite 116
grains
   wheat (*Triticum aestivum* L.) 31
Gramineae 30
gramineous crops 226
grapefruits 226
grapevine 235
green alga 12, 40
GRID force field 212
growth of pollen tubes 149
growth of rice 229
GUS activity 107
GUS expression 167
gymnosperms 29
Gypsophila perfoliata 35

H

Hamamelidaceae 36
haploid seeds 149
heat shock proteins 152
Helianthus annuus L. 35
Helianthus tuberosus 148, 168
hemicelluloses (primarily xyloglucans) 167
herbicide safening 153, 227
homobrassinolide 173
28-homobrassinolide (28-homoBL) 85, 152
   (22S,23S)- 4, 10, 11, 199, 223
28-homobrassinosteroids (28-homoBRs)
   (22S,23S)- 153
28-homocastasterone (28-homoCS) 7, 35
   (22S,23S)- 206
28-homodolicholide (28-homoDL) 7, 22, 37
28-homodolichosterone (28-homoDS) 7, 22, 37
hormone-deficient mutants 163
hormone-insensitive mutants 163
28-homoteasterone (28-homoTE) 30, 35
28-homotyphasterol (28-homoTY) 30
Horner-Wadsworth-Emmons reaction 74
HPLC with ODS columns 52
Hydrodictyon reticulatum 12, 40
hydrogen bonding (H-bonding) 211
6α-hydroxycampestanol 103
22α-hydroxycampesterol 172, 173, 177
6α-hydroxycampesterol
   $[26,28-^2H_6]$- 59
1β-hydroxycastasterone (1β-OH-CS) 36
6α-hydroxycathasterone (6α-hydroxyCT)
   $[26,28-^2H_6]$- 60
20-hydroxyecdysterone 153
25-hydroxy-24-epibrassinolide (25-hydroxy-
   24-epiBL) 132
26-hydroxy-24-epibrassinolide (26-hydroxy-
   24-epiBL) 134
25-hydroxy-3,24-diepicastasterone (25-
   hydroxy-3,24-diepiCS) 129
hypocotyls
   bean 150
   Brassica chinensis 167

Cucurbita maxima 167
sunflower 150

I

immobilized phenylboronic acid gel 62
immunocytochemistry 50
indole-3-acetic acid (IAA) 49
inhibition of root growth 141
insect control 227
insect galls 5, 6, 36
   chestnut (*Castanea crenata*) 36
isofucosterol 101

J

Japanese black pine 38
Japanese cedar 39

K

KM-01 218
Kocienski olefinic synthesis 72
Kovats retention time indices 55

L

Lablab purpreus 37
lamina inclination in rice 141
late C6-oxidation pathway 12, 105
lateral expansion of the pea stem 145
lateral roots 141
LC-MS 63
le (*lepida*) 172
leaf explants
   Rumex 150
   Xanthium 150
leaves
   Distylium racemosum 36, 105
   tea 7, 38
   Thea sinensis 38
   wheat 152
Leguminosae 36
leguminous crops 226
lettuce 151

leucine-rich receptor kinases 178
LeXET gene 164
    tomato 164
Liliaceae 31
Lilium elegans 31
Lilium longiflorum 32, 105
lily 32, 106
limiting step in BR biosynthesis 107
*lip*1 151
lipid peroxidation 150
lipoxygenase 153
LK 14, 174
lk 14, 59, 99, 102, 173, 174
lka 14, 151, 173
LKB 14
lkb 14, 37, 59, 99, 101, 142, 151, 173
long-distance transport 139
long-lasting activity 11
Lotus japonicus 141
low temperature conditions 227
Luffa cylindrica 149, 226
Lycopersicon esculentum 38, 57, 123

M

maize 141, 150, 175
maize root segments 114
male sterile 149
malondialdehyde 150
mango 236
membrane hyperpolarization 145
membrane stability 152
meristems 148
mesophyll cells
    Zinnia elegans 148, 165, 168
methaneboronate-trimethylsilyl (MB-TMS) 54
24-methylcholestan-3-one 101
    $[26,28-^2H_6](24R)$- 59
24-methylcholesterol 95
24-methylcholest-4-en-3β-ol
    $[26,28-^2H_6](24R)$- 59
24-methylcholest-4-en-3-one 101
    $[26,28-^2H_6](24R)$- 59
24-methylcholest-4-ene-3-one
    (24R)- 174

24-methyldesmosterol 101, 173
25-methyldolichosterone (25-Me-DS) 36
24-methylenecholesterol 101, 171, 172, 173
    $[25,26,27-^2H_7]$- 59
24-methylenecycloartanol 101
microtubule organization 146
molecular electrostatic potential (MEP) 208
molecular overlay volume (MOV) 197
monocots 29
Moraceae 37
moth bean 152
mulberry 149
mung bean 116, 141, 147, 150, 164
mustard 140
mutants
    BR-deficient 164
    BR-insensitive 164

N

negative feedback regulation 108
negative ion fast atom bombardment (FAB)-
    MS 63
Nicotiana tabacum 105, 106, 118
NMR (^{13}C, ^{1}H) 52, 115
nonglycosidic metabolites 116
28-norbrassinolide (28-norBL) 7, 35, 36
28-norcastasterone (28-norCS) 7, 35, 36, 37

O

Ornithopus sativus 8, 37, 57, 123
orthoester Claisen rearrangement 85
Oryza sativa 30, 105, 106, 116, 118
osmotic potential 146
6-oxabrassinosteroid 216
oxidative lipid degradation 153
oxido reductase 173
6-oxobrassinosteroids (6-oxoBRs) 12, 108
6-oxocampestanol 103
6-oxocampesterol
    $[26,28-^2H_6]$ 59

P

panicles
   Rheum rhabarbarum L. 38
   rhubarb 38
parenchyma cells 142
pathogenesis-related genes 153
pea 142, 150, 151, 153
pear 237
Perilla 149
Perilla frutescence 226
peroxidase 150
Pharbitis purpurea 35
Phaseolus vulgaris 7, 36, 106
Phaseolus vulgaris L 50
PHD finger proteins 181
9-phenanthreneboronic acid 61
phenylalanine ammonia lyase 165, 168
phloem 142
photomorphogenesis 9, 151
photosynthetic capacity 140, 227
phytochrome signalling 138
phytochrome-controlled phenomena 9
phytosterol acyl conjugates 127
Picea sitchensis 39
Pinaceae 38
pinto 147
Pinus radiata 153
Pinus silverstris 39
Pinus thunbergii 38
Pisum sativum 37, 99
plant receptor kinases 179
pollen 50
   Alnus glutinosa L. 35
   Arizona cypress 38
   Brassica napus 21, 35
   buckwheat 38
   cat-tail 7
   cattail 32
   Citrus sinensis 38
   Citrus unshiu 38
   Cupressus arizonica 12, 38, 105
   Erythronium japonicum 32
   European alder 35
   Fagopyrum esculentum 38
   Helianthus annuus 35
   Japanese black pine 38
   Lilium elegans 31
   Lilium longiflorum 32
   lily 32
   Pinus thunbergii 38
   rape 35
   Robinia pseudo-acacia 37
   sunflower 35
   tulip (*Tulipa gesneriana*) 32
   Typha latifolia 32
   Zea mays 31
Polygonaceae 38
post-column fluorescence detection 62
posttranscriptional regulation of *BRU*1 166
potato 233
pre-labeling reagents 61
pregnanes 124
pregnenolone 70, 77
n-propyl dihydrojasmonate (PDJ) 11, 233
proton ATPase 149
proton extrusion 142, 145
proton secretion 142
protoplasts
   Chinese cabbage 147
Pteridophyte 29, 39

Q

quantitative structure-activity relationship (QSAR) 196

R

Radioimmunoassays (RIAs) 49
radiolabeled BRs 56
radish 35, 140, 141, 142, 147
radish leaf expansion test 8
Ramberg-Backlund rearrangement 83
rape 35
rape pollen 2
rapeseed 226, 233
Raphanus 149
Raphanus sativus 35, 226

Raphanus test 218
red pepper 236
Δ⁷-reductase 172
Rheum rhabarbarum L. 38
rhubarb 38
rice 30, 115, 116, 148, 150, 152, 233, 235
rice lamina inclination test (RLIT)
    5, 48, 105, 116, 132, 192, 194, 218
RLIT using intact rice seedlings 197
RMS similarity index 202
RNA and DNA polymerases 146
Robinia pseudo-acacia L. 37
root mean square (RMS) 197
Rumex 150
Rutaceae 38
rye 30

S

salt stress 152
salt tolerance 227
satsuma mandarin 226
Scots pine 39
Secale cereale 7, 30
secasterone (SE) 7, 30
seed filling 150
seed germination 147
seed treatment 233
seed- and fruit-setting 227
seed-setting 235
seedlings
    Arabidopsis 106
    C. roseus 105, 106, 118
    cucumber (*Cucumis sativus*) 115, 118
    mung bean 150
    Nicotiana tabacum 105, 118, 150
    rice (*Oryza sativa*)
        105, 106, 115, 116, 118
    soybean 165
    wheat (*Triticum aestivum*) 115, 118
seeds 35
    Apium graveolens 38
    Beta vulgaris L. 35
    Brassica campestris 35
    broad bean 37

Cannabis sativa L 37, 63
Cassia tora 37
celery 38
Chinese cabbage 35
Dolichos lablab 7, 37
Eucalyptus camaldulensis 147
garden pea 37
Gypsophila perfoliata L. 35
Lablab purpreus 37
Ornithopus sativus 37
Pharbitis purpurea 35
Phaseolus vulgaris 7, 36, 50, 106
Pisum sativum 37
radish 35, 147
Raphanus sativus 35
rye 30
Secale cereale 30
sugar beet 35
Vicia faba 37
self-incompatibility 227
senescence 150
Sephadex LH-20 chromatography 50
serradella 123, 124
sex expression 149
sex modification 226
Sharpless epoxidation 70, 73
sheaths
    Brassica campestris 35
    Chinese cabbage 35
shoots 35, 37
    Lycopersicon esculentum 38
    rice (*Oryza sativa*) 30
    Sitka spruce (*Picea sitchensis*) 39
    tomato 38
    Zea mays L. 31
short-distance effects of BRs 139
silica gel chromatography 50
silica gel HPLC 52
siliques 35
Sitka spruce 39
sitosterol 101
Solanaceae 38
solid-phase extraction 62
solvent partitionings 50
source/sink 148

soybean 164, 165, 167
soybean elongation assay 204
soybean field experiments 4
split sections from dwarf pea stems 143
spray treatment 233
SQR similarity index 202
square residue (SQR) 197
squash 142
*ste*1 171
stems
   Phaseolus vulgaris L 50
steroid 5α-reductases 97, 173
steroid hormones 1
Δ^7-sterol-C5-desaturase (STE1) 172
stigmasterol 70, 71, 75, 85, 101
strawberry 149, 226, 236
strobilus
   Equisetum arvense L. 39
structure-activity relationships (SAR) 191
sucrose synthase 167
sugar beet 35, 152, 233
sunflower 35, 150
superoxide 150
suspended cells
   Lycopersicon esculentum 123
   Ornithopus sativus 123
   serradella 123
   tomato 123
sweet corn 31
synergism between BL and auxin 138
synergistic interactions 9

T

T-DNA insertion mutant 175
T-DNA-tagged mutant 172
Taxodiaceae 39
*TCH*4 gene 164, 166
*TCH*4 promoter 180
*TCH*4:GUS 166
*TCH*4-BF1 181
TE-3-myristate 63
tea 7, 38
teasterone (TE) 30, 103
   $[26,28-^2H_6]$- 57

teasterone-3-laurate (TE-3-La) 32
teasterone-3-myristate (TE-3-My) 32, 63, 106
temperature stress 152
Thea sinensis 38
Theaceae 38
three-dimensional (3D) 196
TNZ303 11, 233
tobacco pith callus test 8
tomato
   38, 120, 123, 128, 141, 163, 164, 175, 229
tomato mutant *cu-3* 180
tomato mutants 14
tracheary element differentiation 165
tracheary elements 168
tracheid differentiation 148
transformed tobacco cells 147
translocation of photosynthates 227
transplantation stress 153
transplanting 235
transport 114
transverse growth and twining 150
22,23,24-triepiBL 114, 120
22-,23-,24-triepiBL 151
2α,3β,6β-trihydroxy-5α-pregnane-20-one 124
Triticum aestivum 31, 118
TS303 11, 232
tuber explants
   Helianthus tuberosus 168
tulip (*Tulipa gesneriana*) 32
Typha latifolia 32
Typhaceae 32
typhasterol (TY) 7, 22, 30, 105, 176
   $[26,28-^2H_6]$- 57
   $[^2H_6]$- 96

U

Umbellifererae 38
uniconazole 145, 168
uptake 114
uptake of heavy metals 149
uptake of sucrose 148

V

vegetative growth 139, 226
Vicia faba 37, 148

W

wall loosening 167
water potential 142
water stress 152
wheat
      31, 115, 118, 140, 141, 142, 152, 233
wheat leaf unrolling test 49, 105
witchweed 147

X

Xanthium 150
xylem 142, 148
xylem differentiation 9, 168
xylogenesis 148, 168
xyloglucan 165
xyloglucan endotransglycosylase (XET) 164

Y

Yield increases 226
young bean seedlings 137

Z

Zea mays 31
Zea mays pollen 2
zeatin 49
zinc-finger-like transcription factors 181
Zinnia elegans 148, 165, 168